International Journal on Advances in Life Sciences

IARIA

2016 vol. 8 nr. 1&2

Printed from e-media with permission by:

Curran Associates, Inc.
57 Morehouse Lane
Red Hook, NY 12571

Some format issues inherent in the e-media version may also appear in this print version.

Printed by Curran Associates, Inc. (2018)

International Academy, Research, and Industry Association (IARIA)
412 Derby Way
Wilmington, DE 19810

Phone: (408) 893-6407
Fax: (408) 527-6351

petre@iaria.org

Additional copies of this publication are available from:

Curran Associates, Inc.
57 Morehouse Lane
Red Hook, NY 12571 USA
Phone: 845-758-0400
Fax: 845-758-2633
Email: curran@proceedings.com
Web: www.proceedings.com

The *International Journal on Advances in Life Sciences* is published by IARIA.
ISSN: 1942-2660
journals site: http://www.iariajournals.org
contact: petre@iaria.org

Sponsored by IARIA
www.iaria.org

International Journal on Advances in Life Sciences
Volume 8, Number 1 & 2, 2016

International Journal on Advances in Life Sciences
Volume 8, Numbers 1 & 2, 2016

CONTENTS

pages: 1 - 15
A Novel Approach for Healthcare Equipments Lifespan Assessment
Mona Aridi, Lebanese International University, Lebanon
Bassam Hussein, Lebanese International University, Lebanon
Mohamad Hajj-Hassan, Lebanese International University, Lebanon
Hassan M. Khachfe, Lebanese International University, Lebanon

pages: 16 - 25
Electronic Communication as a Tool to Reduce Elective Surgery Cancellations A case study from Norway
Conceição Granja, Norwegian Centre for eHealth Research, University Hospital of North Norway, Norway
Kari Dyb, Norwegian Centre for eHealth Research, University Hospital of North Norway, Norway
Stein Roald Bolle, Norwegian Centre for eHealth Research, University Hospital of North Norway, Norway
Gunnar Hartvigsen, Norwegian Centre for eHealth Research, University Hospital of North Norway; Department of Computer Science, UIT – The Arctic University of Norway, Norway

pages: 26 - 38
Investigation of the Mechanical Behaviour of Porous Silicon Neural Microprobes
Rayan Fayad, Lebanese International University, Lebanon
Soumaya Berro, Lebanese International University, Lebanon
Bakri AbdulHay, Lebanese International University, Lebanon
Hussein Hajj-Hassan, Lebanese International University, Lebanon
Hassan M. Khachfe, Lebanese International University, Lebanon
Mohamad Hajj-Hassan, Lebanese International University, Lebanon

pages: 39 - 49
Digital Users in Pre-Digital Hospital Organisations? An Analysis on the Readiness for Electronic Communication Between a Hospital and Surgical Patients
Kari Dyb, Norwegian Centre for eHealth Research, Norway
Conceição Granja, Norwegian Centre for eHealth Research, Norway
Stein Roald Bolle, Norwegian Centre for eHealth Research, Norway
Gunnar Hartvigsen, Norwegian Centre for eHealth Research and Department of Computer Science UIT – The Arctic University of Norway, Norway

pages: 50 - 64
Assessing Visitor Engagement in Science Centres and Museums
Wolfgang Leister, Norsk Regnesentral, Norway
Ingvar Tjøstheim, Norsk Regnesentral, Norway
Trenton Schulz, Norsk Regnesentral, Norway
Göran Joryd, Expology, Norway
Andreas Larssen, Expology, Norway
Michel de Brisis, Expology, Norway

pages: 65 - 75
Multiscale Cancer Modelling in Terms of Copyright
Iryna Lishchuk, Leibniz Universität Hannover, Germany
Marc Stauch, Leibniz Universität Hannover, Germany

pages: 76 - 86
Optimization of Lead Design and Electrode Configuration in Deep Brain Stimulation
Ruben Cubo, Department of Information Technology, Uppsala University, Sweden
Mattias Åström, Department of Biomedical Engineering, Linköping University, Sweden
Alexander Medvedev, Department of Information Technology, Uppsala University, Sweden

pages: 87 - 102
Reverse Paternalism in Medical and Clinical Engineering Practice
Sarah Hammoud, Lebanese International University, Lebanon
Ali Hage-Diab, Lebanese International University, Lebanon
Hassan Khachfe, Lebanese International University, Lebanon
Bassam Hussein, Lebanese International University, Lebanon

pages: 103 - 111
Monitoring the use of impaired hand by a new low cost device during daily life activities with a real-time visual feedback
Nour El Hoda Saleh, Lebanese University, Lebanon
Ali Hage-Diab, Lebanese International University, Lebanon
Ghadir Salhab, Lebanese University, Lebanon
Bilal Debs, Lebanese International University, Lebanon
Mohamad Hajj-Hassan, Lebanese International University, Lebanon
Hassan Khachfe, Lebanese International University, Lebanon
Ahmad Rifaii Sarraj, Lebanese University, Lebanon
Soha Saleh, Kessler Foundation, USA

pages: 112 - 121
Comparability, Availability and Use of Medication eHealth Services in the Nordic Countries
Heidi Gilstad, Health Informatics Research Group, Department of Neuroscience, Faculty of Medicine, The Norwegian University of Science and Technology (NTNU), Norway
Berit Brattheim, Health Informatics Research Group, Department of Neuroscience, NTNU The Norwegian University of Science and Technology, Norway
Arild Faxvaag, Health Informatics Research Group, Department of Neuroscience, NTNU The Norwegian University of Science and Technology, Norway
Christian Nøhr, Dept. of Development and Planning Aalborg University, Denmark
Sidsel Vilumsen, Dept. of Development and Planning Aalborg University, Denmark
Jarmo Reponen, FinnTelemedicum, Research Unit of Medical Imaging, Physics and Technology (MIPT) University of Oulu. and Raahe Hospital, Finland
Hege Andreassen, Norwegian Centre for eHealth (Norwegian Centre for Integrated Care & Telemedicine), Norway
Maarit Kangas, FinnTelemedicum, Research Unit of Medical Imaging, Physics and Technology (MIPT), Finland
Guðrún Audur Harðardóttir, Dept. of Health Information Management Directorate of Health, Iceland
Sabine Koch, Health Informatics Centre, Dept. of Learning, Informatics, Management and Ethics Karolinska Institutet, Sweden
Hannele Hypponen, Dept. of Information National Institute for Health and Welfare, Finland

pages: 122 - 132
Toward Usable and Trustworthy Online Monitoring on e-Health Applications
Youna Jung, Virginia Military Institute, United States

pages: 133 - 142
Agile Medical Device Software Development: Introducing Agile Practices into MDevSPICE®
Fergal McCaffery, Dundalk Institute of Technology, Ireland
Marion Lepmets, Dundalk Institute of Technology, Ireland

Kitija Trektere, Dundalk Institute of Technology, Ireland
Özden Özcan-Top, Middle East Technical University, Turkey
Minna Pikkarainen, University of Oulu, VTT, Technical Research Centre of Finland, Finland

pages: 143 - 153
The MedITNet Assessment Method – Development and Validation using Action Design Research Self Assessment against IEC 80001-1
Silvana Togneri MacMahon, Dundalk Institute of Technology, Ireland
Fergal McCaffery, Dundalk Institute of Technology, Ireland
Frank Keenan, Dundalk Institute of Technology, Ireland

pages: 154 - 163
Construction of the emotional environment and media
Treton Cécile, DICEN IDF/ – UPEM (Paris East University), France
Bourret Christian, DICEN IDF/ – UPEM (Paris East University), France

pages: 164 - 174
The Mental Organization of Air Traffic and its Implications to an Emotion Sensitive Assistance System
Linda Pfeiffer, Visual Computing Laboratory, Technische Universität Chemnitz, Germany
Georg Valtin, Institute for Media Research, Technische Universität Chemnitz, Germany
Nicholas Hugo Müller, Institute for Media Research, Technische Universität Chemnitz, Germany
Paul Rosenthal, Visual Computing Laboratory, Technische Universität Chemnitz, Germany

A Novel Approach for Healthcare Equipments Lifespan Assessment

Mona Aridi, Bassam Hussein, Mohamad Hajj-Hassan, Hassan M. Khachfe
Lebanese International University
Beirut, Lebanon
{mona.aridi@gmail.com}
{bassam.hussein, mohamad.hajjhassan, hassan.khachfe}@liu.edu.lb

Abstract — **Medical equipments contribute to the quality of healthcare services on several levels. They play a key role in the diagnosis, the treatment, and the rehabilitation of the medical impairment and diseases. However, as any operating machine, medical equipments have a definite lifespan that expires after a period of time. Theoretically, studies specified ten years as the lifespan of medical equipments. In fact, the status of the medical equipments defines its age. This status should be addressed according to a list of criteria that evaluate the efficiency and the performance of these equipments. The purpose of this study is to develop a well-designed plan for evaluating medical equipments. According to this evaluation, the equipments that should be replaced can be ranked in the descending order of urgency, taking into account many criteria and sub-criteria.**

Keywords – efficiency; healthcare; lifespan; medical equipments; performance.

I. LITERATURE REVIEW

Assessment of medical equipment is increasingly becoming the concern of healthcare institutions [1]. For several years, great effort has been devoted to the study of reliability and maintenance of medical technology and the investigation of their malfunctions. In the early 1990s, the world raised the attention to the device-related activities and many regional offices were opened all over Europe, the Middle East, and Asia Pacific [2]. Moreover, the International Medical Device Regulators Forum (IMDRF) discussed the future directions in medical device regulatory harmonization [3]. Furthermore, the International Organization for Standards (ISO) defined ISO 13485 as a standard for assessing and maintaining the efficacy of medical equipments. It deals with the specifications of medical technology to meet healthcare requirements for healthier outcomes [4]. In addition, the US Food and Drug Administration (FDA) generated a Device Evaluation Intern Program (DEIP) to monitor the efficiency, safety, and degree of risk to public health of the medical equipments [5].

Many researchers paid considerable attention to the criticality of medical devices and the significance of the stringent environment surrounding them, so they dedicated their research to the classification of medical equipments and analyzed their preventive maintenance data using Failure Mode and Effects Analysis (FMEA) [6]. Similar studies measured maintenance effectiveness with failure codes as an evidence-based maintenance, where they compared different maintenance strategies adopted for seven types of medical equipments [7], [8]. Other studies focused on the importance of managing the regular maintenance process in hospitals, and proposed programs to increase the efficiency of the utilization of the medical equipments through a Medical Equipment Management Program (MEMP) [9], [10]. Kirisits and Redekop highlighted the economical evaluation as a critical key point that stands behind the decision making for an equipment-upgrading program [11]. Khalaf proposed a maintenance model for minimizing the risk and optimizing the cost-effectiveness of medical equipments [12]. Another study dealt with the problem from another perspective, where it shed the light on the relationship between the reliability of critical medical equipment (CME) and the effectiveness of CME maintenance management in relation to patient outcomes [13]. The clinical investigation of medical devices in Europe focuses on outlining the risks that may threaten both the patient and the staff [14].

All the above studies discussed the importance of preventive maintenance and its effect on the lifespan of medical devices. However, the most interesting approach in this issue has been proposed by a new Canadian systematic study for preventive maintenance prioritization of medical equipments. This study classifies the medical equipments into five levels of prioritization for preventive maintenance. However, this study is limited to the metering of the risks on the medical equipments using the quality function deployment (QFD) as a new concept in preventive maintenance classification [15]. Nevertheless, among all the calls regarding the evaluation of medical equipment, a study done by Sharareh Taghipour in 2011 assigned six main criteria in which some of them are branched into sub-criteria [14]. Taghipour focused on the recalls and hazard alerts that may occur for medical equipments. Moreover, concerning the risks, a great deal of attention was given to the failure frequency, the possible redetect of the risk, and the failure consequences, where we investigated the safety and environment effect of the device.

On the other hand, Taghipour raised the attention to the operational and the non-operational consequences of a failure, to inspect the cost of repair. This inspection covers the 'manpower' and the 'spare parts' costs to fix a defect. Besides, the Canadian study boosted the attention to the out of service periods and the number of waiting patients due to those failures, defined as the downtime of the device.

Here, a new evaluation technique, similar to the Canadian one, which will be highlighted later in the paper, is proposed but with less required data. In our model, we tried to make the investigation simple and direct so we focused on the function and the age of the medical equipment, as well as we focused on the mission criticality, the risks, and the maintenance requirements. Actually, collecting data for each criterion is very hard and requires a long questionnaire, so we designed a checklist questionnaire to gather the required data about each equipment. As a case study, we applied this model on a Lebanese public hospital and we came back with a list of equipments that should be replaced after a period of time as defined by the hospital.

In this paper, we propose the methodology of the study in Section II. Then, we show the way to derive the weights and the intensities of the tested criteria in Section III. After that, we present the missions to accomplish the assessment plan through Section IV. In Section V, we analyze the obtained results and make decisions accordingly. This is followed by a "Case Study" in Section VI to test the validity of the presented technique. After that, we move to the professionals' evaluation in Section VII, where we re-assess the medical equipments from the professionals' perspective. In Section VIII, we go through the budgetary quotation for the procurement process for purchasing the nominated medical equipments. Finally, we end up with a conclusion and our further expectations through Section IX.

II. PROPOSED METHODOLGY

Medical devices play a significant role in providing healthcare, as they affect the patient and the care providers directly. Besides, the design of the medical equipments gives a share in the safety of the environment [16]. The excessive use of the medical equipments is directly proportional to its performance with time, which will shorten its expected lifespan. The clinical evaluation of medical technology should be based on a comprehensive analysis that covers relevant criteria and parameters to appraise the efficiency of the equipment.

This paper proposes a model to evaluate the medical equipments according to measurable criteria and quantitative parameters that identify the time after which this equipment should be replaced. To start, we are going to identify some main criteria in which some of them are branched into sub-criteria. To make our work measurable, we assigned each criterion and sub-criterion to a specific weight that defines its criticality.

Many methods can be used to appraise and weigh clinical data. In our study, we take into account five main criteria in which some of them are divided into sub-criteria. The main criteria are: function, mission criticality, age, risks, and the maintenance requirements of the medical equipment. Among those main criteria, mission criticality is evaluated by two sub-criteria, the utilization of the equipment and the availability of alternative devices. Besides, the risks on the equipment are evaluated through three sub-criteria related to risks: the failure frequency, the detectability of failure, and the failure consequences. Each criterion has a certain weight that specifies its weight in the study. Moreover, each criterion is limited to a certain range of choices, where every choice is assigned to certain intensity. For a clear top view, we summarized the main criteria with their sub-criteria in Table I below.

Table I. OVERVIEW OF THE MEDICAL EQUIPMENT ASSESSMENT CRITERIA.

Main Criteria	Sub-criteria
Function	---
Mission criticality	Utilization
	Availability of alternative devices
Age	---
Risks	Failure frequency
	Detectability
	Frequency consequences
Maintenance Requirements	---

After defining the grades and intensities for all criteria, the model will be ready for use to assess the devices. To compute the final score, we need to calculate the total score that is the summation of the product of intensities and weights for each criterion. After that, we should calculate the Normalized Score Value (NSV) that indicates the relative importance of each device in comparison with other devices, from which we generated the Transformed Score Value (TSV). The transformed score value is the value that allows us to rank the medical device according to its importance. In order to better understanding the whole process, we illustrated the main steps in Figure 1 below.

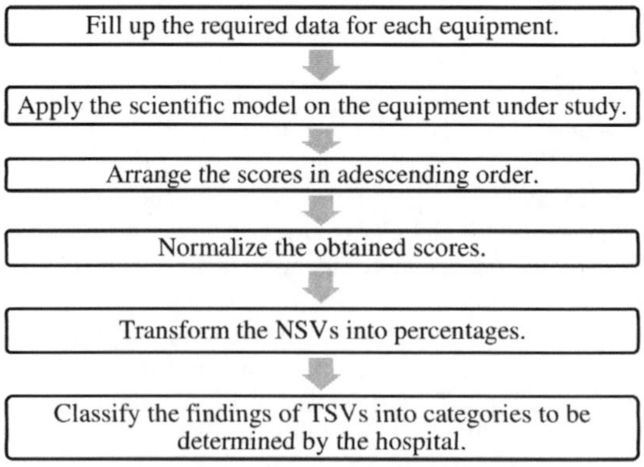

Figure 1. Flow chart of the scientific evaluation.

The above flow chart summarizes the required steps to accomplish the scientific evaluation. Going through such assessment requires a lot of parameters and equations. The derivation of all the scientific relations is explained in the following section.

Before going through the steps of the study in details, we are going to define some key terms in order not to mix up between them.

- Missions: the steps taken by our study to apply the quantitative part of the model. We assigned five consecutive missions; each mission deals with a main criterion at a time.
- Criteria: the standards and norms of evaluation. We considered five main criteria to assess the medical equipments: function, mission criticality, age, risk, and maintenance requirements.
- Parameters: the measurable factors used in our study, such as the weights and the intensities.

III. PARAMETERS

For reliable measurements on medical devices, some grades known as intensities and weights for each criterion and sub-criterion should be introduced. The grades may encounter several classes for one criterion. For example, the maintenance requirements of a device may be high, medium, or low. The definition of each class differs from one hospital to another depending on the decision makers at each hospital. Consequently, the term 'low' for maintenance requirements differs from hospital to another.

If the criterion of a device contributes with its maximum capacity to the upper-level of this criterion, then its intensity should record a value of 1.

According to Sharareh, the intensities and the weights are obtained from a pairwise comparison matrix of qualitative grades, which is built using expert opinion [16], [17].

The weight of each grade is obtained in (1):

$$v_i = \frac{(\prod_{j=1}^{5} a_{ij})^{\frac{1}{5}}}{\sum_{i=1}^{5}(\prod_{j=1}^{5} a_{ij})^{\frac{1}{5}}} \quad (1)$$

where i = 1 to 5 and j = 1 to 5.

The intensity of each grade is obtained in (2):

$$Intensity = \frac{v_i}{\max(v_i)} \quad (2)$$

where i = 1 to 5.

TABLE II. PAIRWISE COMPARISON MATRIX FOR THE GRADE OF THE CRITERION 'FUNCTION'.

	Life saving	Therapeutic	Diagnostic	Analytic	Misc.
Life saving	1.00	5.00	6.00	8.00	9.00
Therapeutic	0.20	1.00	1.60	1.40	1.80
Diagnostic	0.17	0.63	1.00	1.25	1.50
Analytic	0.13	0.71	0.80	1.00	1.29
Misc.	0.11	0.56	0.67	0.78	1.00

Table II shows the pairwise comparison matrix for the grades of the first criterion, 'Function', as assigned by expert opinion. Using the above table and formulas, we can calculate the intensities and the weight for the criterion 'Function'. We listed the results in Table III, using (3) and (4):

$$a = (\prod_{j=1}^{5} a_{ij}) \quad (3)$$

$$b = (\prod_{j=1}^{5} a_{ij})^{\frac{1}{5}} \quad (4)$$

Table III. CALCULATING THE INTENSITIES OF THE CRITERION 'FUNCTION'.

	a	b	v_i	Intensity
Life saving	2160.00	4.64	0.62	1.00
Therapeutic	0.81	0.96	0.13	0.21
Diagnostic	0.20	0.72	0.10	0.16
Analytic	0.09	0.62	0.08	0.13
Miscellaneous	0.03	0.50	0.07	0.11
$\sum_{i=1}^{5} b = 7.45$				

In our model, we discarded the sixth criterion, which is "Recalls and Hazards" from the study, as it is not available in the hospital where the study was done. Hence, we distributed 0.16, the weight of recalls and hazards, equally on the other criteria by adding 0.032 on each of the five criteria ($0.16 \div 5 = 0.032$). For example, the weight of the criterion "Function" was 0.45. After adding 0.032 it becomes 0.482.

IV. MISSIONS

Assessment of medical equipments requires five consecutive missions, where each one deals with a criterion. In the first mission, we classify the function of the equipment. In the second mission, we specify the mission criticality of the equipment through its rate of utilization and availability of alternative devices. In the third mission, we identify the age of the equipment. In the fourth mission, we investigate the risks on the equipment by looking into its failure frequency, detectability of the failure, and the failure consequences. Finally, in the fifth mission, which is the last one, we study the maintenance requirements of the equipment. The core of each mission is gathering data. Before going through any of the missions, we made up an identity card for each equipment by filling up its name, its serial number, its brand, and its manufacturer. This information will not affect our study, but the aim is rather to identify each equipment to make sure that there is no overlapping in case the equipment is shared among the units and departments.

The intensities are obtained from a pair-wise comparison of grades; experts construct these grades as elaborated by the work of Taghipour.

First Mission: In the first mission, we classified the function of each medical equipment into five categories: lifesaving, therapeutic, diagnostic, analytic, and miscellaneous according to the classification developed by Fennigkoh, Smith, and Dhillion [18]. The weight of the

TABLE IV. THE INTENSITIES OF THE FUNCTION OF THE EQUIPMENT.

Function (0.482)				
Life saving	Therapy	Diagnostic	Analytic	Miscellaneous
1.00	0.21	0.16	0.13	0.11

function and the intensity of each category are shown in Table IV.

Second Mission: This mission accomplishes the second criterion; mission criticality of weight (0.132) is divided into two sub-criteria: the utilization and the availability of alternative devices, as shown in Figure 2.

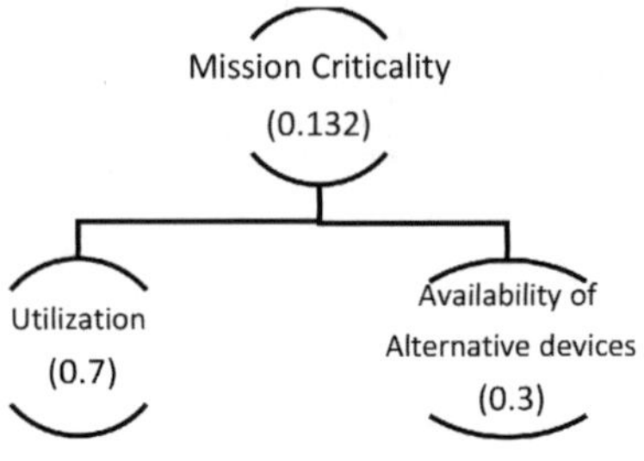

Figure 2. Hierarchy for mission criticality.

The usage of the device and its back-up devices identify the load of work on that device. Moreover, using the equipment excessively will increase the failure on the equipment [19]. In the first sub-criterion, utilization of a device is the total hours the device is used on average in a hospital (the unit can be defined as hours per day or days per week or weeks per year). In our proposed model, we considered the 'average hours a device is used per week' for the utilization criterion divided into three classes as shown in Table V.

TABLE V. THE WEIGHT AND INTENSITIES OF THE USAGE OF MEDICAL EQUIPMENT.

Usage hour/week (0.70)		
24≤	12≤x<24	<12
1.00	0.34	0.15

On the other hand, the availability of alternatives affects the mission criticality as it represents the number of similar or backup devices for one equipment. However, as the number of similar devices at hand becomes fewer because of lack of backup of the medical equipment, the risks on that equipment will increase. Furthermore, having several similar devices with low demand may also harm the device by affecting its performance from one side, and by costing the hospital regular preventive maintenance from the other side. The weight and the intensities of the availability of alternatives are shown in Table VI.

TABLE VI. THE WEIGHT AND THE INTENSITIES OF THE ALTERNATIVES.

Alternatives (0.30)		
≤1	1< x≤4	>4
1.00	0.34	0.20

Third Mission: The third mission deals with the third criterion, which is the age of the equipment. The age of the medical device is based on the actual age of a device and its predictable lifespan. In general, 10 years is the average lifespan for a medical device. The equipments are divided into five categories according to the actual age of the equipment divided by the lifespan as shown in Table VII. As the ratio approaches 1, the equipment is considered as old; otherwise, it is considered to be new as the ratio approaches zero. The age ratio is expressed in equation (5):

$$Age\ Ratio = \frac{Actual\ Age}{LifeSpan} \quad (5)$$

TABLE VII. THE WIGHT AND THE INTENSITIES OF THE AGE OF THE MEDICAL EQUIPMENT.

Age (0.092)				
>1	0.75< x≤1	0.5<x≤0.75	0.25<x≤0.5	0≤x≤0.25
1.00	0.67	0.43	0.17	0.12

Fourth Mission: The fourth mission addresses the fourth criterion, which is the risk of a device (of weight 0.192). In a patient-centric environment, managing risk is the top priority that occupies a worthy space under the umbrella of healthcare [20]. The risk of a device is the summation of all risks threatening patients. These risks can be estimated from the actual failures, which have occurred in that device, and are shown in the figure below.

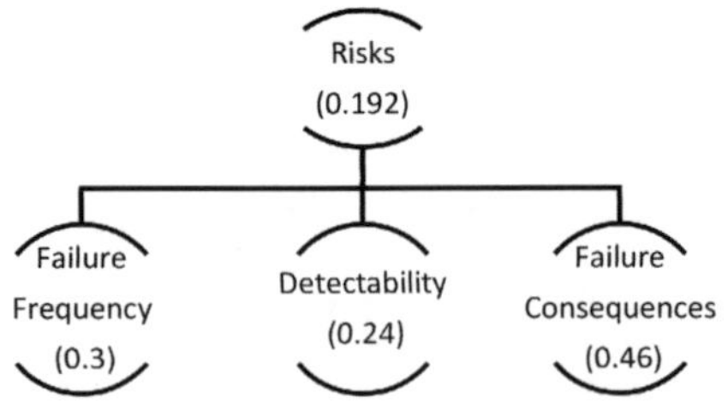

Figure 3. Hierarchy for risks on the medical equipment

Figure 3 illustrates the three sub-criteria of risks. The consequences associated with the risks of a device are assigned by the failure frequency, the detectability, and the failure consequences. These should be extracted or estimated from historical data and device maintenance archives [21].

The frequency of failure indicates how often the failure occurs. In order to capture this dimension, we considered four levels for the frequency of failure as outlined in Table VIII. If the failure is frequent, it means that the failure is likely to occur (several occurrences in 1 year). On the other hand, if the failure is occasional, it means that it probably will occur (several occurrences in 1 to 2 years). Then, if it is uncommon, this means that there is a possibility of occurrence (one occurrence in 2 to 5 years). Finally, if it is remote, it means that it is unlikely to occur (one occurrence in 5 to 10 years).

TABLE VIII. THE WEIGHT AND INTENSITIES OF THE FREQUENCY OF FAILURE.

Frequency of Failure (0.3)			
Frequent	Occasional	Uncommon	Remote
1.00	0.33	0.20	0.15

Failure detectability is the ability to detect a failure when it occurs. This is the most important criterion to assess harm [20]. We can detect the failure at many different levels. In our model, we used four levels of detectability. The failure maybe detected by error, that is when the equipment stops working, or by inspection during the regular preventive maintenance rounds, it might be visible by naked eye or it can be detected by self-announcement, as summarized below in Table IX.

TABLE IX. THE WEIGHT AND INTENSITIES OF THE DETECTABILITY.

Detectability (0.24)			
Error	Inspection	Visible	Self-announcement
1.00	0.33	0.20	0.13

The failure consequences, of weight (0.46) deals with the safety and the environment where we discuss the effect of the failure on the patient and the staff [22]. The failure of the medical equipment may harm the patient at different levels. It may cause death in extreme cases, injury in which it may disable the patient, inappropriate therapy, misdiagnosis, which makes the situation worse or failure, which may cause a delay in the treatment. Finally, in some other situations, it may cause nothing. The intensities of those failures are summarized in Table X.

TABLE X.THE WEIGHT AND INTENSITIES OF THE FAILURE CONSEQUENCES.

Failure Consequences (0.46)				
Death	Injury	Inapp. Therapy or misdiagnosis	Delay in treatment or diagnosis	Non
1.00	0.34	0.21	0.14	0.09

The risk value can then be estimated as a function of frequency, consequence, and detectability for each failure mode. As a result, the risk of the device is the total risk of all its failure modes.

Fifth Mission: The last criterion, which is the fifth one where we studied the maintenance requirement for every medical equipment, is covered in the fifth mission. The availability of the medical equipments should be based on maintenance history and the maintenance requirements [23]. According to Fennigkoh and Smith [24], equipment that is predominantly mechanical, pneumatic, or fluidic often requires the most expensive maintenance. A device is considered to have an average maintenance requirement if it requires only performance verification and safety testing.

Equipment that receives only visual inspection, a basic performance check, and safety testing is classified as having minimal maintenance requirements. We defined each of these classes as high, medium, and low with their corresponding intensities as shown in Table XI.

TABLE XI. THE WEIGHT AND INTENSITIES OF THE MAINTENANCE REQUIREMENTS.

Maintenance Requirements (0.102)		
High	Medium	Low
1.00	0.50	0.17

Identifying the main and the sub-criteria of each equipment allows us to determine their relative importance according to their goal or their upper level criterion using Saaty's eigenvector technique – a mathematical technique that assigns a total score value for each medical device under study. This technique is used in multi-criteria decision-making missions [25]. This total score is generated from the weights and the intensities of those medical devices from the matrix of criteria and sub-criteria observed [26], [27]. Figure 4 shows a schematic diagram of the main and sub criteria of the evaluation test.

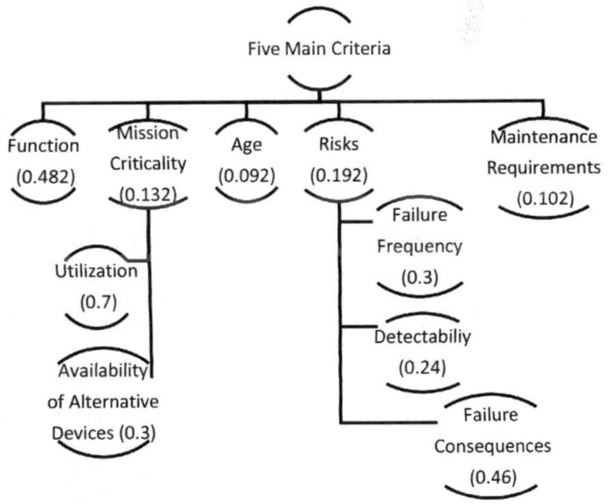

Figure 4. Hierarchy for the five main criteria.

After filling the questionnaire for each equipment, we can compute the scores using the assigned weights and intensities. The total score of each equipment is the

$$Total\ Score = w_{function} \times i_{function} + w_{age} \times i_{age} + w_{mission\ criticality} \times [w_{usage} \times i_{usage} + w_{back-up} \times i_{back-up}] + w_{risks} \times [w_{failure\ consequences} \times i_{failure\ consequences} + w_{detectability} \times i_{detectability} + w_{frequency} \times i_{frequency}] + w_{maintenance\ requirements} \times i_{maintenance\ requirements}$$

summation of the weight × intensity for the five criteria, which is illustrated in (6) below:

$$Total\ Score = \sum_{j=1}^{5} w_j s_{ij} \qquad (6)$$

where "w" is the weight of each criterion "j" = 1, 2 … 5 and "i" is the intensity of each class.

At this stage, the total score for the equipments is listed in descending order from the highest score to the lowest score. This rank helps in calculating the normalized score value that indicates the relative criticality of a device compared to other devices. Therefore, the Normalized Score Value of each equipment is expressed in equation (7):

$$NSV = \frac{Total\ score\ of\ each\ device}{Maximum\ total\ score} \qquad (7)$$

The aim of this study is to prioritize the medical devices according to their criticality. To do so, we have to calculate the transformed score value from the above procedure, which can be used for prioritizing or ranking of devices. The TSV depends on the NSV of each device involved in the model, and on the minimum and the maximum scores that could be achieved. The TSV plays an important role in assessing the medical equipments according to a percentage. In our proposed model, devices can have a total score between (0.1257592, 1.0) where score 1.0 is for a device, which gets the highest intensity when assessed against every single criterion, and 0.1257592 is obtained when the device gets the lowest intensity from all criteria. The calculation is shown below using (6):

$$\begin{aligned} &MinimumTotalScoreValue \\ &= (0.482 \times 0.11) \\ &+ 0.132[(0.7 \times 0.15) + (0.3 \times 0.2)] \\ &+ (0.092 \times 0.12) \\ &+ 0.192[(0.3 \times 0.15) + (0.24 \times 0.13) \\ &+ (0.46 \times 0.09) + (0.102 \times 0.17) \\ &= 0.1257592 \end{aligned}$$

Similarly, we can calculate the maximum value using equation (6):

$$\begin{aligned} &MaximumTotalScoreValue \\ &= (0.482 \times 1) \\ &+ 0.132[(0.7 \times 1) + (0.3 \times 1)] \\ &+ (0.092 \times 1) \\ &+ 0.192[(0.3 \times 1) + (0.24 \times 1) + (0.46 \\ &\times 1) + (0.102 \times 1) = 1 \end{aligned}$$

As it was expected to be, the maximum score value recorded a total score 1. This is because the intensity of each criterion and sub-criterion is the highest.

However, the total scores of devices can be used as absolute measurements for classification. The ranking of the medical devices can be done according to the normalized score value, however, for a better reading we can express the results in percentage, and so the normalized score value can then be mapped to (0, 100%) Transformed Score Value using the following equation:

$$TSV = \frac{Score\ value - Minimum}{Maximum - Minimum} \times 100 \qquad (8)$$

The whole process of doing these calculations is summarized in Table XII.

TABLE XII. THE TRANSFORMED SCORE VALUE.

Equip.	Total Score	NSV	TSV
↓	Descending order	$\frac{(Total\ score\ of\ each\ device)}{max.\ Total\ score}$	$\frac{Score\ value - minimum}{maximum - minimum} \times 100$
			$\frac{Score\ value - 0.1257592}{1.00 - 0.1257592} \times 100$
			$\frac{Score\ value - 0.1257592}{0.87311} \times 100$

As an example, let us apply our model on the monitors of ICU. Starting with the first mission, the monitor is classified as diagnostic equipment (intensity = 0.16), then the score of the function can be calculated as follows: $w_{function} \times i_{function}$= 0.482 × 0.16=0.07712

In the second mission, we found that the usage of the monitor is more than 24 hours per week (intensity = 1). Besides, our investigation showed that there are more than four back-up monitors in the ICU (intensity = 0.2) and hence the score of the mission criticality can be calculated as follows:

$$w_{mission\ criticality} \times [w_{usage} \times i_{usage} + w_{back-up} \times i_{back-up}] = 0.132 \times [0.7 \times 1 + 0.3 \times 0.2] = 0.10032$$

In the third mission, we checked for the age of the monitor and the result is obtained below:

$w_{age} \times i_{age}$= 0.092 × 0.67=0.06164

In the fourth mission, we examined the risks on the monitor of the ICU through the three sub-criteria of risks; frequency of failure, detectability, and failure consequences. The failure on the monitor is frequent so its intensity is high (intensity = 1), this failure is detected by error (intensity = 1), and the consequences of that failure results an inappropriate therapy or misdiagnosis (intensity = 0.21). Hence, the risks on the ICU monitor scores:

$w_{risk} \times [w_{failure\ frequency} \times i_{failure\ frequency} + w_{detectability} \times i_{detectability} + w_{failure\ consequences} \times i_{failure\ consequences}]$=0.192 × [0.3 × 1 + 0.24 × 1 + 0.46 × 0.21] = 0.1222272

Finally, coming to the fifth mission, the maintenance requirements on the monitor of the ICU is low (intensity = 0.17). The score of the maintenance requirements is:

$w_{maintenance\ req.} \times i_{maintenance\ req.}$
= 0.102 × 0.17 = 0.01734

Using equation (6), we can substitute the intensities and weights for the monitor of ICU, as follows:

TSV= 0.07712 + 0.10032 + 0.06164 + 0.1222272 + 0.01734 = 0.3786472

As illustrated in Table XII, in order to determine the normalized score value of the monitor in the ICU, we should compute all the total scores of the devices under study to find the maximum total score.

Similarly, we computed the total score for all the equipments under study in which we obtained a list of total scores. Among those scores, the defibrillator scored the highest value (Total score = 0.470601297). Using the total score of the defibrillator as the maximum score, we got the normalized scores for all other equipments. Since the defibrillator has the maximum total score, its NSV is 1.

$$\text{NSV (defibrillator)} = \frac{0.49337955}{0.49337955} = 1$$

$$\text{NSV (ICU Monitor)} = \frac{Total\ Score(ICU\ monitor)}{Max.Total\ Score} = \frac{0.3786472}{0.49337955} = 0.76745621$$

$$\text{TSV (defibrillator)} = \frac{NSV(Monitor\ ICU) - 0.1257592}{0.8742408} = \frac{0.76745621 - 0.1257592}{0.8742408} = \frac{0.64169701}{0.8742408} = 0.734004876$$

Following the same procedure, we obtained a long list of medical equipments with their transformed score values (TSV).

The obtained list of medical equipments can be classified into many categories according to the prioritizing plan of the hospital, which is related to the budget assigned by the decision makers. In our study, the criticality of a device is classified into three categories in which a transformed score value should belong. The first category is for those which should be replaced urgently. The second one for those which should be replaced after a year and a half (their replacement can be limited to a deadline defined by the hospital according to their budget). The third one is for those which are still functioning normally and can work for several years to come. Using the transformed score value, we can sort the medical equipments according to their urgency using Table XIII.

TABLE XIII. THE CRITICALITY OF A DEVICE FROM THE TRANSFORMED SCORE VALUE.

Criticality class	Transformed Score Value	Maintenance Strategy
High	70% <TSV≤ 100%	To be changed urgently
Medium	30% < TSV ≤ 70%	To be changed after a year and a half
Low	0% ≤ TSV ≤ 30%	To be changed after three years

Before knowing the final scores of the medical devices under study, we cannot assign the suitable thresholds for the evaluation classes. Many factors contribute to the classification of the evaluation classes. One of these factors is the result obtained in the TSV list, the load of work in the hospital, and the rate of in-patient. On the other hand, thresholds should be adjusted after applying the model of inventory of a hospital [28] and studying the obtained transformed score values. The classes in the table below are suggested by Taghipour[16].

Generally speaking, we can classify equally the equipments of a hospital in the order of their urgent need for replacement. If the equipment's score is between 70% and 100%, it means that the equipment should be replaced immediately. If its score ranges between 30% and 70%, then the equipment should be replaced after a while. Finally, if its score is less than 30%, this means that the replacement of the equipments does not need to happen in the near future. This was an example on how to classify the results in a hospital. Keep in mind that we can consider other intervals to sort the tested devices according to the hospital's financial contribution.

The decision makers at the hospital, where the study was applied, set the interval of criticality to be between 65% and 100%. Therefore, referring to our example, the monitor in the ICU scores 73.4%, this belongs to the first class of criticality and should be replaced immediately.

VI. CASE STUDY

In this section, we are going to apply the assessment model on the medical equipments found in some units of a Lebanese hospital in order to evaluate them for an updating program.

The professional work hours needed to apply such a model varies from hospital to another. It depends on the number of units running in the hospital, which implies the variety of sections and fields we are dealing with, and the rate of in-patients in the hospital, which implies the load of work on the professionals and so their availability to cooperate with the ongoing study.

In this case, the study was launched in a public university hospital that includes 430 beds (in-patient treatment), 14 operation rooms, 15 units, and over than 1200 medical items. Besides, the medical staff was busy all around the clock and so scheduling appointments was barely possible. Over and above, because hospitals operate 24 /7 all over the year, emergencies might take place at any time forcing us to reschedule for another appointment with the concerned doctor. On the other hand, the team was built up of four full-timers who dedicated four months working six days a week, six hours a day. This ended up with dedicating about 860 working hours.

All the above factors contribute to the achievement of the evaluation model in a specific duration. Consequently, we cannot define a common timeline for the application of this assessment mission, but we can set definite milestones for the process of whole project.

The whole process is depicted in Figure 5, where firstly, it is very important to start the study by getting introduced to the environment that we are going to work in, to know the units, the technicians, and most importantly, the medical equipments – the core of our study. Based on this step, we can build up our team, and assign the missions for the team members.

Once this stage is attained, the team should be ready to launch the investigation officially, by assigning an opening session that should be held with the presence of the chief of the Biomedical Engineering Department, the biomedical engineers, all the people in charge in all units, the technicians, and every person who work in contact with the medical equipments. The purpose of this session is to introduce the medical staff to the aim of our study and the importance of their cooperation and contribution in every single information they might offer. After the opening session, we need to check on the equipments by launching excessive rounds on floors.

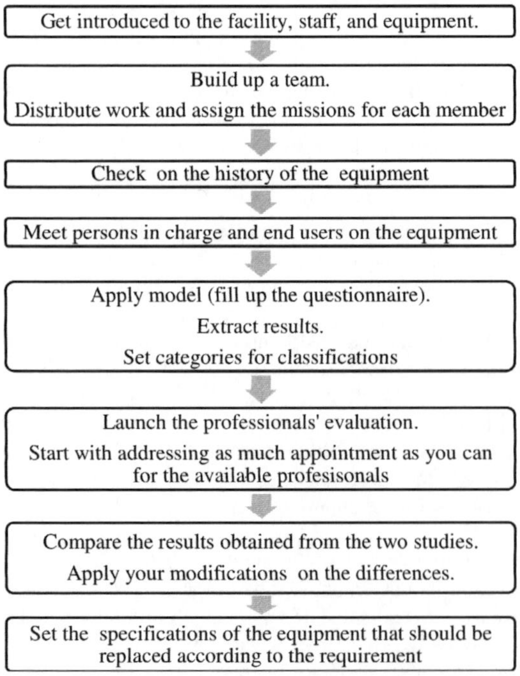

Figure 5. Flow chart showing the steps of the whole study.

This step lead to start filling up the data for the scientific evaluation by questioning the end users on floors. After gathering the required information, we can apply the scientific model theoretically to end up with a list of TSVs. With the help of the Biomedical Engineering Department, the suitable categories for the replacement plan can be classified. According to the list of results, scheduling appointments for the professionals' evaluation can be started.

At this stage, it is extremely important to respect the nature of work of the professionals that we are dealing with, as their job might require a lot of emergencies. Considering that they are on-call workers, rescheduling several times might occur.

At this level, two results emerge: the match between these two findings indicates the precision of the work, one being the user of the equipments and the other being the calculated total score value (TSV) based on the model. However, it is normal to highlight some differences in the ranking of the two results. We can align those expected differences by applying some modifications with the help of the chief of Biomedical Engineering Department, so that to end up with one unified list of equipments nominated for replacement according the classification defined before.

Finally, the specifications of the ordered equipments according to the requirements of the professionals can be addressed. This declares the end of the study, but they went further, by contacting medical companies for the best of their offers. Besides, they gathered the received offers and handled them to the biomedical engineering department and they took it from there. The whole process is illustrated in Figure 5.

In this study, in order to do the scientific evaluation, the researchers have chosen the Dialysis and the Critical Care Units as a sample study. These units normally have the Intensive Care Unit (ICU), which is dedicated to treat patients, who are seriously ill. Besides, we have the Coronary Care Unit (CCU), where patients with a pacemaker, intra-aortic balloon pump, or with cardiac telemetry are treated. Moreover, there is the Cardiac Surgical Unit (CSU), where patients having open-heart, lung, or vascular surgery are recovered. In addition, the Neonatal Intensive Care Unit (NICU) is the unit that monitors the neonates, who are facing newborn problems. Finally, the Pediatric Intensive Care Unit (PICU) is the intensive care specialized for pediatrics.

In these units, the team dealt only with the medical equipments that are in direct contact with the patient and that might affect the patients' safety. The equipments that are related to the ward medical equipments, housekeeping equipments, mortuary equipments, general furniture and accessories, are considered as not urgent at all so they are kept away from the study with "to be replaced after a determined period of time" as a general status. The team gathered the required data for 324 equipments distributed over 35 different items by questioning 24 of the end users, most of them are physicians, nurses, and technicians, and five biomedical engineers who are responsible for the regular maintenance of those equipments. After the feedback of the above professionals, the results listed in Table XIV were compiled. As one can notice from the obtained results, the same item may record different grades when used in different units. For example, the ECG in the ICU records a grade of 57.35 whereas the ECG in the NICU recorded a grade of 42.88. These two different grades for the same item reflect the different mode of use and different urgency of that equipment at its unit.

TABLE XIV. SCORES AND GRADES FOR EACH ITEM.

Nb.	Name	Normalized Score	Transformed Score (%)
1	Defibrillator	1	100
2	Blood Gas system	0.84776143	82.563644
3	Pulse Oximeters	0.83167396	80.721096
4	Infusion pump (CCU)	0.80675075	77.866563
5	Monitor (ICU)	0.76745621	73.400487
6	Oximeters	0.76154527	72.689039
7	Syringe pump (ICU)	0.75121329	71.505686
8	Dialysis	0.74112659	70.350424
9	Monitor (CCU)	0.69471468	65.034724
10	**Monitor (Endoscopy)**	**0.68912238**	**64.394221**
11	**Syringe pump (PICU)**	**0.68817904**	**64.286177**
12	Refrigerator Pharmacy)	0.68541098	63.969142
13	Monitor (Dialysis)	0.68198543	63.576804
14	Incubator(PICU)	0.67304633	62.552981
15	Refrigerator (NICU)	0.66928522	62.122209
16	Refrigerator (PICU)	0.66928522	62.122209
17	Incubator (mobile)	0.66080627	61.151088
18	Syringe pump (floors)	0.65123852	60.055265
19	Incubator (Therapeutic)	0.64902559	59.801811
20	ECG (ICU)	0.62764624	57.353167
21	Fetal Monitor	0.62328606	56.853783
22	x-ray (ICU)	0.60806278	55.110213
23	Ultrasound Unit	0.59212049	53.284293
24	Reanimation & warming table	0.58451048	52.412695
25	ECG (CCU)	0.57761153	51.622536
26 *	**ECG (Dialysis)**	**0.56972919**	**50.719747**
27	**Infusion Pump (NICU)**	**0.56173091**	**49.80368**
28	**Infusion Pump (floors)**	**0.54855002**	**48.294032**
29	Lactina Electric pulse	0.52413795	45.498041
30	CPR	0.52413795	45.498041
31	ECG (NICU)	0.50133159	42.885958
32	Fetal Doppler	0.48090312	40.546222
33	Incubator (Delivery Unit)	0.43585151	35.386322
34	Otoscope	0.39837394	31.093899
35	Bair Hugger	0.24364574	13.372397

At this time, the team is able to make an educated decision. According to the hospital's budget, and with the help of head of Biomedical Department, three consecutive categories were set, each bounded within an interval of grades that matches the updating strategic plan of the hospital. In this case study, the three categories were assigned based on a strategic updating plan set by the hospital. The decision makers at that hospital were planning to spend a certain budget after the results of the study, and another amount after a year and a half and finally another amount after three years. Consequently, the coming three missions, as seen in Table XV, were set; the equipments with grades between 65% and 100% should be replaced directly. Those with grades between 50% and 65% can be replaced after a year and a half, and finally, those with grades below than 50% can be replaced after three years from the first updating plan.

Table XV. THE CRITICALITY OF A DEVICE FROM THE TRANSFORMED SCORE VALUE - CASE STUDY

Criticality class	Transformed Score Value	Maintenance Strategy
High	65% $<$TSV$\leq$ 100%	To be changed urgently
Medium	50% $<$ TSV $\leq$ 65%	To be changed after a year and a half
Low	0% $\leq$ TSV $\leq$ 50%	To be changed after three years

Based on the above three ranges of grades, one can summarize the three groups of medical equipments as shown in Table XVI.

TABLE XVI. RESULTS FOR THE UPDATING PLAN.

To be changed urgently	**To be changed after a year and a half**	**To be changed after three years**
High 70% <TSV≤ 100%	**Medium** 50% < TSV ≤ 70%	**Low** 0% ≤ TSV ≤ 50%
Defibrillator	Monitor (Endoscopy) *	Infusion Pump (NICU) *
Blood Gas System	Syringe pump (PICU)	Infusion Pump (floors) *
Pulse Oximeter	Refrigerator (Pharmacy)	Lactina Electric pulse
Infusion pump	Monitor (Dialysis)	CPR
Monitor (ICU)	Incubator(PICU)	ECG (NICU)
Oximeters	Refrigerator (NICU)	Fetal Doppler
Syringe pump (ICU)	Refrigerator (PICU)	Incubator (Delivery Unit)
Dialysis	Incubator (mobile)	Otoscope
Monitor (CCU)	Syringe pump (floors)	Bair Hugger
	Incubator (Therapeutic)	
	ECG (ICU)	
	Fetal Monitor	
	x-ray (ICU)	
	Ultrasound Unit	
	Reanimation & warming table	
	ECG (CCU)	
	ECG (Dialysis)**	

From the above table, the hospital can conduct a plan of three phases for upgrading its medical equipments. Each phase would be set along a period of time, according to the procurement process and the installation program that should be launched for each equipment based on its requirements.

VII. PROFESSIONALS' EVALUATION

To make sure that the obtained results are correct and the devices that are changed meet the hospital's requirements, a survey was designed that questions the physicians, the technicians, and the nurses, where the questions were about the equipments that should be replaced directly. Interestingly, a list that matched the above one, which was achieved by the scientific study, was obtained.

Since the aim of this evaluation is to check the validity of the scientific approach used, it was assumed that there would be some error within a short interval $\pm \varepsilon$. Consequently, the team expected to make some fine-tuning on the results obtained, especially for the equipments with grades close to the boundaries chosen. Hence, according to the evaluation of the professionals and of the persons on charge, one can add or remove an epsilon ($\pm \varepsilon$) to the grade of the equipments whose score is close to the boundaries of the intervals chosen, where epsilon is the discrepancy between the end user recommendation and the findings of our model. For example, the endoscopy monitor recorded 64.39%, so it should belong to the second category in the updating program. As we can notice here, even though 64.39% is very close to 65 but we cannot move it to the first category in the program. However, if the professionals, who work on the endoscopy monitor, recommended an urgent replacement for this monitor for specific reasons to be discussed, we can move it to the first category and add it to the equipments to be replaced directly. This will not be considered an error since the endoscopy monitor is on the boundary so it may belong to both categories.

On the other hand, this step played an important role on checking the accuracy of the results obtained. It also served on checking if the professionals recommend any additional new equipment that was not available at the hospital, and consequently not included in the study done.

Moreover, questioning the end user helped in setting out the desired specifications and requirements of the equipments to be replaced. To do so, two forms were designed: the first one is a general form to check if the units being questioned needed any equipment to be replaced, or if they recommended any new medical technology that they saw might raise the level of the medical care at their unit. The second form is a specific one to highlight the requirements and specifications of the desired equipments.

1) The General Form:

The aim of this form is to specify the list of equipments needed in each unit as suggested by the end users. In this form, the name of the requesting person and his/her position is identified, to make sure that his/her job description empowers him/her to suggest the medical technology used in the unit. The requester should provide his/her extension number so that the team can refer to him/her any time a clarification or further information is needed.

The main part of this form contains a table of three columns, the first column to list the name of the new equipment, the second column to specify if the named equipment is replacing an old one already found in the unit, and the third one to specify some details concerning the old equipment that is being replaced. The general form is shown in Appendix A (CED-F-03).

2) The Specific Form:

This form includes detailed information about each equipment named in the first form. In this form, care is taken about some other information related to the replacing decision, such as the clinical application of the named device, the accessories that should be provided with that device. Attention is also paid to underline some suggested brands and models for that device with some external notes if required. Identifying a specific brand for each equipment serves in the purchasing process while making the decision among many several offers and budgetary quotations.

If the equipment under discussion is replacing an old one, then the third part of the form titled: "Old Equipment Identification" will be filled.

Finally, in the last part of this form that has many sections, some specifications concerning the new equipments are highlighted. The end user is asked to clarify the status of the old equipments that should be replaced; whether it is of old technology, obsolete, out of order, or affects patients' safety.

The other section of this part is designed to make sure that the nominated equipment satisfies the suitable conditions to meet the international standards for medical equipments. One of these conditions is that the equipment should comply with the actual clinical standards. Moreover, the requester should mention whether his/her recommendation is cost effective or not. Besides, some other specifications will increase the acceptance of the proposal, such as whether the equipment of the new brand results in a better patient care.

Following the above section, the frequency of use of the named equipment as well as we asked for the requirements of replacing it was probed. In this manner, a check is made whether this replacement requires new installation or whether it requires training of the staff.

On the other hand, the professionals have to justify the choice of the new suggested equipment. At this stage, it is preferable to list two to three other accredited hospitals using the proposed technology; this will empower the suggestion in hand. Finally, the value of each request according to the emergency status should be specified:

i. High – so that the equipment should be replaced immediately without any delay
ii. Medium – means the replacement is critical but can be delayed for a short period of time
iii. Low – means that the equipment has no harm on patient and the replacement can be postponed for a longer period of time.

The "Specific Form" is shown in Appendix B (CED-F-04).

3) Experimental Versus Theoretical Evaluation:

The "Professionals' Evaluation" served in checking the accuracy of the results obtained from the scientific study. Although this step made a slight change in the three categories obtained, it did not induce a fundamental change in the list of equipments chosen. As discussed in the example above, the endoscopy machine recorded a score close to the lower boundary of the first category, and hence can be kept in the second category, or can be move to the first one based to the evaluation of the main users of this machine.

After consulting with the chief of the endoscopy unit, and in the presence of the physician and technician working there, the above-mentioned two forms were filled. The results showed that the endoscopy monitor is seriously facing some technical problems and some unexpected failures form time to time, and hence needed to be replaced immediately.

Based on that evaluation, the grade of the endoscopy machine requires an ε upward, so that the grade becomes $64.39 + \varepsilon$, and hence can be included in the first category, raising the status of the endoscopy to an urgent call. The same thing is applied on the other equipments with grades close to the upper and lower boundaries.

At the end of this evaluation, some modifications were deemed necessary. Actually, these modifications were expected, as they were related to the medical equipments with grades close to the upper and lower boundaries of the three categories.

The endoscopy monitor and the syringe pump recorded 64.39 and 64.28, respectively, so they should belong to the second category. However, according to the evaluation of the end users and the technicians working on them, it was found that they constitute risks and should be replaced directly. Considering their grades, an ε can be added to each one and move them to the most critical category. Similarly, the professionals recommended a very soon replacement for the infusion pumps for the floors and the Neonates Intensive care Unit. Consequently, the infusion pump was moved to the second category with those medical equipments to be replaced after a year and a half.

On the other hand, the person in charge of the dialysis unit found that the ECG at their unit is functioning normally and there is no load on it, so the replacement of this equipment can be postponed for a longer period of time. However, the ECG recorded 50.71, which places it in the second category, but the recommendation of the second evaluation moves it to the third category of replacement.

It is worth noting that the questionnaire was filled by thirteen professionals in charge of the seven units under study, as shown in table XVII.

Most of those professionals' suggestions were the same as those obtained theoretically by the study. However, there was a slight mismatch between the results obtained by the

Table XVII: NUMBER OF PROFESSIONALS WHO ANSWERED THE QUESTIONNAIRE.

Unit	Number of professionals
Delivery Unit	2
Dialysis Unit	3
ICU	2
CCU	2
CSU	2
NICU	1
PICU	1

model and the recommendation of the persons in charge and this refers back to the reason that the user tends to recommend new technologies when it comes to his/her choice, whereas the quantitative results show reasonable real values. To better assess the correlation between the scientific and the professional's evaluation, let's denote by R the non-zero ratio of the scientific results to the professional one.

$$R = \frac{X+1}{Y+1} \quad (9)$$

where X is the number of equipments that should be replaced as obtained from the scientific model, and Y is the number of equipments that should be replaced as recommended by the professionals' evaluation. Getting R=1 means that there is a match between the equipments obtained by the scientific evaluation and those nominated by the professionals.

Figure 6 shows the ratio R for the seven units under study in the three assigned classes. As it is clearly shown in the graph below, there is a good match between the equipments ordered upon the scientific evaluation and those ordered by the professionals' evaluation and this is the case in the three critical units, ICU, CCU, and CSU. On the other hand, the criticality of some medical equipments was increased, as recommended by the professionals, such as the endoscopy monitors and the syringe pump for the PICU, which were classified in middle class by the model, yet were recommended by the professionals for an urgent replacement, so their criticality was boosted to the first class. Similarly, the infusion pump was classified theoretically, as third class in criticality, however, the professionals recommended an earlier replacement.

On the contrary, the ECG was moved from the second to the third class, as the professionals showed an acceptable satisfaction of its work as compared to other devices.

Referring to Table XIII, one can notice the four medical equipments that required a $+\varepsilon$ to their obtained grades, marked in bold face. Besides, the ECG that required a $-\varepsilon$ to its grade is also bolded.

At the end of the evaluation, the four equipments that required displacement from their category to an upper one were designated with a single asterisk (*), and those requiring moving to a lower category with double asterisks (**), as shown in Table XVI. The professionals' evaluation did not fundamentally change the results obtained from the scientific evaluation done before; yet, it introduced some little modifications – or fine tuning – to the equipments at the boundaries of the categories chosen. We cannot consider this change an error since it was expected to occur. Consequently, the professionals' evaluation served as an experimental tool to test the validity of the theoretical evaluation.

VIII. Budgetary Qoutations

After updating the results, a list of equipments that should be replaced directly was compiled. At this phase, several companies were contacted asking for a budgetary quotation for each ordered equipment.

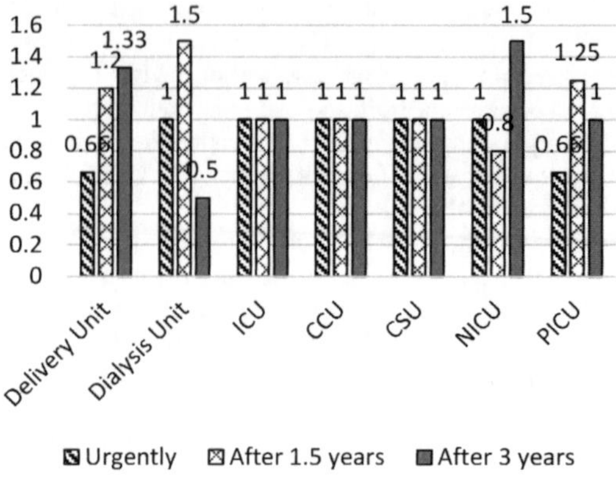

Figure 6. Ratio of the scientific evaluation to the professionals' evaluation

In the request, the specifications of the equipments needed to be replaced were identified. The description of that equipment and the unit to which it belongs were mentioned, as well as the minimal requirements and the quantity needed. Actually, mentioning the number of the ordered item was very effective from a budget aspect. Offers streamed positively with higher quantities of each item.

Besides, the ordered equipment should comply with the international standards for medical equipments, such as FDA approval or CE certificate. In addition, the electrical specifications were matched with the British standards, as recommended by the hospital. Finally, a budgetary price for the equipment and all the spare parts that may function with it were sought, as depicted in Appendix C.

After collecting the budgetary quotations received from several companies, the team listed them according to the best specifications and offers provided. Several offers from different companies were received for each piece of medical equipment, and that required some excessive meetings for the decision-taking committee at the hospital to come up with a verdict. At this level, the list was raised to that committee to decide on the equipments that best suit their demand.

Whenever the decision is made, the biomedical engineering department can launch the procurement process for purchasing the chosen medical equipments.

IX. Conclusion

Medical equipment is a critical interface between the patient and the diagnosis, the treatment, or the rehabilitation process. It provides an opportunity for a better medical service. Consequently, medical devices are expected to operate in the required way providing the ultimate results of accuracy, safety, and reliability for an efficient and healthy contribution. As such, this study provides a new model for assessing the life of medical equipment based on its actual usage, and not only speculated based on its suppositional lifespan. This method would result in a more accurate scheme that would most probably extend the life and usage of the equipments thus resulting in substantial savings to the healthcare institution from one side, and would serve as an assessment tool based on a multi criteria decision-making approach from the other side.

Using such a model of evaluation, the wheel of change in the assessment of medical equipments can be turned to overreach several sectors in the world of machinery. Moreover, adapting an automated management system to monitor the evaluation of the medical equipments will be revolutionary move towards safety and efficiency. Furthermore, the proposal assessment approach would be further enhanced by using information technology, where the lifespan of the equipments may be monitored in real time. This can be addressed by integrating the equipments with information technology software and hardware through the usage of the internet. When done, precise and up-dated reports may be generated anytime and anywhere, to assess the present status of the equipments' lifespan.

X. APPENDICES

Appendix A:

Here is the general form for listing the medical equipment that should be replaced.

Evaluation of Medical Technology Platform And Updates at RHUH	CED-F- 03		
Medical Equipment Suggestion Form	Edition 1	Page 1/1	

This form is to be filled upon suggesting equipments. Kindly, fill in the details and attach any additional documents if needed.

Requester Identification	
Department/ unit:	Name:
Extension Number:	Position:
Date:	

New Equipments	Replacing an Old Equipment(Yes/No)	Old Equipments

Signature: ______________

Appendix B:

Here is the specific form for interpreting the reason of replacement and the specifications of the new recommended medical equipments.

Evaluation of Medical Technology Platform And Updates at RHUH	CED-F- 04	
Medical Equipment Suggestion Form	Edition 1	Page 1/1

This form is to be filled upon suggesting equipments. Kindly, fill in the details and attach any additional documents if needed.

End User Identification	
Department/ unit:	Name:
Extension Number:	Position:
Date:	

New Equipment Identification	
Equipment name:	Quantity Requested:
Clinical Application:	Sample brands and models suggested:
Accessories:	Notes:

Old Equipment Identification (if replacing)	
Brand:	Quantity Available:
Date of Purchase:	

Suggestion Justification
☐ The used equip. is of old technology/ obsolete ☐ The used equip. is out of order ☐ The used equip. affects the patient's safety
☐ The suggested equip. is cost effective ☐ The suggested equip. complies with actual clinical standards ☐ The suggested equip. is of better quality ☐ The suggested equip. is better for patient care ☐ Other reason:
Frequency of use:
Requirements: ☐ Training ☐ New Installation
Name of other hospitals using this new equipment: ______________
Who can work on this equipment? ______________

State of request: ☐ urgent		☐ normal
	☐ can be postponed	

Appendix C:

Here is the request as emailed to the companies for a budgetary quotation.

	Evaluation of Medical Technology Platform And Updates at RHUH	CED-F-05	
	Request for Budgetary Quotation and Specifications	Edition 1	Page 1/1

Item Identification	
Item Number	
Needed Item	
Description	
Department/Unit to be used in	
Minimal Requirement	
Quantity Needed	
Electrical Standards	B.S.
International Standards	FDA/CE or others to be specified
Special Requirements	Optional accessories shall be quoted separately

Remark: The supplier is kindly recommended to provide us with a budgetary price quotation, in addition to the technical specifications and details in hard soft copy (when possible).

Clinical Engineering Department

REFERENCES

[1] M. Aridi, B. Hussein, M. Hajj-Hassan, and H. M. Khachfe, "An Objective Program for the Evaluation of Medical Equipment for Updates and Maintenance," in GLOBAL HEALTH 2014: The Third International Conference on Global Health Challenges, Rome, Italy, August, pp. 24-28, 2014.

[2] J. C. Lerner, "https://www.ecri.org/About/Pages/history.aspx," ECRI Institute, 2014. [Online]. [Accessed 2014].

[3] IMDRF Medical Device Single Audit Program Work Group, "http://www.imdrf.org/about/about.asp#man," International Medical Device Regulators Forum (IMDRF), 19 February 2014. [Online]. Available: http://www.imdrf.org/docs/imdrf/final/consultations/imdrf-cons-assessment-outcomes-140402.pdf.

[4] ISO 13485, Medical devices. Quality management systems. Requirements for regulatory purposes, second edition ed., Switzerland: ISO copyright office, 2003.

[5] "http://www.fda.gov/AboutFDA/WorkingatFDA/FellowshipInternshipGraduateFacultyPrograms/DeviceEvaluationInternProgramCDRH/," 26 February 2013. [Online].

[6] M. Ridgway, "Analyzing Plannes Maintenance (PM) Inspection Data by Failure Mode Effect Analysis Methodology," Biomed Instrum Technol, vol. 37, no. 3, pp. 167-179, 2003.

[7] Wang et al., "Evidence-Based maintenance: Part I: Measuring Maintenance Effectiveness With Failure Codes," Journal of Clinical Engineering, vol. 35, no. 3, pp. 132-144, July/ August/ September 2010.

[8] B. Wang et al., "Evidence-Based maintenance: Part II: Comparing Maintenance Strategies Using Failure Codes," Journal of Clinical Engineering, vol. 35, no. 4, pp. 223-230, 2010.

[9] D. Mutia, J. Kihiu, S. Maranga, "Maintenance Management of Medical Equipment in Hospitals," Industrial Engineering Letters, vol. 2, no. 3, pp. 9-20, 2012.

[10] A. Jamshidi, S. A. Rahimi, D Ait-kadi, and A. Ruiz Bartolome, "Medical Devices Inspection and MAintenance: A Literature Review," in Proceedings of the 2014 Industrial and Systems Engineering Research Conference, Canada, 2014.

[11] A. Kirisits and W. K. Redekop, "The Economic Evaluation of Medical Devices," Applied Health Economics and Health Policy, vol. 11, no. 1, pp. 15 - 26, Febraury 2013.

[12] A. Khalaf, "Maintenance Model For Minimizing And Optimizing Cost-Effectiveness of Medical Equipment In Palestine," in World Congress , vol. 14, Springer Berlin Heidelberg, 2007, pp. 3649-3653.

[13] K. A. Mkalaf, P. Gibson, and J. Flanagan, "A Study of Current Maintenance Strategies and the Reliability of Critical Medical Equipment in Hospitals in Relation to Patient Outcomes," International Journal of Social, Education, Economics, and Management Engineering, vol. 7, no. 10, pp. 1383-1390, 2013.

[14] N. J. Stark, "Clinical Evaluation Report: The New Requirement," MD+DI, Chicago, 2010.

[15] N. Saleh et al., "A New Approach for Preventive Maintenance Prioritization of Medical Equipment," in XII Mediterranean Conference on Medical and Biological Engineering and Computing 2013, Seville, 2014.

[16] S. Taghipour, "Prioritization of Medical Devices," in Reliability and Maintenance of Medical Devices,

Toronto, 2011, pp. 24 - 59.

[17] T. L. Saaty, "How to make a Decision: The Analytic Hierarchy Process," European Journal of Operational Research, vol. 48, no. 1, pp. 9-26, 5 September 1990.

[18] B. S. Dhillion, "Medical Device Maintenance and Maintainability," in Medical Device Reliability and Associated Areas, Florida, CRC Press: Boca Raton, FL., 2000, pp. 179 - 195.

[19] L. Pruitt and J. Furmanski, "Polymeric Biomaterials for Load-bearing Medical Devices," Journal of Minerals, Metals, and Materials Society (JOM), vol. 61, no. 9, pp. 14 - 20.

[20] M. Viswanathan et al., "Assessing the Risk of Bias of Individual Studies in Systematic Reviews of Health Care Interventions," in Methods Guide for Effectiveness and Comparative Effectiveness Reviews, AHRQ Publication No. 10(12)-EHC063-EF, April 2012, pp. 69 - 97.

[21] J. Robson, P. Yeo, M. Riches, T. Carlisle, and N.l Kitto, "Risk Management and Biomedical Devices," in Engineering in Medicine and Biology 27th Annual Conference, Shanghai, China, 2006.

[22] D. V. Agarkhed, "healthcare.financialexpress.com," Tuesday 13 November 2012. [Online]. Available: http://healthcare.financialexpress.com/strategy/949-hazard-identification-and-risk-analysis. [Accessed Wednesday 2 April 2014].

[23] A.B. Khalaf, Y. Hamam, Y. Alayli, and K. Djouani, "The effect of maintenance on the survival of medical equipment," Journal of Engineering, Design and Technology, vol. 11, no. 2, p. 112, 2013.

[24] L. Fennigkoh and B. Smith, Clinical equipment management, JCAHO Plant, Technology, and Safety Management Series 2, pp. 5-14, 1989.

[25] T. L. Saaty, "How to make a Decision: The Analytic Hierarchy Process," European Journal of Operational Research, pp. 9 - 26, 1990.

[26] T. L. Saaty, "Decision-making with the AHP: Why is the principal eigenvector necessary," Europe Journal of Operational Research, vol. 145, no. 1, pp. 85-91, February 16, 2003.

[27] I. P. Smirnov, "Organizing function of medical equipment in public health," Biomedical Engineering, vol. 6, no. 6, pp. 343 - 346, 1972.

[28] M. D. Rossetti, D. Marek, S. Prabhu, A. Bhonsle, S. Sharp, and Y. Liu, "Inventory Mangement Issues in Health Care Supply Chains," Fayetteville, AR 72701, June 5, 2008.

Electronic Communication as a Tool to Reduce Elective Surgery Cancellations
A Case Study from Norway

Conceição Granja
Norwegian Centre for eHealth Research
University Hospital of North Norway
Tromsø, Norway
conceicao.granja@telemed.no

Kari Dyb
Norwegian Centre for eHealth Research
University Hospital of North Norway
Tromsø, Norway
kari.dyb@telemed.no

Stein Roald Bolle
Norwegian Centre for eHealth Research
University Hospital of North Norway
Tromsø, Norway
stein.roald.bolle@telemed.no

Gunnar Hartvingsen
Norwegian Centre for eHealth Research
University Hospital of North Norway
Tromsø, Norway
Department of Computer Science
University of Tromsø
Tromsø, Norway
gunnar.hartvigsen@telemed.no

***Abstract*—Surgery cancellations are undesirable in hospital settings as they increase costs, reduce productivity and efficiency, increase waiting lists, and directly affect patients. The elective surgery cancellation problem in a northern Norwegian university hospital is addressed. Based on observations and interviews conducted at the hospital, lack of information during pre-operative planning was identified as the main cause for elective surgery cancellations. The problems with the existing pre-operative process were identified, and a new process is proposed. By studying the pre-operative planning at the hospital, we have determined that part of the information flow can be moved to the patient at home through electronic communication. From the work presented herein, we conclude that the assessment information required during the pre-operative planning can be compiled in a personal health assessment questionnaire, and requested from the patient, at an earlier stage.**

Keywords-elective surgery cancellations; pre-operative planning; electronic communication; clinical process; context-awareness

I. Introduction

Surgical departments are simultaneously the major source of investment, and the greatest source of revenue for most hospitals [1][2][3] . However, it is known that between 10 and 40 % of elective surgeries are cancelled [2][4][5][6]. In western countries, up to 20 % of elective surgeries are cancelled on the day of surgery [7][8][9]. Furthermore, it has been reported that 50 % of these cancellations might be avoided [2][10][11].

Surgery cancellations are undesirable in hospital settings as they increase costs, reduce productivity and efficiency, increase waiting lists, and directly affect the patient [4][9][12]. Considerable resources are invested in maintaining operating theatres, and having surgeons and theatre staff available on an agreed schedule [2][13]. In spite of this, the cancellation rate of elective surgeries is high, especially in the public sector [10][14]. Cancellations can significantly inconvenience patients and their families [15][16]. It is also reported that patients may suffer psychological stress, and/or financial hardships [10]. Accordingly, cancellations are stressful and costly, with a high level of emotional involvement before surgery [2].

The causes for elective surgery cancellation are diverse and may be divided in two major categories: (a) hospital, and (b) patient related reasons, when considering who took the underlying decision to cancel. Hospital related reasons are the most frequent and encompass causes such as the unavailability of the surgical team [4][8][9], incomplete pre-operative study/preparation [8][17], lack of surgical/anaesthetic readiness [8][9], and lack of theatre time due to extended duration of scheduled surgeries [8]. On the other hand, patient related causes are mostly due to patient no-shows and refusal to undergo surgery [8][9][17]. It is argued that the majority of cancellations are due to information that existed prior to the day of surgery, but was not available when required [10][14][18][19][20][21].

In line with what is reported in literature, our site of research, the University Hospital of North Norway (UNN), has identified inadequate planning due to lack of information as a main cause for cancellations (Figure 1). The hospital has reported that more than 50 % of all cancellations at UNN are

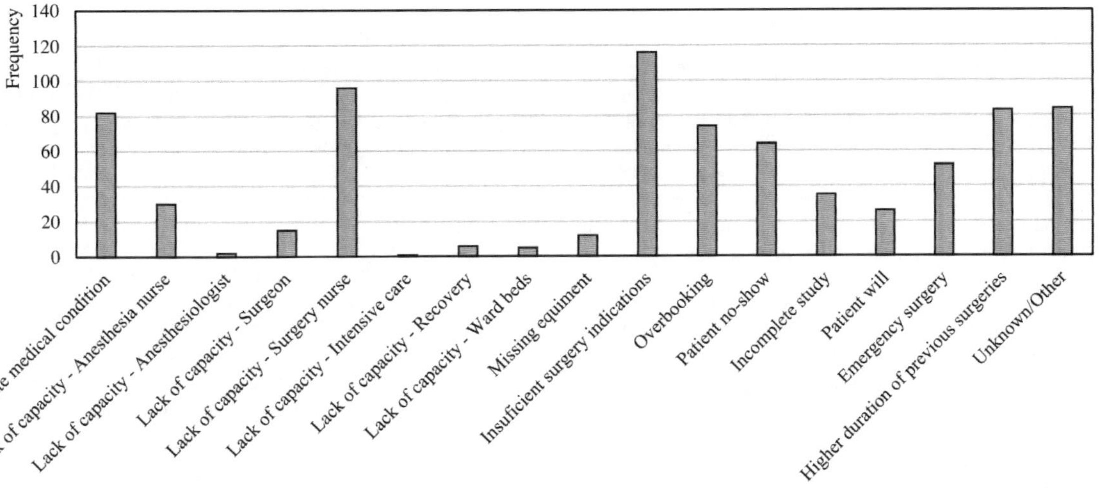

Figure 1. Causes for day of surgery cancellations distribution at UNN from January to June 2011.

related to inadequate pre-operative planning [18]. It is anticipated that the pre-operative planning process may be improved if adequate patient information is gathered at an earlier stage, before the patient is admitted at the hospital. The cancellation problem at UNN is addressed by the eTeam-Surgery project group. This group consists of an interdisciplinary research team, who is studying if and how electronic communication can involve the patient in the pre-operative planning process to provide the missing information.

In this paper, the elective surgery cancellation problem caused by inadequate pre-operative planning, in a university hospital in Norway, is addressed. We started by studying and evaluating the pre-operative planning process at UNN. The aim was to explore a system for gathering information from patients on their condition through a personal health assessment questionnaire. It is suggested an architecture for a two-way electronic communication tool to support the pre-operative planning process.

The paper is divided in six sections. In the first section the problem object of the study is described and classified according to its causes. In the second section a brief review of the state of the art is presented. Data collection methodologies, with which the results were obtained, are presented and explained in the third section. The results are disclosed in section four. In the fifth section, the results are interpreted and discussed in relation to the context-awareness methodology. In the last section conclusions about the results are drawn, and some indicators of future work in the area foreseen.

II. BACKGROUND

In literature, pre-operative planning is reported to be approached in several different ways. A brief state-of-the-art on how information is gathered from the patient prior to surgery is presented below.

A widely studied approach to the elective surgery cancellation problem is the establishment of pre-operative assessment clinics (POACs). The aim of a POAC is to prepare patients for the administration of anaesthesia and for surgery. The implementation of POACs may take different configurations relating to the health worker leading the appointment. Doctor-led POACs were implemented by [17][22][23][24] in an attempt to solve elective surgery cancellations due to lacking information. In this settings, patients are referred to the POAC either from the ward or the outpatient clinic. It was concluded that the number of cancellations was reduced but considered not significant [17]. In nurse-led POACs [25][26][27] the role of the physicians is transferred to the nurse. Thus, in such environments, the pre-operative assessment is undertaken by nurses, with overall supervision of a consultant anaesthesiologist. Nurse-led pre-operative assessment systems POACs do not address the hypothesis that the pre-operative assessment information may be collected from the patient at home.

A different, but still similar, approach to improve pre-operative planning is to re-evaluate the role of health workers in the pre-operative process, and create tools that enable the transfer of responsibilities from physicians to nurses. It is advocated that the pre-operative assessment of elective surgical patients may be undertaken by trained nurses [19][28][29]. Following this hypothesis, nurse-led pre-operative assessment systems have been implemented [19][28][29], using protocols to guide nurses in the decision making process. Nurse-led pre-operative systems do not address the hypothesis that the pre-operative assessment information may be collected from the patient at home.

Searches on the major academic literature databases (i.e., PubMed, Web of Science, Inspec, SCOPUS), on pre-operative planning that use electronic communication with the patient at home, did not retrieve any relevant result. Following, an approach to the problem of elective surgery cancellations by contacting the patient at home is presented.

Telephone calls are being studied as a solution to reduce elective surgery cancellations, due to patient no-shows, on the day-of-surgery [30][31]. Such studies propose a communication channel between the patient and the provider to enable the confirmation of the patient's intention to attend surgery, or simply address patient's questions and concerns. Information exchange between health personnel and patients, while the patients are still at home, may solve some of today's challenges with late pre-operative planning and, consequently, cancellations of surgical procedures.

In developed countries, like Norway, where the population is well prepared and able to use ICT (e-readiness), a new approach is possible [32] to promote patient-centred health care [33][34]. Many patients [35], including elderly or less-educated [36], are strongly motivated to use electronic services [37]. In order to address the potential of electronic communication in the healthcare sector, a brief state-of-the-art on the use of electronic communication in other health care settings was conducted, and is presented hereafter.

A promising application of health information and communication technology (ICT) is the facilitation of web-based communication between patients and health workers. Such tools are expected to improve health care services by promoting streamlined communication, improving resource usage, facilitating shared decision-making, and patient self-management [38][39][40][41][42].

A web-based questionnaire of patient symptoms in primary care has been implemented at the Mayo Clinic, with a 40 % decrease of office visits [43].

Zhou et al. [44] reported on the use of secure e-mail between physicians and patients at the Kaiser Permanente, for a period of two months. The patient portal was integrated with the EHR, and 35 % of the hospital patients were registered in the portal. The study focus on patients with diabetes and hypertension. During the study period, the authors registered 556 339 e-mails with a total of 630 807 messages, and 85 % of the threads were initiated by the patient. In this study was shown that the use of secure e-mail between the patients and physicians was associated with higher performance of quality measures.

In Rosen and Kwoh's [39] study, a consecutive series of patients' families in paediatric care were offered e-mail access over a 2 years period. The authors reported that 5.7 % of patients' e-mails were urgent (i.e., notification of disease flare or new symptoms) and only 0.002 % of the e-mails required physicians' emergent attention. After 1 year of enrolment in the patient-physician e-mail service, the majority of families agreed that service increased access to the physician and improved the quality of care.

A web-based collaborative care management tool was presented by Ralston et al. [45]. The tool targeted patients diagnosed with type 2 diabetes, and the aim was to support the patient at home in the management of their disease. Features such as patient access to the EHR, and secure e-mail with health workers were included in the tool. During the period from August 2002 to May 2004, 83 patients were randomized to receive care through the web-based tool as an addition to the established care procedures. The authors reported that the use of secure e-mail between the patient and the physician improved glycaemic control in type 2 diabetes.

Increased collaboration with patients, as active participants, through ICT solutions, are also defined as a priority area, as stated in the Norwegian Ministry of Health and Care Services' Coordination Reform [46].

At the same time, an extensive ICT investment is taking place in the northern health region of Norway, including at the UNN hospital, our site of research. Helse-Nord, the Northern Norway Regional Health Authority, is investing €62.5 million in the FIKS (from the Norwegian *Felles innføring kliniske systemer*) project to develop the electronic health record (EHR) for the future – a fundamental tool for high-quality patient treatment [47]. The planning tool on the surgical module in the current EHR system has been recognized as an unused resource by the FIKS project of the Northern Norway Regional Health Authority and the Lean Project [47]. The described health care trends in Norway open new possibilities to approach the elective surgery cancellation problem.

The aim of our research is to reduce the elective surgery cancellations at UNN, by studying pre-operative planning and determine if it may be moved from the hospital to the patient at home. We will explore if surgical patients and health personnel can collaborate in a team while the patient is still at home, through an electronic communication tool, and if this reduces elective surgery cancellations, by better preparing hospitals and patients for surgical procedures.

III. Materials and Methods

To develop an efficient and functional web-based tool for hospital-patient collaboration is not an easy task, and it has not always been successful. As the development of health ICT grows, there is also an increasing number of reports on unsuccessful implementation projects, challenges and unforeseen consequences of ICT in health care, particularly in hospitals [48][49][50][51][52][53][54][55][56][57][58][59]. A contributing factor to such results may be found on the focus of health ICT on improving individual tasks rather than supporting value added care processes. By supporting individual tasks, ICT is focusing on the provider. This is a significant contribution to a lower quality and high cost health care. On the other hand, process focused care is centred on the patient. It integrates the team work (e.g., patients, physicians, nurses, caregivers, managers, and administrative personnel) to provide high quality and efficient care throughout the full process. Value added care processes are the goal of the patient centred health care. However, few health care processes have been modelled comprehensively enough to provide a basis for specifying software requirements to health ICT designers. Thus, health ICT designers have focused on supporting the work of individual care team members by taking existing paper-based tools, as their models. The result is that most health ICT systems do little to support care teams. Hence, prior to development, eTeam-Surgery carried out an in-depth study of the pre-operative planning at UNN.

The management at UNN, our site of research, is determined to reduce the cancellation rate at the hospital. Resources have been allocated, and a Lean process for elective surgical patient pathways at the Operation and Intensive care clinic has been initiated at UNN. Lean projects are commonly used to transform healthcare organizations for improvements in patient care through the development of a quality driven culture [12]. At UNN, Lean is defined to concern the right things at the right place, time and amount, with a minimum of waste while, at the same time, being flexible and prepared for changes. The Lean process at UNN is organized as a project team, including a project manager, a Lean consultant, a Lean mentor, an economics and an IT-consultant. In addition, the Lean project has an executive board, a project group and a focus group. At the start of the Lean Project, the focus group, which is the actual working group, consisted of; one anaesthetist nurse, one theatre nurse, two anaesthesiologists, three surgeons, one member of the staff responsible for sterilization of surgical equipment, three staff members responsible for elective surgery planning and waiting lists in the surgery ward, one paediatric nurse, two ICT consultants (one of them responsible for the EHR), one employee representative, and one user (patient).

Two researchers from our research team have followed the Lean process since the initial group meeting in April 2012. One has participated solely as a researcher, conducting observations during Lean meetings, while the other had an active role and contributed as an anaesthesiologist in the Lean process. The researchers observed and participated in more than twenty meetings.

The data on the causes for elective surgery cancellation, presented in Figure 1, is a result of the work carried out by the Lean project. In addition to following the Lean process, we have accomplish three weeks of fieldwork at the Operation and Intensive care clinic, conducting observations and unstructured interviews while following an anaesthesiologist and an anaesthetist nurse in their daily work. We have also conducted thirteen structured interviews with physicians, nurses and administrative personnel. Based on the knowledge from the quantitative and qualitative inquiries, it was identified the need to proceed with further analysis of the data on the causes for cancellation provided by the Lean project, as explained in the results section.

Data collected through observations and interviews was analysed together with observational data from the Lean project. Our analytical qualitative approach focuses on the interaction between technical and social factors that produces particular outcomes [60]. The preliminary results are limited to the identification of the information needed for pre-operative assessment from the anaesthetists and surgeons' point of view.

An empirical inquiry of the reported causes for cancellations in the hospital's EHR, and on the pre-operative planning process at the hospital, was conducted. The study was carried out using mixed methods, involving both a quantitative and a qualitative approach. A quantitative approach was used to map the causes for cancellation reported in the hospital's EHR, while a quantitative approach was applied in the study of the current preoperative planning process at the hospital.

The aim of the quantitative approach was to quantify and map the different causes for cancellations at the hospital in order to determine if extended communication between the patient and the hospital is an adequate initiative to reduce surgical cancellations. In order to make an analysis on how such interaction can be organized and integrated in the existing work practices at the hospital, it was identified the need for a qualitative study of the preoperative planning at the hospital. The qualitative approach consisted of an in-depth study of the current preoperative planning process, including observations and interviews at the hospital. The aim was to acquire in-depth knowledge on the information flow and workflow during the pre-operative planning process, and identify bottlenecks and/or challenges that lead to cancellations.

IV. RESULTS

A process model facilitates a systematic description of the events permitting the identification of decision activities, and the health workers responsible for each of them. In addition, it allows us to learn about the information flow, and to identify the underlying process issues that are causing the patient assessment information not to be available when required. The observations and interviews, described in Section III, allowed the identification of the activities involved in the pre-operative planning as it is done today at UNN.

The data on elective surgery cancellation, presented in Figure 1, is categorized according to the causes reported in the EHR system. The data presented in Table I evidences that the identified causes for elective surgery cancellation inside the same category are not all related to the same context. Thus, these causes required further classification according to the decision context for the cancellation: management, medical, or patient. Decisions within the patient context are related to the patient, as patient no-show and will. Decisions in the medical context are related to clinical issues and are, therefore, taken by surgeons and anaesthesiologists. Finally, the decisions taken in the management context relate to planning and operational issues and are taken by secretaries, nurses, and physicians. This allowed us to identify the responsible (e.g., health workers) for the activities required for the process modelling. The mapping of the existing pre-operative process model is shown in Figure 2.

TABLE I. CATEGORIZATION OF THE CAUSES FOR ELECTIVE SURGERY CANCELLATION AT UNN, ACCORDING TO WHO TOOK THE UNDERLYING DECISION TO CANCEL AND THE RELATED ACTOR.

	Cause for elective surgery cancellation		Decision context
Hospital	Lack of capacity	Anesthesia nurse	Management
		Anestheseologist	
		Surgeon	
		Surgery nurse	
		Intensive care	
		Recovery	
		Ward beds	
	Missing Equipment		
	Overbooking		
	Emergency surgery		
	Higher duration of previous surgery		
	Insufficient surgery indications		Medical
	Incomplete study		
Patient	Other acute medical condition		
	Patient no-show		Patient
	Patient will		

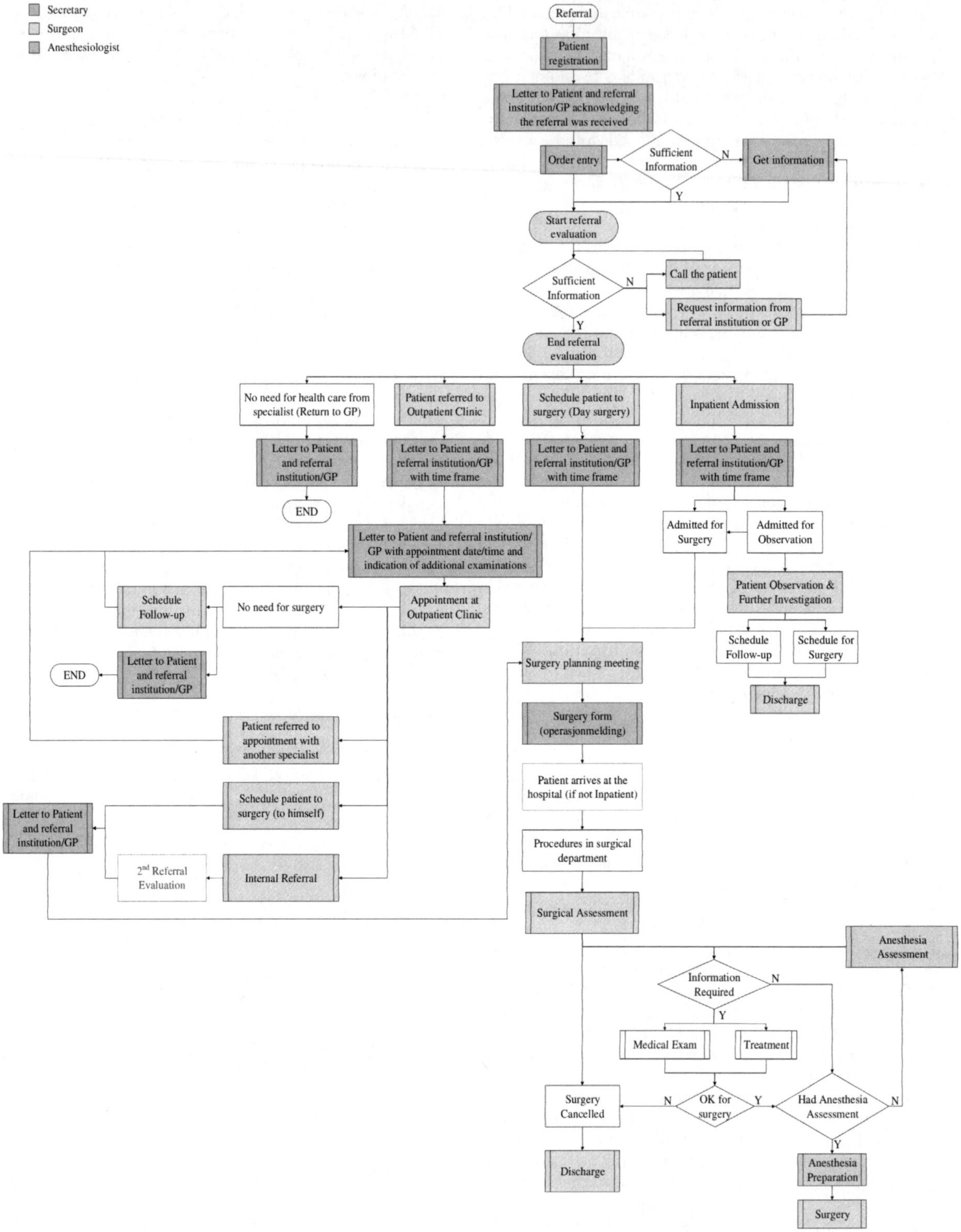

Figure 2. Scheme of the surgery process at UNN. Assessment activities after patient arrival (Box with green border), may contribute to late cancellations (box with red border) while there are many possibilities for hospital-patient interaction at earlier stages (Letters to patient and patient at the hospital).

At UNN, as seen in Figure 2, the final pre-operative planning is often done after the patient has arrived at the hospital for the scheduled surgery. Which means, the final pre-operative planning might be done the day before, or even on the day of surgery. During this final planning process, new information is gathered from patients which may lead to cancellations. Considering the data collected during the observations and interviews, and the analysis of the existing pre-operative process, all the decision activities were identified and characterized. Based on the information requirements on each of those activities, a new pre-operative process was proposed. In the new pre-operative process the assessment information is requested to the patient at an earlier stage and while the patient is still at home. The assessment information identified as required might be included in the personal health assessment questionnaire which some departments ask the patients to fill out and bring to the hospital when hospitalized for surgery.

V. Discussion

This paper addresses the elective surgery cancellations problem at UNN. Observations and interviews were conducted at UNN, and lack of information during the pre-operative planning was identified as the main cause of elective surgery cancellations. The problems with the existing pre-operative process were identified and a new process was proposed. In the new process, the assessment information is systematized in a personal health assessment questionnaire, and provided by the patient at an earlier stage, while the patient is still at home.

The mapping of a generic pre-operative process model facilitated the identification of the decision activities, and the health workers responsible. The identification of activities, and their responsible health worker, allowed us to carry out semi-structured interviews to determine the information required to complete the pre-operative assessment.

Surgeons and anaesthesiologists at UNN considered that the identified information may be provided by the patient. Some departments ask the patients to fill out a personal health assessment questionnaire and bring it to the hospital when hospitalized for surgery. The information classified as required might be included in this questionnaire. Such questionnaires can be sent to the patient through the postal system, and the patient can fill it out at home.

At this stage, the collaboration between surgical patients and health personnel cannot do much for the elective surgery cancellations related to management context. These are mostly due to inadequate planning and should be approached within the adequate research field. On the other hand, the causes for elective surgery cancellation related to the remaining decision contexts, patient and medical, may be positively influenced by such collaboration, either by gathering information on the patient health status based in the afore mentioned personal health assessment questionnaire or, by improving the dialogue through electronic communication.

The aim of the research project "eTeam-Surgery" is to reduce the number of elective surgery cancellations at UNN. In today's surgical process the information required for anaesthetic evaluation is gathered after the patient is hospitalized, as shown in Figure 2. The aim of eTeam-Surgery is to provide a tool for two-way electronic communication, Figure 3, between the hospital and the patient prior to hospital admission. Such tool will enable the hospital to collect the lacking information at an earlier stage in the pre-operative planning process, while the patient is still at home. The tool proposed by the eTeam-Surgery project, will provide two communication channels: (1) Collect the lacking information based on the Patient Health Assessment Questionnaire described in Section I. To guarantee that the information collected through this channel can be shared by the health workers involved in the patient episode, it should be structured data that can be included in the EHR system. However, by collecting the information while the patient is still at home means that patients loose the support from health workers while completing it. Therefore, eTeam-Surgery will support (2) two-way electronic communication. This part of the tool will provide an asynchronous messaging service that will enable the patient to pose questions to the hospital, and vice-versa.

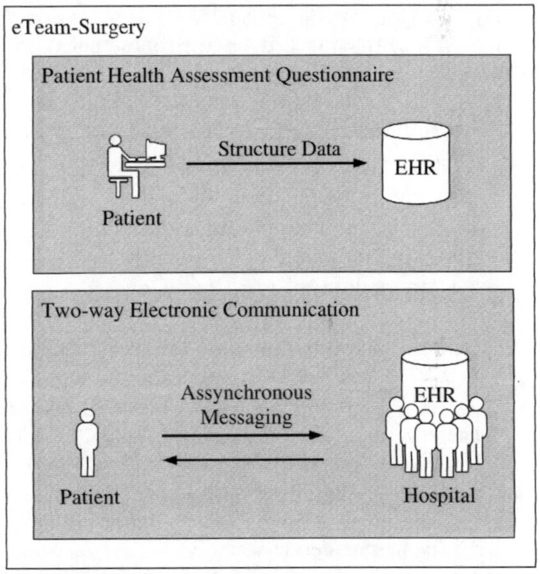

Figure 3. Ilustration of the eTeam-Surgery solution architecture.

The implementation of such tool implies that the pre-operative planning process, described in Figure 2, needs to be restructured. Which means that some of the identified activities might be removed from the process, and new ones be created. It might also involve the re-evaluation of the health workers responsible for the activities. This knowledge, combined with the reports in literature that patients often evidence "selfish" communication practices which could result in an overload of conversations that would not occur if the tools were not available [61], leads to the question: "How to balance the need to gather information from the patient, and the requirements for electronic communication, with the reported consequences of communication tools on clinical processes?"

A possible solution may be found in the field of context-awareness. Let us first define context. Abowd et al. [62] defined context as "*any information that can be used to characterize the situation of an entity. An entity is a person, place, or object that is considered relevant for the interaction between a user and an application, including the user and application themselves*". This definition shows the importance of which information is relevant or not in a context-aware system. A context-aware system could, therefore, be defined as a system allowing interactions between multiple entities using relevant information. In [62] is state that: "*A system is context-aware if it uses context to provide relevant information and/or services to the user, where relevancy depends on the user's task*". This definition shows that a context-aware system can change its behaviour and send some relevant information according to the context, which reflects our view.

The trend in the health IT field has been to push as much information as possible to the users, in order to provide more sophisticated and useful services while, at the same time, making users more available. During a preliminary research study on the Aware Media system [63], they suggested a classification that splits the above listed information along three main axes:

- Social awareness: `where a person is', `activity in which a person is engaged on', `self-reported status';
- Spatial awareness: 'what kind of operation is taking place in a ward', 'level of activity', 'status of operation and people present in the room';
- Temporal awareness: 'past activities', 'present and future activities' that is significant for a person.

A context-aware system, comprises two main modules:

- Context engine: This module interfaces with other information systems and devices to collect raw data. These are then fed to an analyzer to classify raw data and generate context data;
- Rules engine: This module acts as filter between the data and the user. By applying a set of pre-defined conditions that define what, when, and to who the information must be presented. Such rules can be defined manual or automatically.

The adoption of context-aware systems based on these definitions is growing in a variety of domains such as, smart homes, airports, travel/entertainment/shopping, museum, and offices, as mentioned in [64].

In the scope of the eTeam-Surgery project, context information may be used to decide, e.g., who should communicate with the patient, the urgency level of the communication, and time frame when the communication should be available. In this way, it would be possible to balance the eagerness of the patients to communicate with hospital with the fact that communication tools make health care workers "fatally" available.

VI. CONCLUSIONS

By studying the pre-operative planning at UNN, we have determined that parts of the information flow can be moved to the patient at home. From the work presented herein, we conclude that the assessment information required during the pre-operative planning can be compiled in a personal health assessment questionnaire, and requested from the patient, at an earlier stage.

The authors acknowledge that the paper-based pre-operative planning process proposed is not in line with the best practices suggested in literature. When using the postal system the information flow between the patient and the hospital is time consuming, and it is not possible for the hospital to confirm the reception and submission of the personal health assessment questionnaire. At the same time, due to: (a) the patient prioritization rules in Norway, (b) waiting list, (c) and emergency surgeries, surgeries can be delayed and the patient might be requested to complete the personal health assessment questionnaire more than once. On the other hand, when asking the patient to answer a personal health assessment questionnaire from home, the patient might require support from health workers when interpreting the questions, and selecting the relevant information.

The international healthcare trends on paperless and patient focused clinical processes, combined with the e-readiness in Norwegian society, point to new possibilities on how to gather assessment information from the patient at home. To access this information, low-cost communication with patients and their families has been recommended [13]. Thereby improving pre-operative planning, and reducing the number of cancelations, due to lack of information. In order to enable the communication between the patient and the hospital, the interaction with patients should take place through a variety of synchronous and asynchronous secure communication channels, including phone, messaging systems, email, and web-pages. Considering the impact of health ICT in clinical processes, the authors suggest the adoption of the context-aware methodology in the development of electronic communication tools to reduce elective surgery cancellations.

ACKNOWLEDGMENT

The authors would like to thank the regional health authority Helse-Nord for funding the research project HST 1119-13 and HST 1125-13. We would also like to thank the personnel at UNN and the Lean project, especially Tonje Drecker.

REFERENCES

[1] C. Granja, K. Dyb, S. R. Bolle, and G. Hartvingsen, "Reduced Elective Surgery Cancellations through Patient Involvement in Pre-Operative Planning in Norway," eTELEMED 2014, The Sixth International Conference on eHealth, Telemedicine, and Social Medicine, 2014, pp. 164-169.

[2] W. N. Schofield, G. L. Rubin, M. Piza, Y. Y. Lai, D. Sindhusake, M. R. Fearnside, and P. L. Klineberg, "Cancellation of Operations on the Day of Intended Surgery at

a Major Australian Referral Hospital," Medical Journal of Australia, vol. 182, pp. 612-615, 2005.

[3] B. Denton, J. Viapiano, and A. Vogl, "Optimization of Surgery Sequencing and Scheduling Decisions under Uncertainty," Health Care Management Science, vol. 10, pp. 13-24, 2007.

[4] A. Foster, "Operating Theatres: Review of National Findings," Audit Commission for Local Authorities and the National Health Service, London 2003.

[5] M. J. Lacqua and J. T. Evans, "Canceled Elective Surgery: An Evaluation," American Surgeon, vol. 60, pp. 809-811, 1994.

[6] B. Ivarsson, P. O. Kimblad, T. Sjoberg, and S. Larsson, "Patient Reactions to Cancelled or Postponed Heart Operations," Journal of Nursing Management, vol. 10, pp. 75-81, 2002.

[7] M. Aaserud, M. Trommald, and J. Boynton, "Elektiv Kirurgi - Strykninger, Skjerming Og Effektivitet," Tidsskr Nor Lægeforen vol. 21, pp. 2516-2519, 2001 (in Norwegian).

[8] A. González-Arévalo, J. I. Gómez-Arnau, F. J. DelaCruz, J. M. Marzal, S. Ramírez, E. M. Corral, and S. García-del-Valle, "Causes for Cancellation of Elective Surgical Procedures in a Spanish General Hospital," Anaesthesia, vol. 64, pp. 487-493, 2009.

[9] S. Z. Yoon, S. I. Lees, H. W. Lee, H. J. Lim, S. M. Yoon, and S. H. Chang, The Effect of Increasing Operating Room Capacity on Day-of-Surgery Cancellation vol. 37. Edgecliff, AUSTRALIE: Anaesthesia Society of Anaesthetists, 2009.

[10] T. L. Trentman, J. T. Mueller, S. L. Fassett, C. L. Dormer, and K. P. Weinmeister, "Day of Surgery Cancellations in a Tertiary Care Hospital: A One Year Review," Journal of Anesthesia & Clinical Research, vol. 1, 2010.

[11] P. Sanjay, A. Dodds, E. Miller, P. J. Arumugam, and A. Woodward, "Cancelled Elective Operations: An Observational Study from a District General Hospital," Journal of Health Organisation and Management, vol. 21, pp. 54-58, 2007.

[12] R. R. Cima, M. J. Brown, J. R. Hebl, R. Moore, J. C. Rogers, A. Kollengode, G. J. Amstutz, C. A. Weisbrod, B. J. Narr, and C. Deschamps, "Use of Lean and Six Sigma Methodology to Improve Operating Room Efficiency in a High-Volume Tertiary-Care Academic Medical Center," Journal of the American College of Surgeons, vol. 213, pp. 83-92, 2011.

[13] V. Haana, K. Sethuraman, L. Stephens, H. Rosen, and J. G. Meara, "Case Cancellations on the Day of Surgery: An Investigation in an Australian Paediatric Hospital," ANZ Journal of Surgery, vol. 79, pp. 636-640, 2009.

[14] M. B. Ferschl, A. Tung, B. Sweitzer, D. Huo, and D. B. Glick, "Preoperative Clinic Visits Reduce Operating Room Cancellations and Delays," Anesthesiology, vol. 103, pp. 855-859, 2005.

[15] M. G. Perroca, M. d. C. Jericó, and S. D. Facundin, "Surgery Cancelling at a Teaching Hospital: Implications for Cost Management," Revista latino-americana de enfermagem, vol. 15, pp. 1018-1024, 2007.

[16] A. R. Tait, T. Voepel-Lewis, H. M. Munro, H. B. Gutstein, and P. I. Reynolds, "Cancellation of Pediatric Outpatient Surgery: Economic and Emotional Implications for Patients and Their Families," Journal of clinical anesthesia, vol. 9, pp. 213-219, 1997.

[17] M. Knox, E. Myers, I. Wilson, and M. Hurley, "The Impact of Pre-Operative Assessment Clinics on Elective Surgical Case Cancellations," Surgeon-Journal of the Royal Colleges of Surgeons of Edinburgh and Ireland, vol. 7, pp. 76-78, 2009.

[18] R. Busund, "Rapport Fra Prosjekt: Optimal Ressursutnyttelse Av Opperasjonskapasiteten I Unn," Universitetssykehuset Nord-Norge, Norway 2008 (in Norwegian).

[19] M. R. Rai and J. J. Pandit, "Day of Surgery Cancellations after Nurse-Led Pre-Assessment in an Elective Surgical Centre: The First 2 Years," Anaesthesia, vol. 58, pp. 692-699, 2003.

[20] A. R. Seim, T. Fagerhaug, S. M. Ryen, P. Curran, O. D. Sæther, H. O. Myhre, and W. S. Sandberg, "Causes of Cancellations on the Day of Surgery at Two Major University Hospitals," Surgical Innovation, vol. 16, pp. 173-180, 2009.

[21] W. A. van Klei, C. L. Rutten, K. G. Moons, B. Lo, J. T. Knape, and D. E. Grobbee, "Limited Effect of Health Council Guideline on Outpatient Preoperative Evaluation Clinics in the Netherlands: An Inventory," Nederlands Tijdschrift voor Geneeskunde, vol. 145, pp. 25-29, 2001 (in Dutch).

[22] M. A. Starsnic, D. M. Guarnieri, and M. C. Norris, "Efficacy and Financial Benefit of an Anesthesiologist-Directed University Preadmission Evaluation Center," Journal of clinical anesthesia, vol. 9, pp. 299-305, 1997.

[23] M. Agah and B. Radpay, "Efficacy of Anesthesia Clinic on Preoperative Evaluation, Preparation and Related Costs," Tanaffos, vol. 3, pp. 63-68, 2004.

[24] R. Odijk, "Process Optimization of the Preoperative Anesthesia Clinics at Zgt Almelo and Hengelo," 2012.

[25] M. Reed, S. Wright, and F. Armitage, "Nurse-Led General Surgical Pre-Operative Assessment Clinic," Journal of the Royal College of Surgeons of Edinburgh, vol. 42, pp. 310-313, 1997.

[26] B. J. Kirkwood, K. Pesudovs, P. Latimer, and D. J. Coster, "The Efficacy of a Nurse-Led Preoperative Cataract Assessment and Postoperative Care Clinic," Medical journal of Australia, vol. 184, pp. 278-281, 2006.

[27] H. Walsgrove, "Piloting a Nurse-Led Gynaecology Preoperative-Assessment Clinic," Nursing times, vol. 100, pp. 38-41, 2004.

[28] AmandaBeck, "Nurse-Led Pre-Operative Assessment for Elective Surgical Patients," Nursing Standard, vol. 21, pp. 35-38, 2007.

[29] H. Kinley, C. Czoski-Murray, S. George, C. McCabe, J. Primrose, C. Reilly, R. Wood, P. Nicolson, C. Healy, S. Read, J. Norman, E. Janke, H. Alhameed, N. Fernandes, and E. Thomas, "Effectiveness of Appropriately Trained Nurses in Preoperative Assessment: Randomised Controlled Equivalence/Non-Inferiority Trial," BMJ (Clinical research ed), vol. 325, p. 1323, 2002.

[30] M. A. G. de Avila and S. C. M. Bocchi, "Confirmação De Presença De Usuário À Cirurgia Eletiva Por Telefone Como Estratégia Para Reduzir Absenteísmo," Revista da Escola de Enfermagem da USP, vol. 47, pp. 193-197, 2013 (in Portuguese).

[31] K. Haufler and M. Harrington, "Using Nurse-to-Patient Telephone Calls to Reduce Day-of-Surgery Cancellations," AORN Journal, vol. 94, pp. 19-26, 2011.

[32] H. K. Andreassen, M. M. Bujnowska-Fedak, C. E. Chronaki, R. C. Dumitru, I. Pudule, S. Santana, H. Voss, and R. Wynn, "European Citizens' Use of E-Health Services: A Study of Seven Countries," BMC Public Health, vol. 7, p. 53, 2007.

[33] B. Kaplan and P. F. Brennan, "Consumer Informatics Supporting Patients as Co-Producers of Quality," Journal of the American Medical Informatics Association, vol. 8, pp. 309-316, 2001.

[34] E. Randeree, "Personal Health Records: Patients in Control," in *Health Information Systems: Concepts, Methodologies, Tools, and Applications*, J. Rodrigues, Ed., ed: IGI Global, 2010, pp. 2111-2124.

[35] L. Millsopp, S. Frackleton, D. Lowe, and S. N. Rogers, "A Feasibility Study of Computer-Assisted Health-Related Quality of Life Data Collection in Patients with Oral and Oropharyngeal Cancer," International journal of oral and maxillofacial surgery, vol. 35, pp. 761-764, 2006.

[36] P. E. Kummervold, D. Gammon, S. Bergvik, J. A. Johnsen, T. Hasvold, and J. H. Rosenvinge, "Social Support in a Wired World: Use of Online Mental Health Forums in Norway," Nordic Journal of Psychiatry, vol. 56, pp. 59-65, 2002.

[37] P. M. Webb, G. D. Zimet, J. D. Fortenberry, and M. J. Blythe, "Comparability of a Computer-Assisted Versus Written Method for Collecting Health Behavior Information from Adolescent Patients," Journal of Adolescent Health, vol. 24, pp. 383-388, 1999.

[38] J. S. Brown and P. Duguid, The Social Life of Information: Harvard Business Press, 2000.

[39] P. Rosen and C. K. Kwoh, "Patient-Physician E-Mail: An Opportunity to Transform Pediatric Health Care Delivery," Pediatrics, vol. 120, pp. 701-706, 2007.

[40] E. M. Liederman, J. C. Lee, V. H. Baquero, and P. G. Seites, "The Impact of Patient-Physician Web Messaging on Provider Productivity," Journal of Healthcare Information Management, vol. 19, pp. 81-86, 2005.

[41] T. Delbanco and D. Z. Sands, "Electrons in Flight-E-Mail between Doctors and Patients," New England Journal of Medicine, vol. 350, pp. 1705-1707, 2004.

[42] J. D. Ralston, C. M. Rutter, D. Carrell, J. Hecht, D. Rubanowice, and G. E. Simon, "Patient Use of Secure Electronic Messaging within a Shared Medical Record: A Cross-Sectional Study," Journal of general internal medicine, vol. 24, pp. 349-355, 2009.

[43] S. C. Adamson and J. W. Bachman, "Pilot Study of Providing Online Care in a Primary Care Setting," Mayo Clinic Proceedings, vol. 85, pp. 704-710, 2010.

[44] Y. Y. Zhou, M. H. Kanter, J. J. Wang, and T. Garrido, "Improved Quality at Kaiser Permanente through E-Mail between Physicians and Patients," Health affairs, vol. 29, pp. 1370-1375, 2010.

[45] J. D. Ralston, I. B. Hirsch, J. Hoath, M. Mullen, A. Cheadle, and H. I. Goldberg, "Web-Based Collaborative Care for Type 2 Diabetes a Pilot Randomized Trial," Diabetes care, vol. 32, pp. 234-239, 2009.

[46] Helse- og omsorgsdepartement, "Samhandlingsreformen," Det Kongelige helse- og omsorgsdepartement 2009 (in Norwegian).

[47] Helse-Nord. Dette Er De Nye Systemene. Available: http://www.helse-nord.no/helse-nord-nytt/dette-er-de-nye-systemene-article90645-1526.html 4 November 2013 (in Norwegian).

[48] J. Starling and S. Foley, "From Pilot to Permanent Service: Ten Years of Paediatric Telepsychiatry," Journal of Telemedicine and Telecare, vol. 12, pp. 80-82, 2006.

[49] P. Whitten, B. Holtz, and L. Nguyen, "Keys to a Successful and Sustainable Telemedicine Program," International journal of technology assessment in health care, vol. 26, pp. 211-216, 2010.

[50] P. Zanaboni and R. Wootton, "Adoption of Telemedicine: From Pilot Stage to Routine Delivery," BMC Medical Informatics and Decision Making, vol. 12, 2012.

[51] M. Berg, "Implementing Information Systems in Health Care Organizations: Myths and Challenges," International journal of medical informatics, vol. 64, pp. 143-156, 2001.

[52] R. Heeks, "Health Information Systems: Failure, Success and Improvisation," International journal of medical informatics, vol. 75, pp. 125-137, 2006.

[53] C. May, M. Mort, F. S. Mair, and T. Finch, Telemedicine and the Future Patient: Risk, Governance and Innovation: Economic and Social Research Council, 2005.

[54] C. May and N. T. Ellis, "When Protocols Fail: Technical Evaluation, Biomedical Knowledge, and the Social Production of 'Facts' About a Telemedicine Clinic," Social Science & Medicine, vol. 53, pp. 989-1002, Oct 2001.

[55] S. Dünnebeil, A. Sunyaev, I. Blohm, J. M. Leimeister, and H. Krcmar, "Determinants of Physicians' Technology Acceptance for E-Health in Ambulatory Care," International Journal of Medical Informatics, vol. 81, pp. 746-760, 2012.

[56] KS, "Ikt I Helse- Og Omsorg 2008-2012 - Strategi- Og Handlingsplan," Oslo 2008 (in Norwegian).

[57] H. K. Andreassen, "What Does an E-Mail Address Add? Doing Health and Technology at Home," Social Science & Medicine, vol. 72, pp. 521-528, 2011.

[58] N. Schreurs. (2012) Fiasko Eller Fremtid? *Computerworld*. (in Norwegian)

[59] J. C. Wyatt and F. Sullivan, "Ehealth and the Future: Promise or Peril?," BMJ, vol. 331, pp. 1391-1393, 2005.

[60] B. Latour, Science in Action: How to Follow Scientists and Engineers through Society: Harvard university press, 1987.

[61] F. P. Glascoe, "Improving Clinical Communication: A View from Psychology," Ambulatory Child Health, vol. 7, pp. 53-60, 2001.

[62] G. D. Abowd, A. K. Dey, P. J. Brown, N. Davies, M. Smith, and P. Steggles, "Towards a Better Understanding of Context and Context-Awareness," Handheld and ubiquitous computing, Springer, 1999, pp. 304-307.

[63] J. E. Bardram, T. R. Hansen, and M. Soegaard, "Awaremedia: A Shared Interactive Display Supporting Social, Temporal, and Spatial Awareness in Surgery," presented at the 20th anniversary conference on Computer supported cooperative work, Banff, Alberta, Canada, 2006.

[64] A. Hristova, "Conceptualization and Design of a Context-Aware Platform for User Centric Applications," Master of Science in Communication Technology Master, Department of Telematics, Norwegian University of Science and Technology, Trondheim, 2008.

Investigation of the Mechanical Behaviour of Porous Silicon Neural Microprobes

Rayan Fayad, Soumaya Berro, Bakri AbdulHay, Hussein Hajj-Hassan, Hassan M. Khachfe, Mohamad Hajj-Hassan
Dept. of Biomedical Engineering
Lebanese International University
Beirut, Lebanon
rayanfayyad@gmail.com, Soumaya.berro@liu.edu.lb, bakri.abdulhay@liu.edu.lb, hussein.hajjhassan@liu.edu.lb, hassan.khachfe@liu.edu.lb, mohamad.hajjhassan@liu.edu.lb

Abstract—**Porous silicon has become the gold standard when it comes to improving biocompatibility and bioactivity. For this reason, it has become a primary candidate in neural electrodes research and development. Consequently, the purpose of this work was to investigate the mechanical strength of porous silicon neural electrodes. Thus, a finite element model representing the proposed electrode was generated. Mechanical simulation was done on porous and non-porous electrodes using COMSOL® Multiphysics. Results showed that porosity decreased the mechanical strength of the neural electrode without risking the mechanical requirements for neural applications.**

Keywords- biocompatiblity; finite element model; failure analysis; neural microelectrodes; porous silicon.

I. INTRODUCTION

The application of technological advances to cure neurological diseases has long seized the attention of researchers in neural engineering. The first acknowledged use of electrical current in an approach to treat a neural disease goes back to the year 1757. This has gravely evolved nowadays due to progresses in neuroscience and microtechnology, where a wide range of neural electrodes have been fabricated and used in neuroscience and neural prosthetic research (brain machine interfaces) [1], [2], [3], [4].

Neural electrodes, which are micro structures that are implanted in the brain, serve as a communication channel between the electro-active neurons in the brain and an outer electronic circuitry [5], [6], [7].

These electrodes are used in both recording action potentials from neurons and stimulating specific brain regions. The electrical stimulation of nerve tissue and recording of neural electrical activity are the foundation of evolving prostheses and treatments for spinal cord injury, stroke, sensory deficits, and neurological disorders such as seizures, epilepsy, and migraine [7], [8].

As schematized in Figure 1, brain-computer interfaces (BCIs), which incorporate the use of neural electrodes, provide a linkage between the brain and the external world by computer processing the recorded neural signal to extract the subject's command to control an external device. This technology can allow restoring neural functions of patients with severe neurologic impairment [9].

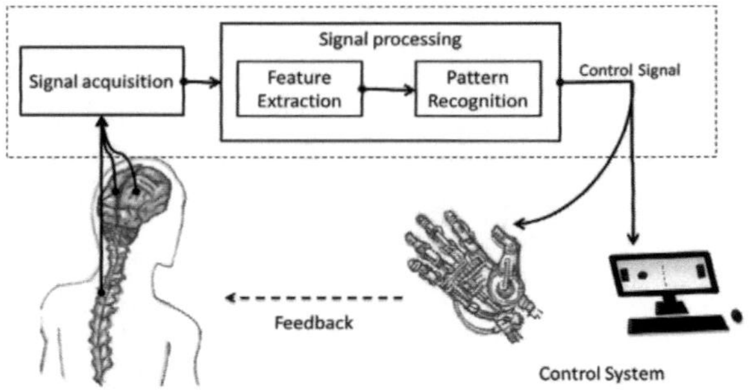

Figure 1. Basic components of a Brain-Computer Interface [10].

When it comes to the general characteristics of a neural electrode, it is notable to mention that the ideal electrode should have a very small cross section in a way that it displaces or damages as little tissue as possible during insertion in order to minimize neural damage. Another reason that makes the small footprint of the neural electrode desirable is that the smaller size allows the selective targeting of the desired neurons, thus improving the signal-to-noise ratio of the recorded activity or targeting a very specific region to be stimulated. On the other hand, the ideal electrode should also be wide enough to incorporate a large number of electrode sites to be able to record and reliably separate different neurons. Regardless of the size, the electrode must be optimized for long term biocompatibility and should have enough mechanical strength to survive the various forces to which it is subjected during insertion and retraction forces while implantation [11], [12]. Several types of neural electrodes have been designed and developed until this day (Figure 2). In the 1950s, the initial use of implantable microelectrodes to record electrical activities in the extracellular environments was traced [13], [14], [15]. These microelectrodes, also known as microwires (Figure 2-A), have the longest history and the widest use in the field. Microwires are wires made of a conducting metal, such as platinum, gold [16], tungsten [17], iridium [18], or stainless steel [19]. Each microelectrode is composed of a metal wire entirely insulated except for its tip; it is left exposed acting as a recording site [20], [21], [22]. Later, the use of microelectrodes advanced into offering the ability of provide multiple metal electrode arrays [23].

Those were constructed by pasting individual microelectrodes together [24], [25], or by gathering several metal wires on a ceramic plate [26]. However, the drawback of this type is that only one recording site is available, which is located at the exposed tip. Hence, any attempt to increase the number of recording sites would include increasing the number of electrodes. This will in turn increase the total size of the electrode, which is a feature that is not usually preferred due to the resulting neural tissue damage it might cause. Furthermore, thought a metal microelectrode has the advantage of simplicity in terms of the process of fabrication, this simplicity is the reason of a major drawback, which is the lack of common standards and automation. Thus, the characteristics of the electrode would vary from one institution to another or from one laboratory to another [11].

This drawback was solved with the introduction of silicon based neural electrodes that emerged with the development of microfabrication techniques and the advancement in microelectronics and microelectrochemical systems (MEMS) [27], [28], [29]. It was in the 1970s when Wise and Angell have published a silicon-based electrode to interface neural tissues [30], [31]. Generally speaking, silicon electrodes offered a superiority over the metal microelectrodes since they allowed an increased number of recording sites without increasing the whole size of the electrode as well as their precise and reproducible fabrication [32]. In other words, by using the photolithography process, the designer would be able to gain control over the recording site size, shape and spacing enabling multiple recording sites to be placed at variable heights on a single electrode shank. Consequently, it would be possible to introduce an increased number of recording sites in a small volume, which is not possible with metal electrode arrays [28]. In addition, silicon offers well-recognized biocompatibility and mechanical properties appropriate for neural electrodes [11]. Well-known examples of silicon neural electrodes are the Utah [33], [34] and the Michigan [6], [35], [36], [37] electrodes. The Utah electrode array (Figure 2-C to the left) is a famous MEMS microelectrode array, which is a widely used type of implantable interface in BCI. The fabrication of the Utah electrode array includes micromachining monocrystalline silicon blocks to form a shape similar to a bed of nails. As for the Michigan electrode (Figure 2-C to the right), it is composed of a boron-diffused silicon substrate, a silicon dioxide and silicon nitride dielectric stack, polysilicon traces, and iridium electrode sites [38].

Another type of electrodes is the polymer-based neural electrode characterized by improved flexibility and biocompatibility as well as the advantage of a simpler fabrication process [39], [40], [41], [42], [43]. Traditionally, in the process of fabrication of a neural silicon electrode, the electrode was usually insulated by a silicon nitride or silicone dioxide layer. This was swapped by the use of polymeric materials in neural electrodes [44], [45], [46]. Serving the purpose of forming a biocompatible interface between the neural electrode and the brain tissue where it is implanted, several biocompatible polymers have been used. These include the use of polyimide and Parylene-C, which play the role of insulating the metal and silicon region of the electrode [46], [47], [48]. However, a significant limitation of this type of electrodes is that they are not stiff enough to penetrate the brain tissue on their own. In other words, these electrodes suffer from lack of rigidity, which leads to less accurate neural targeting due to the buckling of the electrode during the insertion phase [11].

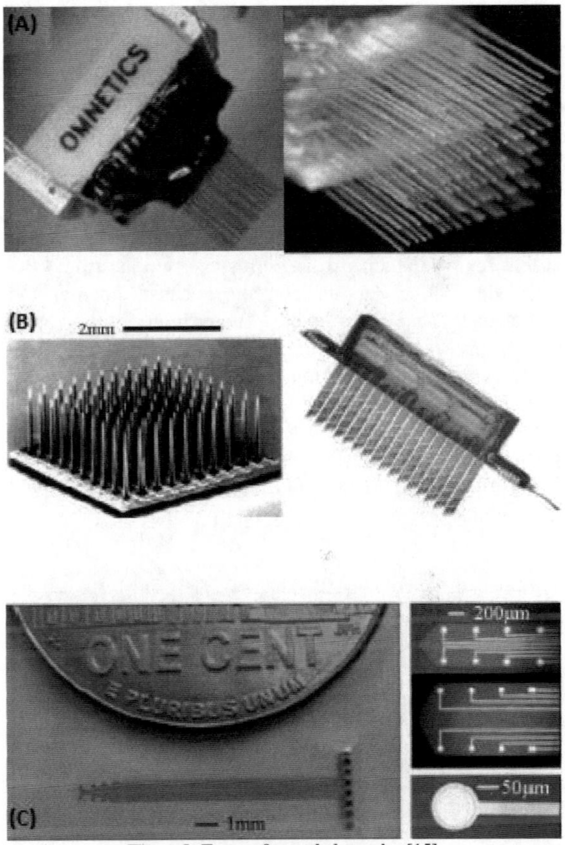

Figure 2. Types of neural electrodes [15].

This article covers the simulation of the mechanical behavior of porous silicon neural electrode. The next section presents the major limitation of present types of chronically implanted neural electrodes due to the resultant brain tissue response. Section III shows how using porous silicon plays a critical role in solving these limitations. Section IV introduces the design of the proposed electrode. Section V discusses the simulation strategy followed. Section VI details the results yielded.

II. LIMITATION OF CHRONICALLY IMPLANTED NEURAL ELECTRODES: THE BRAIN TISSUE RESPONSE

While the previously mentioned systems function well during acute recordings, they frequently do not succeed to operate reliably in clinically relevant chronic settings. The reason why these electrodes fail has been attributed to the brain tissue reaction against these implants. This brain tissue response is provoked by the neural injury upon the implantation of the electrode. Consequently, the resultant tissue response threatens the long-term functioning of the neural electrode. This response includes two major stages known as the acute immune and the chronic immune responses [20], [49], [50], [51] as illustrated in Figure 3.

The implantation of any neural electrode is always a traumatic procedure. When a neural electrode is inserted into the brain, it breaches the vasculature, the extracellular matrix, and destroys neuronal and glial processes in its path. This gives rise to the acute immune response. Forthwith, by misplacing structures along its way, the electrode would cause an alternation in the pressure status in that region inducing a high-pressure region around the electrode. These factors combined cause edema and hemorrhage near the implant. Accordingly, the wound healing process will be commenced as a result. Since the blood vessels are disrupted, this provokes them to discharge erythrocytes, activates platelets, clotting factors, and the complement cascade. This process will assist in macrophage stimulation and the beginning of tissue reconstruction. One day after implementation, activated microglia will show up around the implant.

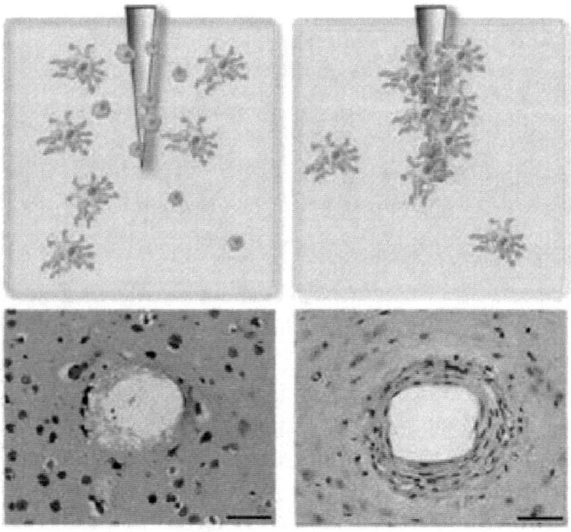

Figure 3. Acute and chronic neural injury caused by the insertion of the neural electrode into the brain cortex [49].

Soon enough, the acute immune response will start to deteriorate, which is directly followed by a chronic immune response. The most influential participants in this phase are reactive astrocytes and activated microglia [52], [53]. This response results in the formation of an encapsulation layer termed the "glial scar" around the electrode. Activated microglia will be engaged in the phagocytosis of the foreign matter for eventual degradation. However, the most significant event that is notable in the long-term response to the chronically implanted electrode is the formation of the encapsulation layer. This glial scar is a reactive glial tissue with reactive astrocytes being its major element. It isolates the implanted neural electrode from the surrounding tissue in a process that resembles fibrotic encapsulation reaction encountered with non-degradable implants in soft tissues. This encourages the inhibition of diffusion and increases the impedance of the tissue-electrode interface. In addition, this also extends the distance between the electrode and the nearest desired neurons. Accordingly, the neurite extensions will find themselves in a non-encouraging environment for growth, the thing that pushes redeveloping neural processes away from the recording sites. This leads to signal deterioration [20], [51], [54], [55], [56].

III. ADVANTAGES OF POROUS SILICON FOR NEURAL ELECTRODES

As stated earlier, interactions between the brain tissue and the electrode are critical in determining the functional performance of the electrode. Various strategies have been investigated and experimented as an attempt to minimize the immune response towards the chronically implanted neural electrode. Some related the degree of the immune response to the size of the electrode [56]. Others thought that the severity of the immune response can be controlled by altering the shape of the electrode [57]. Additionally, it has been proven that tissue response is highly dependent on the surface topography [58]. In the latter approach, it has been tested and proved that rough surfaces are more biocompatible than smooth surfaces [59]. An alternation in the surface topography so that it would be transformed from smooth to rough could be by making the surface porous. In particular, the use of porous silicon as the material that the implants would be made of has shown enhanced biocompatibility and bioactivity [60], [61], [62].

For instance, in a study performed by Hajj-Hassan et al. [62], the biocompatibility and bioactivity of porous silicon wafers (Si) was assessed by examining the survival and replication of mesenchymal stromal cells (MSC) isolated from the bone marrow of wild type mice. These results were compared with that of cells growing in 2D culture on tissue culture plastic (TCP) and on smooth titanium (SmTi), which is well known for its superiority (gold standard) for the manufacture of implants. In the first experiment performed, bone marrow derived MSC were seeded in porous silicon wafers etched to a depth of 20 μm (Si20) in 12 well plates and harvested after 3, 6, and 9 days of culture. Control cells plated at the same density on tissue culture plastic were harvested at 6 days and stained with toluidine blue to visualize the cells. Results showed that the Si20 substrate supported the MSC growth. Additionally, an Alamar Blue metabolic assay was used to analyze the metabolic activity of cells grown porous silicon substrates etched to a depth of 20 μm (Si20) or 30 μm (Si30) and compared with TCP or smooth titanium, which is a common implant material. Representative results of the Alamar Blue assay, shown in Figure 4(A), indicate a small increase in metabolic activity of the cells grown on Si20 and Si30 samples compared to smooth titanium and tissue culture plastic controls. The cell counts indicated a steady increase in numbers that appeared to be dependent on the substrate on which they were grown.

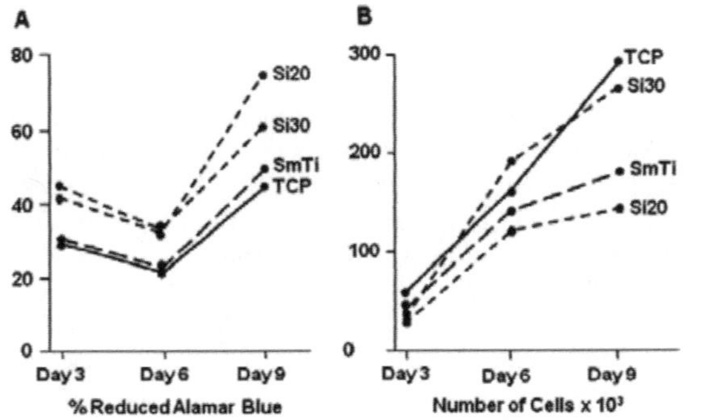

Figure 4. Quantitative analysis of MSC grown on porous silicon etched to a depth of 20 μm (Si20), 30 μm (Si30), commercial grade smooth titanium (SmTi) and tissue culture plastic (TCP) using the Alamar Blue assay to assess the metabolic activity [61].

So, it has been shown that the introduction of the pores improves the biocompatibility and bioactivity. However, a fundamental question imposes itself regarding whether their introduction influences the mechanical strength of the electrode. In other words, we are interested in knowing if the implanted porous electrode will still survive the forces exerted by the brain environment during and after implantation. The solution to this question is demonstrated in the sections that follow.

IV. DESIGN

The following section covers the design of the proposed neural electrode. The developed neural electrode is constructed using a silicon substrate (Young's Modulus = 190 GPa & Poisson's ratio = 0.17) and is considered to be ultra-long with a total length of 10.5 mm. Its overall structure is tapered, which facilitates the penetration. The geometry of the electrode is sectioned into three main regions; a base region, a measuring region incorporating the metal recording sites, and a piercing region. The base region measures 250 μm in length with a width of 350 μm at the base that rapidly reduces to a width of 150 μm. This design aids in diminishing brain tissue damage and displacement. The measuring region that has a length of 10 mm starts with a width of 150 μm at the base and ends up with 50 μm at the other end. Following the measuring region is the piercing region, which has a length of 250 μm and is designed to be of 10 μm width at the end of the probe.

The relative dimensions of these regions are indicated in Figure 5. The electrode was implemented using COMSOL© Multiphysics 4.3 as depicted in Figure 6.

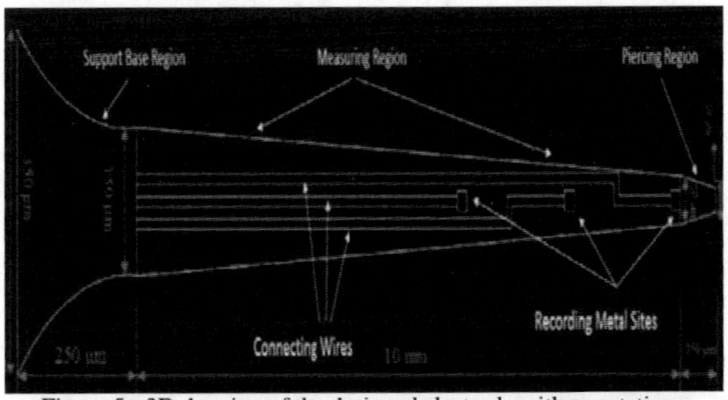

Figure 5. 2D drawing of the designed electrode with annotations.

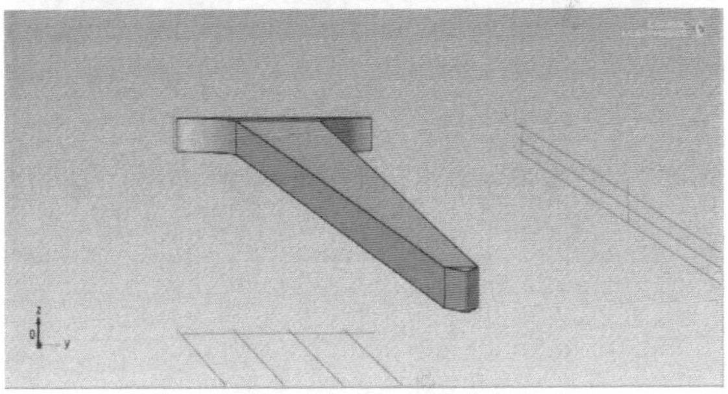

Figure 6. Complete 3D model of electrode.

In the selection of the criteria to develop the pores, we chose the various dimensions of the pores according to the limitations imposed by the standard fabrication processes followed. In other words, our selection should be similar to what is available and applicable in the fabrication world. The medium in which the probe is to be inserted contains features that exhibit micro and nano dimensions. In an attempt to mimic this medium, we select the radius of the pores in the low micro-level and the pore depth was constant at 0.6 μm due to fabrication standards. Our approach included pores with a cylindrical geometry with a radius of 1.5 μm (Figure 7).

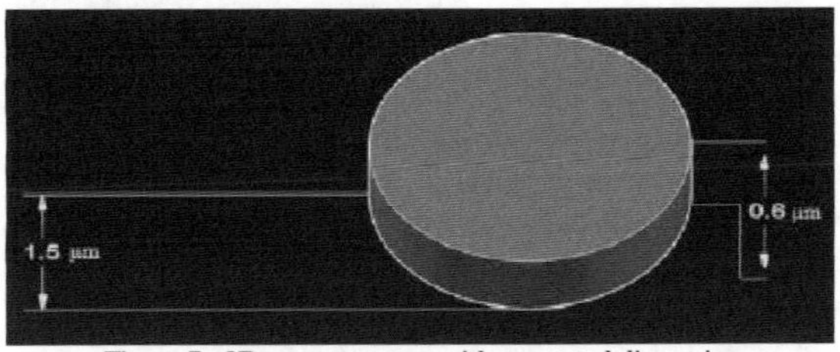

Figure 7. 3D pore geometry with annotated dimensions.

The distance separating the pores in both the x and y directions is approximately 3 μm. As mentioned earlier, each pore had a depth of 0.6 μm. The pores distributed along the entire geometry of the probe except at the regions were the metal sites and connecting wires are placed. This is depicted in Figures 8 and 9, respectively.

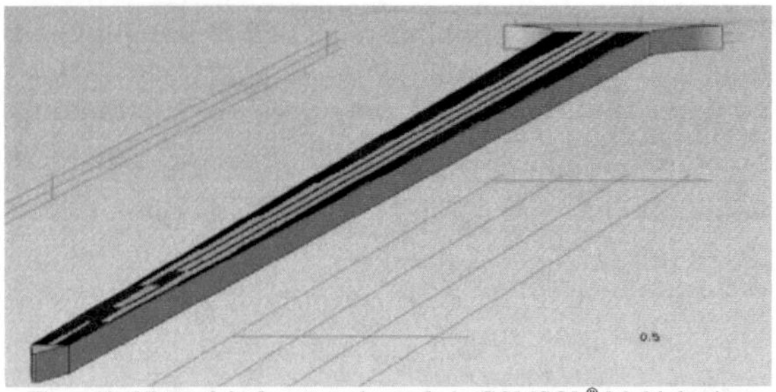

Figure 8. 3D model of porous electrode in COMSOL® Multiphysics.

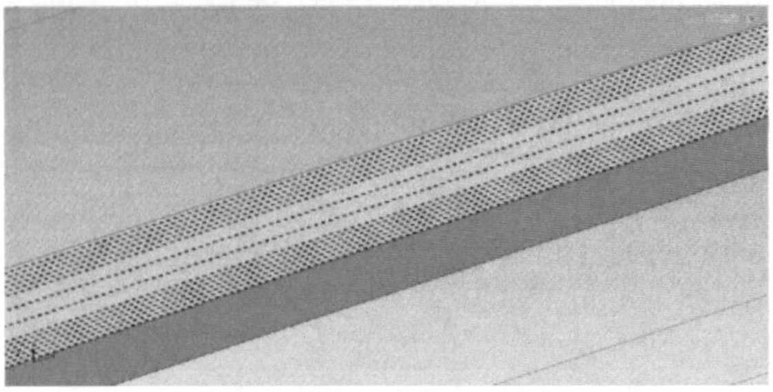

Figure 9. A close up of the arrays of pores.

V. SIMULATION STRATEGY

As an attempt to predict the mechanical behavior of the proposed probe, both a finite element model simulation approach and an analytical calculation approach were performed for both a non-porous and porous electrode. This is targeted to approximate the differences between the two approaches.

The aim is to mimic the forces applied to the electrode during and after insertion into the brain. The naturally imposed forces can be classified into three different cases; case one includes the application of two axial forces, which are imposed during the penetration phase, case two includes the application of a single axial force, which occurs directly after penetration and may cause the buckling of the probe, and case three includes the application of a vertical force, which occurs after the probe implementation and may result in the bending of the probe. Out of plane forces rarely happen because when the probe is implanted it is mounted on a motion controller, which goes in one direction towards the brain (x-axis). Hence, the force along x-axis will be focused upon throughout this paper in the different strategies followed.

Most importantly, in both strategies, the maximum critical stress was to be yielded and this stress was to be compared to the yielding stress of the material. The yielding stress of Silicon is approximately equal to 1GPa [63]. It is essential to mention that the maximum critical stress in both strategies followed is determined using the "Maximum Distortion Energy Theory", also known as the "R. von Mises Theory", which is demonstrated in equation (1) below [64], where σ_e is the effective stress or von Mises stress and $\sigma_{1,2}$ are the principal stresses.

$$\sigma_e=(\sigma_1^2+\sigma_1\sigma_2+\sigma_2^2)^{1/2} \quad (1)$$

The maximum distortion energy theory is one of the famous failure theories for ductile material. This theory states that failure is predicted to occur in the multiaxial state of stress when the distortion energy per unit volume becomes equal to or exceeds the distortion energy per unit volume at the time of failure in a simple uniaxial stress test using a specimen of the same material [65]. In other words, a given structural material is safe as long as the maximum value of the distortion energy per unit volume in that material remains smaller than the maximum distortion energy per unit volume required for causing yield in a tensile test specified of the same material. The simulated effective stress is then compared to the yielding stress of the material.

Regarding the Finite Element Model (FEM) strategy increasing forces were gradually applied on both the porous and the non-porous electrodes until the yield stress of the material is reached. These forces were applied on the front face of the piercing tip while fixing the back face of the support base region as illustrated in Figure 10.

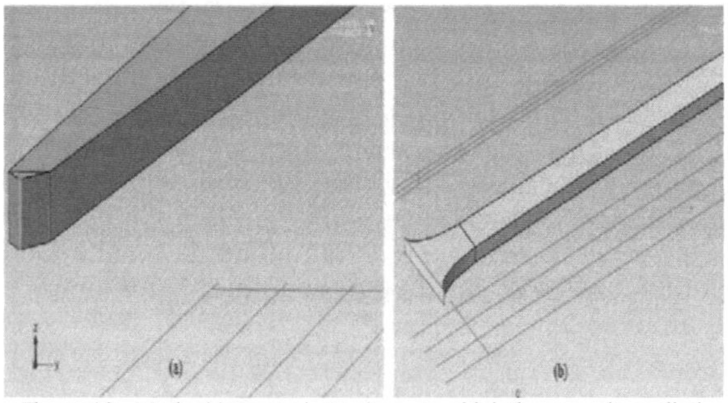

Figure 10. Marked in green is (a) Area on which the stress is applied, (b) area which is a fixed constraint.

It is notable to mention that during the simulation of the porous electrode, pores were restricted to the weakest regions of the electrode as seen in Figure 11.

This was done to reduce the computational complexity. These regions are the middle of the electrode (during axial loads), and the base region of the electrode (during vertical loads) [66].

As for the analytical strategy, the internal forces that generated in the porous and the non-porous probes upon the application of the different combination of loads in the

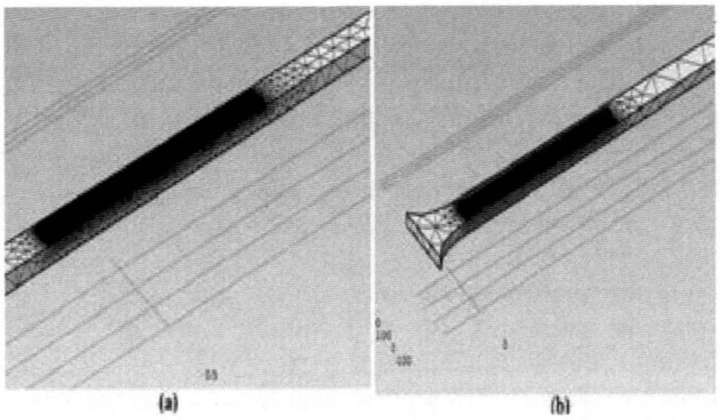

Figure 11. (a) Porous region on the middle of the probe, (b) porous region on the base.

three cases were calculated. These internal forces were used to calculate the resulting principal stresses according to equation (2) below, where σ_x and σ_y are the induced stresses and τ_{xy} is the induced shear stress.

$$\sigma_{1,2} = \frac{\sigma_x + \sigma_y}{2} \pm \sqrt{\left(\frac{\sigma_x - \sigma_y}{2}\right) + \tau_{xy}{}^2} \quad (2)$$

These principal stresses were then employed either to calculate the maximum distortion energy or the von Mises stress in case 1 (x-axis and y-axis axial forces) and case three (z-axis vertical force) or to compare with the critical stress and the elastic stress in case 2 (x-axis axial force). The calculation of the von Mises stresses is done by the previously mentioned equation. Meanwhile, the calculation of the critical stress is done by the calculation of the critical buckling load, which is yielded by the extended Euler's Formula stated below [67] .

$$P_{critical} = \frac{\pi^2 \times E \times I}{(KL)^2} \quad (3)$$

$$\sigma_{critical} = \frac{P_{critical}}{A} \quad (4)$$

where A is the cross-sectional area upon which the load is applied, $P_{critical}$ is the critical load, E is the Young's Modulus of the material, I is the moment of inertia of the cross-section, KL is the effective length of the electrode. In the case of a column fixed at its base K is equal to 2 [67].

Solving the equations will give us an idea about how far the finite element model is from the analytical equations. As mentioned earlier, out of plane forces rarely happen because when the probe is implanted it is mounted on a motion controller, which goes in one direction towards the brain (x-axis). Hence, the force along x-axis will be focused upon in the analytical strategy.

However, before starting with the FEM strategy, and in order to study the stability of the FEM with respect to the results obtained, we tried different element sizes until the results started converging to the same average values. As it will be shown in Table I in the results, the element size is changed, the probe is meshed, and the number of elements is measured.

VI. RESULTS

The following section elaborates on the results obtained from the study done to assess the stability of the FEM and the simulation of both the porous and non-porous electrodes in the FEM strategy and the analytical strategy. For each simulation, a plot of the induced principal stress in MPa versus the length of the electrode in μm as a result of applying the loads is obtained. Subsection 1 will cover the results yielded in from the FEM stability study, subsection 2 will demonstrate the results yielded from the analytical strategy, subsections 3, 4, and 5 will cover the results yielded from the FEM simulation strategy in the three different cases. Subsection 6 will show a comparison between the analytical and the FEM strategies.

1) Results of the FEM Stability Study

As mentioned earlier, different element sizes were chosen until the results starting converging to the same average values. As shown in Table I, the element size is changed, the probe is meshed, and the number of elements is measured.

TABLE I. ELEMENT SIZE VERSUS THE NUMBER OF ELEMENTS FOR THE MODELLED NEURAL PROBE

Element Size (μm)	Number of Elements
2000	265
1000	311
500	426
250	532
125	961
72	2041
36	13712
18	133688
9	1158561
4	13900199

The results obtained are also plotted in a curve as illustrated in Figure 12. One can notice the exponential increase in the number of elements when the element size decreased.

The results were also calculated for different element size. Element size was decreased until the finite element model started converging to a range of close values as shown in Table II.

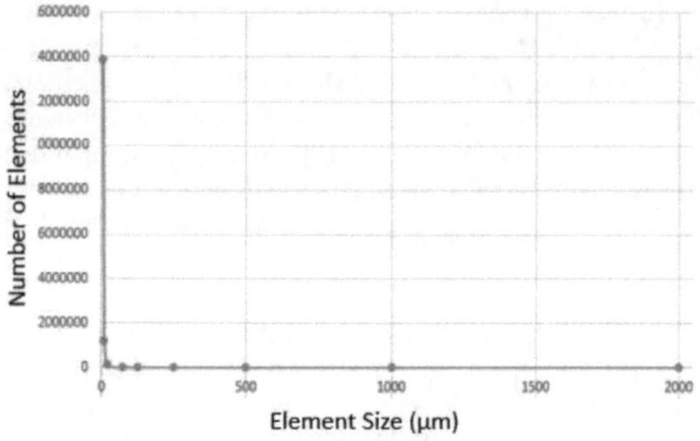

Figure 12. A plot showing the element size versus the number of elements.

The results obtained are plotted on a curve as shown in Figure 13. It is important to note that the y-axis starts from 1000 MPa. Starting from 0 MPa the result was a straight horizontal line, hence it was changed to start from 1000 MPa to shown the fluctuations before the results converged.

TABLE II. ELEMENT SIZE VERSUS VON-MISES STRESS

Element Size (µm)	Result (MPa- Von-Mises)
2000	1053.1
1000	1076.45
500	1066.91
250	1081.88
125	1077.49
72	1082.47
36	1064.15
18	1063.12
9	1064.29
4	1064.76

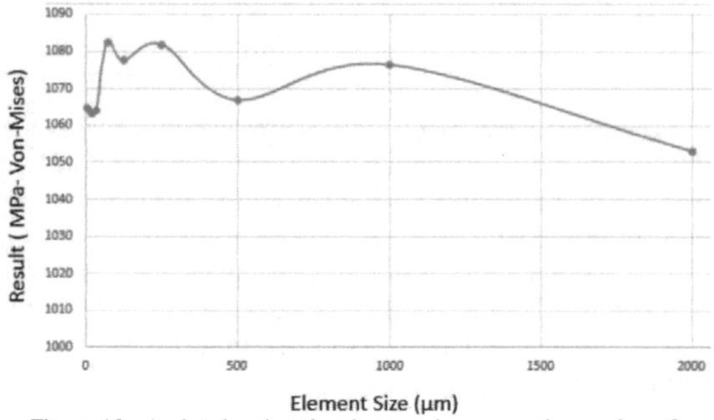

Figure 13. A plot showing the element size versus the number of elements.

2) Results of the Analytical Strategy

It is beneficial to reiterate that the results presented in this section are yielded by performing calculations upon the application of the axial force since is it the most significant.

For the non-porous electrode, upon the application of the maximum force of 527.5 mN the internal stress induced was found to be equal to 1055 MPa. Meanwhile, upon the application of the minimum force of 2.42 mN [68], the internal stress induced was found to be 4.94 MPa. Additionally, the critical force and critical stress that are required for comparison were also calculated according to the previously mentioned equations and were found to be equal to 4.4 mN and 8.8 MPa, respectively.

Regarding the porous electrode, two main parameters must be obtained in order to characterize the mechanical behavior analytically. The parameters are the young's modulus of silicon at the specific porosity and the yield stress at which the material will fail at this porosity. The porosity was calculated by taking into consideration the number of pores and their cylindrical volume with respect to the total volume of the electrode. This is demonstrated in equations (5) and (6).

$$Ratio\ of\ Volume = \frac{V_{pores}}{V_{electrode}} = \frac{8.396\times10^{-14}}{5.35\times10^{-11}} = 1.56 \qquad (5)$$

$$\%\ Porosity = Ratio\ of\ Volume \times 100 = 1.569 \times 10^{-3} \times 100 \cong 0.16\% \qquad (6)$$

In order to obtain the young's modulus of porous silicon at our porosity, different studies were researched. These studies are illustrated in Figure 14. The closest study for the change in young's modulus relative to the percentage of porosity is the one done by Al-Douri et al. [69]. As for the yield stress of our porous silicon material, there are no previous studies in the literature so far. Since the geometry of the porous electrode is the same as the non-porous one, and since both are of the same material we will do the following assumption. We will assume that both will have the same yield strain that will to failure. Having this yield strain and knowing the young's modulus we are able to calculate the yield stress of our porous silicon according to Hook's law given in the equation below.

$$\sigma = E \times \varepsilon_y \qquad (7)$$

Since in the selected literature the closest young's modulus to ours, which is 190 GPa, is 185, we will redo the FEM simulation at 185 GPa in order to compare it to the analytical solution. After repeating the simulation for case 2 only at 185 GPa, result showed that in the porous probe a stress of 970 MPa induced the yield stress of 961.67 MPa. The color map of this FEM simulation is displayed below in Figure 15 only to confirm the yielded result.

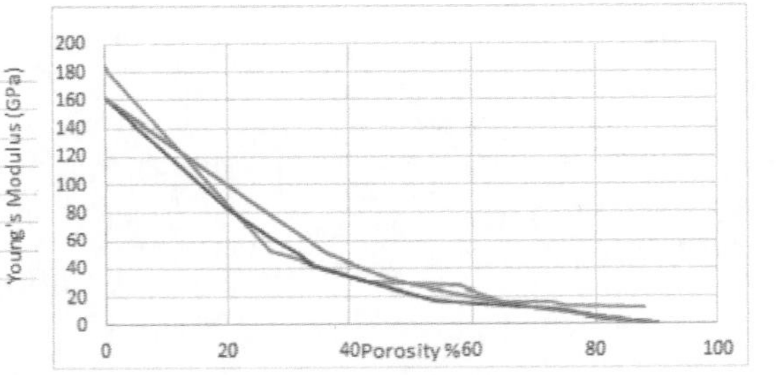

Figure 14. Different Studies on the Effect of Porosity on the Young's Modulus of Porous Silicon. The three curves are extracted from three different studies (In green [70], in blue [69], and in red [71]).

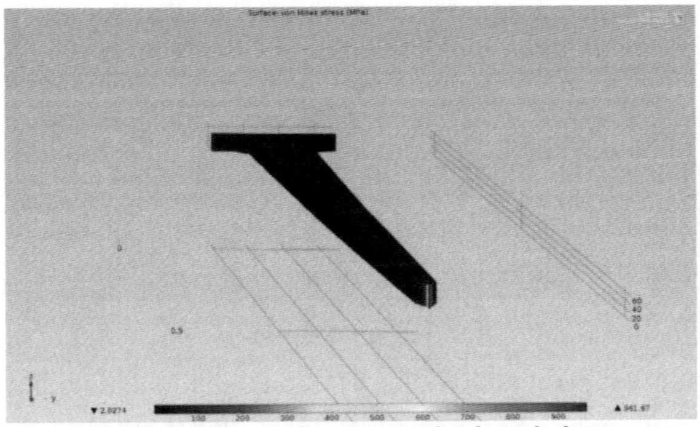

Figure 15. 3D plot of the surface stress on the electrode due to –ve x-axis axial stress at E= 185 (Porous).

That being said, the new young's modulus at our porosity (0.16%) is 184.2 GPa and the yield stress is 969.47 MPa. According to the new E, the new critical stress and force are to be calculated. The critical stress for the porous electrode was found to be equal to 8.5 MPa, and the critical force equal to 4.26 mN.

3) Results of the Application of a Compression Force Along X-axis During Penetration (Case 2)

This section demonstrates the result of the application of a single axial force that occurs directly after penetration. For the non-porous electrode, an axial stress of 1055 MPa that is equivalent to a force of 527.5 mN along the negative x-axis induced stress of around 1GPa at a length of around 10.5 mm, which corresponds to the tip of the electrode. This is depicted in Figure 16. The concentration of the induced stresses is illustrated in Figure 17.

As for the porous electrode, a force of 522.5 mN induced a similar response at a similar location as shown in Figure 19 and illustrated in Figure 18. It is worth noting that the high stress is concentrated at the tip of the electrode that is in direct contact with the brain. This contact induces the most significant stress that may cause the failure of the probe.

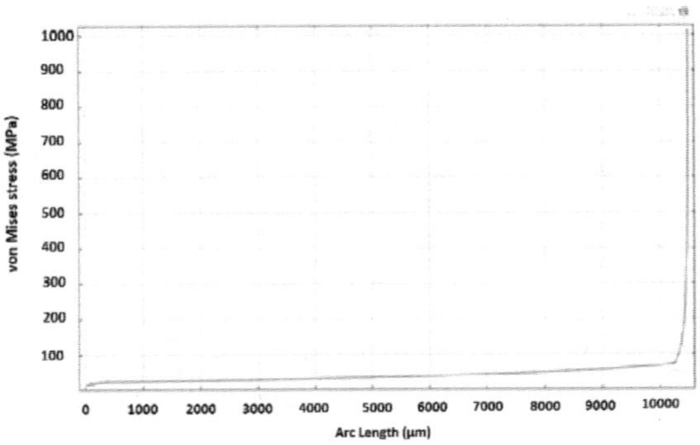

Figure 16. *Von Mises* stress induced upon applying an axial force along the negative *x*-axis versus the electrode length (non-porous).

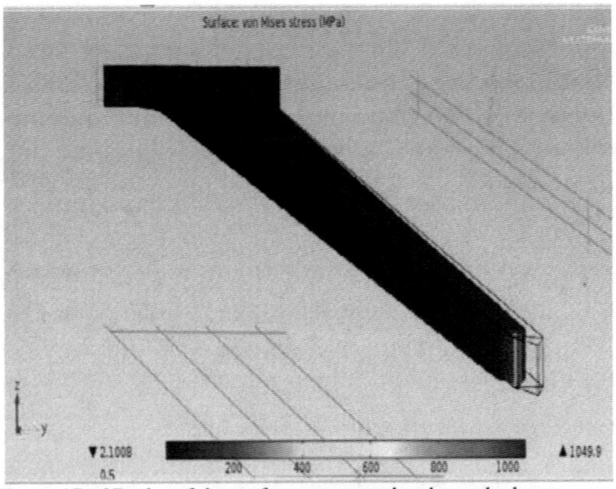

Figure 17. 3D plot of the surface stress on the electrode due to –ve x-axis axial stress (non-porous).

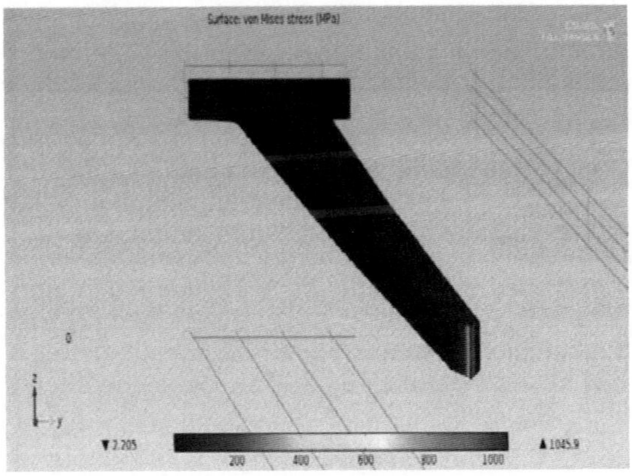

Figure 18. 3D plot of the surface stress on the electrode due to –ve x-axis axial stress (Porous).

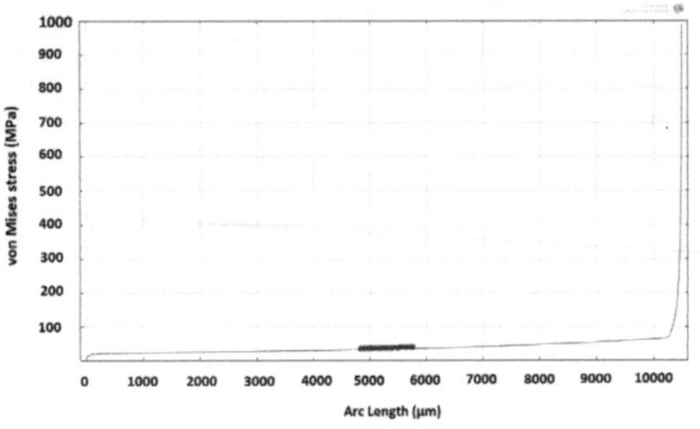

Figure 19. *Von Mises* stress induced upon applying an axial force along the negative *x*-axis versus the electrode length (porous).

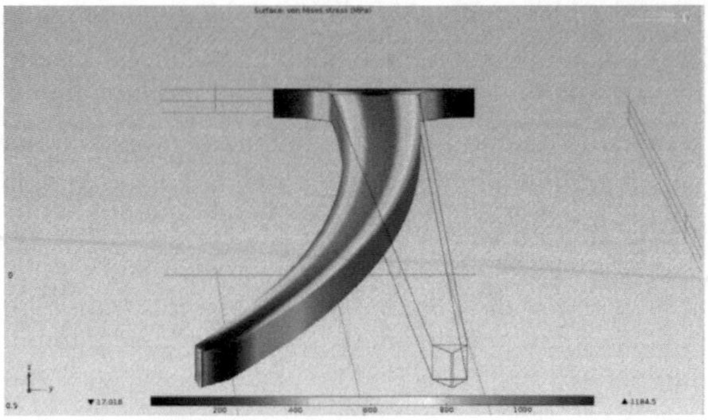

Figure 21. 3D plot of the surface stress on the electrode due a combination of axial stresses on –ve x & y axes (Non-Porous).

4) Results of the Application of the Force Induced Due to Slipping Upon Insertion (Case 1)

This section covers the application of two axial forces (along negative x and y axes), which are imposed during the penetration phase. For the non-porous electrode, a force of 378 mN (x: 375, y: 47.5) induced a stress of ~ 1 GPa at two locations in the electrode; at a distance of ~ 5800 µm corresponding to the middle region (the blue curve in Figure 20), and at a distance of ~ 10.5 mm corresponding to the tip of the electrode (the green curve in Figure 20). The resulting surface stresses are demonstrated in Figure 21.

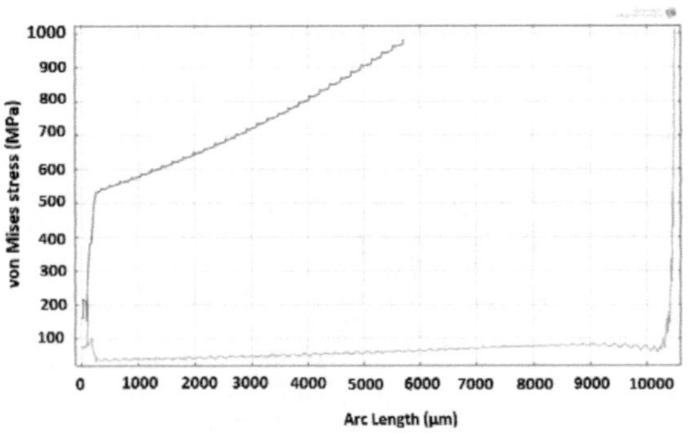

Figure 20. *Von Mises* stress induced upon applying an axial force along the negative y-axis versus the electrode length (non-porous).

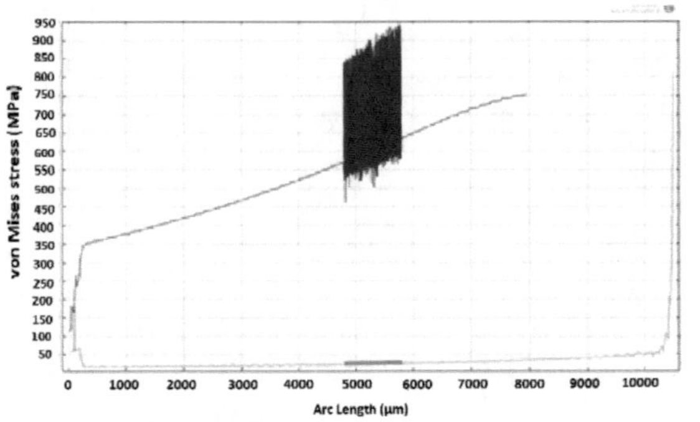

Figure 22. *Von Mises* stress induced upon applying an axial force along the negative y-axis versus the electrode length (porous).

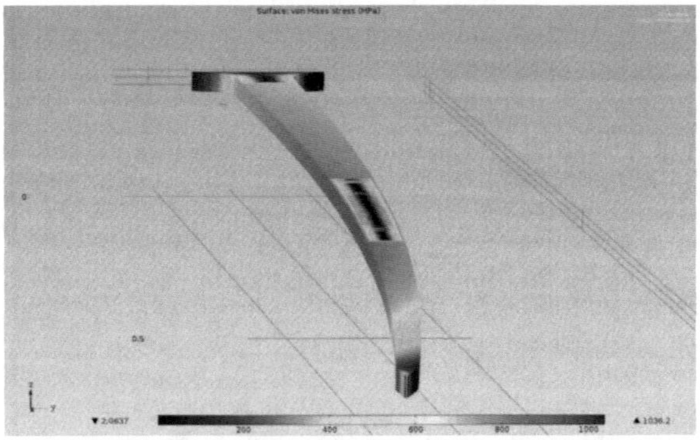

Figure 23. 3D plot of the surface stress on the electrode due a combination of axial stresses on –ve x & y axes (Porous).

Meanwhile, in the porous electrode, a force of 378 mN (x: 375, y: 47.5) induced a stress of ~ 1 GPa at two locations in the electrode; at a distance of ~ 5800 µm corresponding to the middle porous region (the blue curve in Figure 22), and at a distance of ~ 10.5mm corresponding to the tip of the electrode (the green curve in Figure 22). The resulting surface stresses are demonstrated in Figure 23.

5) Results of the Application of the Force Induced Upon Brain Movement After Implantation (Case 3)

Finally, this section covers the application of a vertical force that occurs after the probe implementation. For the non-porous electrode, a stress of 36.5 MPa (18.25m N) applied along the –ve z-axis induced a stress of ~ 1000 MPa (Figure 24) at a length of around 400 µm, which corresponds to the fixed bottom region of the electrode as depicted in Figure 25.

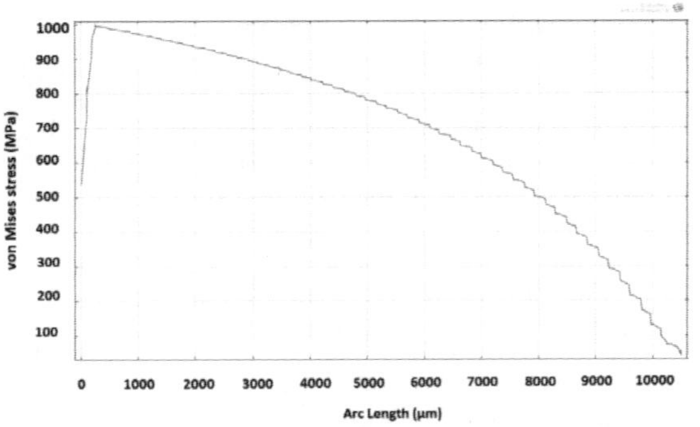

Figure 24. *Von Mises* stress induced upon applying a vertical force along the negative z-axis versus the electrode length (non-porous).

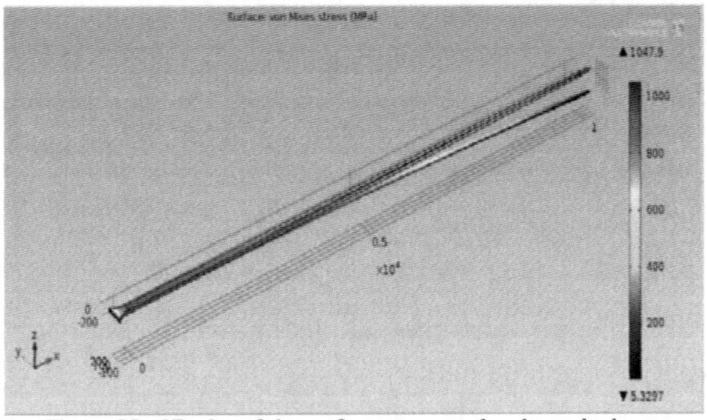

Figure 25. 3D plot of the surface stress on the electrode due to –ve z-axis vertical stress (non-porous).

On the other hand, in the porous electrode, a stress of 23 MPa (11.5 mN) applied along the –ve z-axis induced a stress of ~ 1000 MPa (Figure 26) at a length of around 750 μm, which corresponds to the weakest porous fixed bottom region of the electrode as depicted in Figure 27.

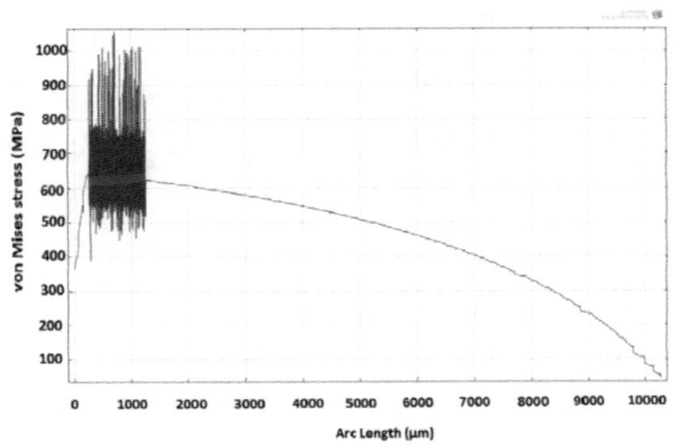

Figure 26. *Von Mises* stress induced upon applying a vertical force along the negative z-axis versus the electrode length (porous).

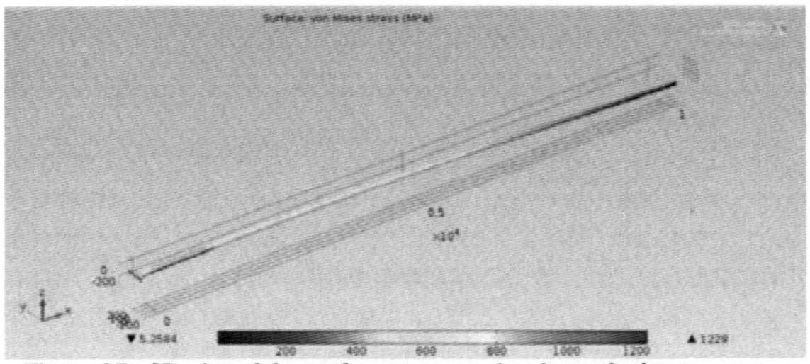

Figure 27. 3D plot of the surface stress on the electrode due to –ve z-axis vertical stress (Porous).

6) Discussion

After the mechanical behavior of the electrodes was analyzed using two different approaches, a FEM and an analytical approach, we were able to estimate the difference between the FEM model simulation and the analytical solution by a comparison of both results (Table III). This comparison showed that the similarity between the two approaches was equal to 96.76%. The minimal difference is possibly due to that FEM is a physical model and the equations cannot fully represent the model.

TABLE III. COMPARISON BETWEEN RESULTS OF POROUS AND NON-POROUS PROBES

Load along –ve x (mN)	σ_x (MPa) Non-porous	Load along –ve x (mN)	σ_x (MPa) Porous
527.5	1055	485	961
2.42	4.94 < $\sigma_{cr=8.8}$	2.42	4.94 < $\sigma'_{cr=8.5}$

On the other hand, regarding the FEM strategy performed on the three different cases, due to the comparison of the forces that induced the maximum yield stress in the three different cases for the non-porous probe (Case 1: 378 mN (x: 375, y: 47.5); Case 2: 527.5 mN; Case 3: 18.25 mN) and the ones for the porous probe (Case 1: 376.8 mN (x: 375, y: 37.5); Case 2: 522.5 mN; Case 3: 11.5 mN), one can notice the mechanical weakening of the porous silicon probe. This is true due to the fact that the values of the forces that induced the maximum yield stress in the porous probe are less than those in the non-porous probe. The weakening was 0.3 %, 0.1%, 37% in case 1, 2 and 3, respectively.

Nonetheless, this weakening will not risk the mechanical integrity of the neural probe used in brain applications. The reason is that the force that induced the maximum yield stress of the porous probe (522.5mN along the x-axis) is still much higher than the minimum force that the probe must withstand during the penetration of the brain tissue (2.42±0.77 mN along the x-axis [66]).

VII. Conclusion

We have presented the novel idea of the mechanical simulation of a porous neural electrode. Even though the introduction of the pores relatively weakened the neural electrode, the electrode was found still capable of surviving the brain environment. Nevertheless, certain limitations were present especially related to the finite element model. The full arrays of pores could not be simulated due to computational complexity and they were restricted to the weakest areas. Moreover, different radii of pores and volume porosity percentages should be tested. The porous electrode is superior to the non-porous electrode due to the improved biocompatibility and bioactivity it offers. Furthermore, the presence of the pores gives an additional advantage where they can behave as scaffolds for entrapping neural growth factors that encourage the re-growth of neurons. This alteration to the electrode's design is able to advance the healthcare services provided to neural diseases' patients all around the world.

References

[1] S. B. R. Fayad, B. AbdulHay, H. Hajj-Hassan, H. Khachfe, M. Hajj-Hassan, "Modelling and Simulation of Mechanical Behaviour of an Ultra-Long Porous Silicon Neural Microelectrode," in Proc. of The Third International Conference on Global Health Challenges, Rome, Italy, pp. 50-54, 2014.

[2] B. He, *Neural Engineering*: Springer London, Limited, 2013.

[3] M. A. L. Nicolelis, "Brain–machine interfaces to restore motor function and probe neural circuits," *Nature Reviews Neuroscience,* vol. 4, pp. 417-422, 2003.

[4] A. B. Schwartz, "Cortical neural prosthetics," *Annual Review of Neuroscience,* vol. 27, pp. 487–507, 2004.

[5] K. C. Cheung, "Implantable microscale neural interfaces," *Biomed Microdevices,* vol. 9, pp. 923–938, 2007.

[6] K. D. Wise, A.M. Sodagar, Y. Yao, M. N. Gulari, G. E. Perlin, and K. Najafi, "Microelectrodes, microelectronics, and implantable neural microsystems," *Proceedings of the IEEE,* vol. 96, pp. 1184-1202, 2008.

[7] R. Jung, *Biohybrid Systems: Nerves, Interfaces and Machines*: Wiley, 2012.

[8] S. F. Cogan, "Neural stimulation and recording electrodes," *Annual Review of Biomedical Engineering,* vol. 10, pp. 275-309, 2008.

[9] T. Yoshimine, "Brain-machine interface (BMI) - application to neurological disorders," *Rinsho Shinkeigaku,* vol. 53, pp. 962-965, 2013.

[10] P. Ponce, A. Molina, D. C. Balderas, and D. Grammatikou, *Brain Computer Interfaces for Cerebral Palsy*, 2014.

[11] V. C. M. Hajj-Hassan, and S. Musallam, "NeuroMEMS: Neural probe microtechnologies," *Sensors,* vol. 6704-6726, pp. 6704-6726, 2008.

[12] T. Kozai, N. Langhals, P. R. Patel, X. Deng, H. Zhang, K. L. Smith, *et al.*, "Ultrasmall implantable composite microelectrodes with bioactive surfaces for chronic neural interfaces," *Nature Materials,* vol. 11, pp. 1065–1073, 2012.

[13] F. Strumwasser, "Long-term recording from single neurons in brain of unrestrained mammals," *Science,* vol. 127, pp. 469-470, 1958.

[14] J. C. M. Jeon, Y. K. Kim, D. Jung, E-S. Yoon, S. Shin, and I-J. Cho, "Partially flexible MEMS neural probe composed of polyimide and sucrose gel for reducing brain damage during and after implantation," *Journal of Micromechanics and Microengineering,* vol. 24, 2014.

[15] B. Ghane-Motlagh and M. Sawan, "Design and implementation challenges of microelectrode arrays: A review " *Materials Sciences and Applications,* vol. 4, pp. 483-495 2013.

[16] T. G. H. Yuen and W. F. Agnew, "Histological evaluation of polyesterimide-insulated gold wires in brain," *Biomaterials,* vol. 16, pp. 951–956, 1995.

[17] J. C. Williams, R. L. Rennaker, and D. R. Kipke, "Stability of chronic multichannel neural recordings: Implications for a long-term neural interface," *Neurocomputing,* vol. 26-27, pp. 1069-1076, 1999.

[18] D. B. M. L. Xindong, R. R. Carter, L.A. Bullara, T. G. H. Yuen, and W. F. Agnew, "Stability of the interface between neural tissue and chronically implanted intracortical microelectrodes," *IEEE Transactions on Neural Systems and Rehabiliation Engineering,* vol. 7, pp. 315-326, 1999.

[19] D. D. M. A. L. Nicolelis, J. M. Carmena, R. Crist, G. Lehew, J. D. Kralik, and S. P. Wise, "Chronic, multisite, multielectrode recordings in macaque monkeys," *Proceedings of the National Academy of Sciences of the United States of America,* vol. 100, pp. 11041–11046, 2003.

[20] P. F. G. Bora Kim, and M. R. Abidian, "Electrode–neural tissue interactions: Immune responses, current technologies, and future directions," in *Biomaterials Surface Science*, J. F. M. A. Taubert, and J. C. Rodriguez-Cabello, Ed., ed: Wiley-VCH, 2013.

[21] D. A. Robinson, "The electrical properties of metal microelectrodes," *Proceedings of the IEEE,* vol. 56, pp. 1065-1071, 1968.

[22] J. Skrzypek and E. Keller, "Manufacture of metal microelectrodes with the scanning electron microscope," *IEEE Transactions on Biomedical Engineering,* vol. 22, pp. 435-437, 1975.

[23] C. A. Terzuolo and T. Araki, "An analysis of intra-versus extracellular potential changes associated with activity of single spinal motoneurons," *Annals of the New York Academy of Sciences,* vol. 94, pp. 547-558, 1961.

[24] M. T. V. Tsytsarev, F. Schottler, S. Tanaka, and M. Hara, "A new planar multielectrode array: recording from a rat auditory cortex," *Journal of Neural Engineering,* vol. 3, pp. 293–298, 2006.

[25] I. F. H. Scherberger, S. Musallam , D. J. Dubowitz, K. A. Bernheim, B. Pesaran, B. D. Corneil, B. Gilliken , and R. A. Andersen "Magnetic resonance image-guided implantation of chronic recording electrodes in the macaque intraparietal sulcus," *Journal of Neuroscience Methods,* vol. 130, pp. 1-8, 2003.

[26] S. Musallam, M. J. Bak, P. R. Troyk, and R. A. Andersen, "A floating metal microelectrode array for chronic implantation," *Journal of Neuroscience Methods,* vol. 160, pp. 122-127, 2007.

[27] Z. Yang, *Neural computation, neural devices, and neural prosthesis*: Springer, 2014.

[28] D. Zhou and E. Greenbaum, *Implantable neural prostheses 2: Techniques and engineering approaches*: Springer, 2010.

[29] R. Brette and A. Destexhe, *Handbook of neural activity measurement*: Cambridge University Press, 2012.

[30] K. D. Wise, J. B. Angell, and A. Starr, "An integrated-circuit approach to extracellular microelectrodes," *IEEE Transactions on Biomedical Engineering,* vol. 17, pp. 238-247, 1970.

[31] K. D. Wise and J. B. Angell, "A low-capacitance multielectrode probe for use in extracellular neurophysiology," *IEEE Transactions on Biomedical Engineering,* vol. 22, pp. 212-219, 1975.

[32] J. K. S. Kisban, P. Janssen, R. V. Metzen, S. Herwik, U. Bartsch, T. Stieglitz, O. Paul, and P. Ruther, "A novel assembly method for silicon-based neural devices," in *World congress on medical physics and biomedical engineering*. vol. 25, O. D. a. W. C. Schlegel, Ed., ed: Springer, 2009, pp. 107-110.

[33] E. M. M. R. A. Normann, P. J. Rousche, and D. J. Warren, "A neural interface for a cortical vision prosthesis," *Vision Research,* vol. 39, pp. 2577-2587, 1999.

[34] P. K. C. K. E. Jones, and R. A. Normann, "A glass silicon composite intracortical electrode array," *Ann Biomed Eng,* vol. 20, 1992.

[35] R. J. V. D. R. Kipke, J. C. Williams, and J. F. Hetke, "Silicon-substrate intracortical microelectrode arrays for long-term recording of neuronal spike activity in cerebral cortex," *IEEE Transactions on Neural Systems and Rehabilitation Engineering,* vol. 11, pp. 151-154, 2003.

[36] K. Najafi, "Solid-state microsensors for cortical nerve recordings," *IEEE Engineering in Medicine and Biology Magazine,* vol. 13, pp. 375-387, 1994.

[37] K. Najafi, "Micromachined systems for neurophysiological applications," in *Handbook of microlithography, micromachining, and microfabrication: Micromachining and microfabrication*, P. Rai-Choudhury, Ed., ed: SPIE Optical Engineering Press, 1997.

[38] J. Wolpaw and E. W. Wolpaw, "Acquiring brain signals from within the brain," in *Brain-computer interfaces: Principles and practice*, ed: Oxford University Press, 2012, pp. 81-103.

[39] D. S. P. P. J. Rousche, D. P. Pivin, J. C. Williams, R. J. Vetter, and D. R. Kipke "Flexible polyimide-based intracortical electrode arrays with bioactive capability," *IEEE Transactions on Biomedical Engineering,* vol. 48, pp. 361-371, 2001.

[40] A. B. S. Metz, D. Bertrand, and P. Renaud, "Flexible polyimide probes with microelectrodes and embedded microfluidic channels for simultaneous drug delivery and multi-channel monitoring of bioelectric activity," *Biosensors and Bioelectronics,* vol. 19, pp. 1309-1318, 2004.

[41] T. Stieglitz and M. Gross "Flexible BIOMEMS with electrode arrangements on front and back side as key component in neural prostheses and biohybrid systems," *Sensors and Actuators B: Chemical,* vol. 83, pp. 8-14, 2002.

[42] T. Stieglitz, "Flexible biomedical microdevices with double-sided electrode arrangements for neural applications," *Sensors and Actuators A: Physical,* vol. 90, pp. 203-211, 2001.

[43] H. B. T. Stieglitz, M. Schuettler, and J. Meyer, "Micromachined, polyimide-based devices for flexible neural interfaces," *Biomedical Microdevices,* vol. 2, pp. 283-294, 2000.

[44] T. C. O. B. L. Seala, and A. Panitch, "Polymeric biomaterials for tissue and organ regeneration," *Materials Science and Engineering: R: Reports,* vol. 34, pp. 147–230, 2001.

[45] D. C. F. J. Rodríguez, M. Schüttler, A. Valero, E. Valderrama, T. Stieglitz, and X. Navarro, "Polyimide cuff electrodes for peripheral nerve stimulation," *Journal of Neuroscience Methods,* vol. 98, pp. 105-118, 2000.

[46] R. R. Richardson, J. A. Miller, and W. M. Reichert "Polyimides as biomaterials: preliminary biocompatibility testing," *Biomaterials,* vol. 14, pp. 627-635, 1993.

[47] K. E. J. P. K. Campbell, R. J. Huber, K. W. Horch, and R. A. Normann, "A silicon-based, three-dimensional neural interface: manufacturing processes for an intracortical electrode array," *IEEE Transactions on Biomedical Engineering,* vol. 38, pp. 758-768, 1991.

[48] W. L. C. Xu, and C. Liu, "Design and fabrication of a high-density metal microelectrode array for neural recording," *Sensors and Actuators A: Physical,* vol. 96, pp. 78-85, 2002.

[49] E. F. Cristina Marin, "Biocompatibility of intracortical microelectrodes: current status and future prospects," *Frontiers in Neuroengineering,* vol. 3, 2010.

[50] B. G. E. Fernández, P. A. House, I. Aranda, C. Botella, J. Albisua, C. Soto-Sánchez , A. Alfaro, and R. A. Normann, "Acute human brain responses to intracortical microelectrode arrays: challenges and future prospects," *Frontiers in Neuroengineering* vol. 7, pp. 1-6, 2014.

[51] P. A. T. V. S. Polikov, and W. M. Reichert, "Response of brain tissue to chronically implanted neural electrodes," *Journal of Neuroscience Methods,* vol. 148, pp. 1–18, 2005.

[52] W. S. J. N. Turner, D. H. Szarowski, M. Andersen, S. Martins, M. Isaacson, and H. Craighead, "Cerebral astrocyte response to micromachined silicon implants," *Experimental Neurology,* vol. 156, pp. 33-49, 1999.

[53] S. S. Stensaas and L. J. Stensaas, "The reaction of the cerebral cortex to chronically implanted plastic needles," *Acta neuropathalogica,* vol. 35, pp. 187-203, 1976.

[54] J. P. Seymour and D. R. Kipke , "Neural probe design for reduced tissue encapsulation in CNS," *Biomaterials,* vol. 28, pp. 3594-3607, 2007.

[55] H. D. R. G. C. McConnell, A. I. Levey, C. A. Gutekunst, R. E. Gross, and R. V. Bellamkonda, "Implanted neural electrodes cause chronic, local inflammation that is correlated with local neurodegeneration," *Journal of Neural Engineering,* vol. 6, p. 056003, 2009.

[56] M. D. A. D. H. Szarowski, S. Retterer, A. J. Spence, M. Isaacson, H. G. Craighead, J. N. Turner, and W. Shain "Brain responses to micro-machined silicon devices," *Brain Research,* vol. 983, pp. 23-35, 2003.

[57] E. J. K. N. K. Wood, and R. J. Oglesby, "The significance of implant shape in experimental testing of biological materials: Disc vs. Rod," *Journal of Biomedical Materials Research,* vol. 4, pp. 1-12, 1970.

[58] S. R. Taylor and D. F. Gibbons, "Effect of surface texture on the soft tissue response to polymer implants," *Journal of Biomedical Materials Research,* vol. 17, pp. 205-227, 1983.
[59] A. Rosengren, L.M. Bjursten, N. Danielsen, H. Persson, and M. Kober, "Tissue reactions to polyethylene implants with different surface topography," *Journal of Materials Science: Materials in Medicine,* vol. 10, pp. 75-82, 1999.
[60] L. W. Agneta Rosengren, Nils Danielsen, Thomas Laurell,and Lars Magnus Bjursten, "Tissue Reactions Evoked by Porous and Plane Surfaces Made Out of Silicon and Titanium," *IEEE Transactions on Biomedical Engineering,* vol. 49, 2002.
[61] S. H. K. A. Moxon, A. Aslani , N. M. Kalkhoran, and P. I. Lelkes, "Bioactive properties of nanostructured porous silicon for enhancing electrode to neuron interfaces," *Journal of Biomaterials Science, Polymer Edition,* vol. 18, pp. 1263–1281, 2007.
[62] M. Hajj-Hassan M, H. Wang, V. Chodavarapu, and J.E. Henderson, "Response of murine bone marrow-derived mesenchymal stromal cells to dry-etched porous silicon scaffolds," *Journal of Biomedical Materials Research Part A,* 2011.
[63] C. J. Wilson and P. A. Beck, "Fracture Testing of Bulk Silicon Microcantiliver Beams Subjected to a Side Load," *Journal of Microelectromechanical Systems,* vol. 5, pp. 142-150, 1996.
[64] A. Blake, *Handbook of Mechanics, Materials, and Structures*: Wiley, 1985.
[65] J. A. Collins, H. R. Busby, and G. H. Staab, *Mechanical Design of Machine Elements and Machines*: Wiley, 2010.
[66]. M. Hajj-Hassan, and S. Musallam, "Reinforced silicon neural microelectrode array fabricated using a commercial MEMS process," *Journal of Micro/Nanolithography, MEMS, and MOEMS,* vol. 8, p. 033011, 2009.
[67] J. Gere and B. Goodno, *Mechanics of Materials*: Cengage Learning, 2008.
[68]. M. Hajj-Hassan, and S. Musallam, "Reinforced silicon neural microelectrode array fabricated using a commercial MEMS process," *Journal of Micro/Nanolithography, MEMS, and MOEMS,* vol. 8, 2009.
[69] Y. Al-Douri, N. M. Ahmed, N. Bouarissa, and A. Bouhemadou, "Investigated optical and elastic properties of Porous silicon: Theoretical study," *Materials and Design,* vol. 32, pp. 4088-4093, 2011.
[70] D. Bellet;, P. Lamagnere;, A. Vincent;, and Y. Bre´chet;, "Nanoindentation investigation of the Young's modulus of porous silicon," *Journal of Applied Physics,* vol. 80, 1996.
[71] "Nanocrystalline Porous Silicon," in *Crystalline Silicon - Properties and Uses*, S. Basu, Ed., ed: InTech, 2011.

Digital Users in Pre-Digital Hospital Organisations?

An Analysis on the Readiness for Electronic Communication Between a Hospital and Surgical Patients

Kari Dyb
Norwegian Centre for eHealth Research
University Hospital of North Norway
Tromsø, Norway
kari.dyb@telemed.no

Conceição Granja
Norwegian Centre for eHealth Research
University Hospital of North Norway
Tromsø, Norway
conceicao.granja@telemed.no

Stein Roald Bolle
Norwegian Centre for eHealth Research
University Hospital of North Norway
Tromsø, Norway
stein.roald.bolle@telemed.no

Gunnar Hartvigsen
Norwegian Centre for eHealth Research
University Hospital of North Norway
Tromsø, Norway
Department of Computer Science
UIT – The Arctic University of Norway
Tromsø, Norway
gunnar.hartvigsen@telemed.no

Abstract—**Are hospitals and surgical patients ready for electronic communication prior to surgery? This paper discusses the readiness for electronic communication between surgical patients and a university hospital in Norway. The first step in our study approach was to map the main actors involved in pre-operative planning and electronic communication. This work mapped six diverse main actors and revealed the need for multiple approaches to address the e-readiness of the different actors. The study approach consists of the following: (1) a study of the most recent health reforms in Norway, focusing on e-readiness from political and policy perspectives; (2) an in-depth empirical observation and interview study of the pre-operative planning process at a university hospital, focusing on the readiness for two-way electronic communication prior to surgery; (3) a qualitative interview study of patients' experiences with surgical cancellations, focusing on the patients' readiness for electronic communication; (4) an inquiry into the readiness of the hospitals' electronic health record to integrate two-way communication and (5) a study of the readiness for electronic patient–hospital communication from the perspective of the regional health authority's ICT operational unit. The results are reported in six analytical categories based on the identification of the main actors in the field. The authors' conclusion is that Norwegian health policy strongly promotes electronic collaboration and that patients and healthcare workers are ready to use new electronic tools. However, the hospital as an entity, together with the electronic health record system and the authority's ICT operational unit—all of which are important actors in the field—are currently not ready for electronic communication between patients and the hospital.**

Keywords—e-readiness; electronic communication; ICT; health policy; hospitals; surgical departments; healthcare workers; patients; electronic health record; health authority; Norway

I. INTRODUCTION

The aim of this paper is to discuss the readiness for electronic communication between surgical patients and a university hospital in Norway. The interest in the subject, e-readiness, stems from recent work in a research and development project, 'eTeam-surgery' [1]. The overarching goal of the eTeam-surgery project was to develop a tool for electronic communication between surgical patients and the University Hospital of North Norway (UNN). The idea of the project was to actively involve the patient in the pre-operative planning process prior to hospitalisation in order to reduce the number of surgery cancellations at the hospital.

For many patients, undergoing surgery is a major life event which involves a high level of anxiety before admission to the hospital [2][3][4][5]. In most hospitals, surgical departments are both a major area of investment and the greatest source of revenue [1][6][7]. Nonetheless, elective surgeries are regularly cancelled, and cancellation rates of 10–40 % have been reported [7][8][9]. In Western countries, up to 20 % of elective surgeries are cancelled on the day of the surgery, and this percentage increases if cancellations within the week of the scheduled surgery are included. Despite the waste of hospital resources, the anxiety and emotional stress placed on patients and the frustrations among healthcare workers, surgical cancellations seem to be commonplace in public hospitals.

The reasons for elective surgery cancellations vary. In the literature, the causes of cancellation are often divided into the following two major categories: (a) hospital-related and (b) patient-related. Hospital-related reasons encompass such issues as the unavailability of the surgical team [10][11][12], incomplete pre-operative study/preparation [13][14], lack of surgical/anaesthetic readiness [11][13] and lack of theatre time due to the extended duration of scheduled surgeries [13]. On the other hand, patient-related causes are mostly due to patient no-shows or to a patient's need to reschedule the assigned surgical date [11][13].

At our research site, 50 % of all surgery cancellations were identified as avoidable [15], which corresponds with the literature [7][8][9]. Avoidable cancellations refer to, among other things, those involving a lack of information; hence, these are cancellations where information existed prior to surgery, but was not available when required [9][15][16][17][18][19]. For further information on the rate of elective surgery cancellations at the research site and the reasons reported for these cancellations, refer to [20].

The eTeam-surgery project group consists of an interdisciplinary research team which is studying if and how electronic communication, prior to hospitalisation, can actively involve the patient in the pre-operative planning process, provide the missing information and thereby reduce cancellations (Figure 1). For further information on the eTeam-surgery project, refer to [21]. A central goal for the eTeam-surgery project was to develop and test a two-way communication tool that would enable patients to communicate with the surgical department by storing the patient's input in the UNN hospital electronic health record (EHR). In this way, patients could inform the surgical department about changes in their health status or other factors that could potentially affect their forthcoming surgery. The eTeam-surgery intervention can be seen as part of what Gale and Sultan [22] label 'a wider trend to move medical technologies from the hospital to home'.

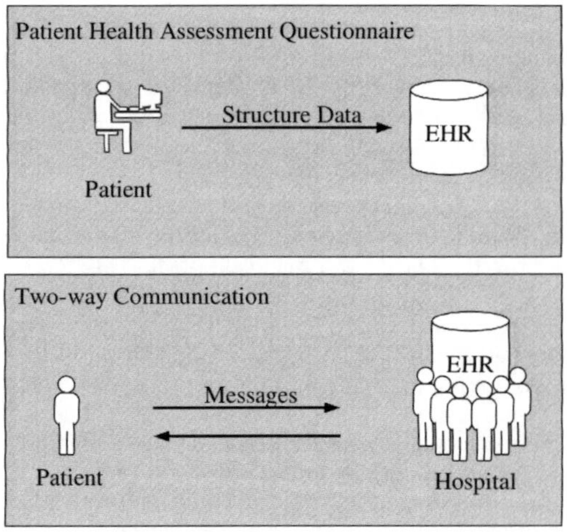

Figure 1. Illustration of the eTeam-Surgery architecture.

Developing a tool for electronic collaboration between the patient and the hospital is not straightforward. Several health information and communication technology (ICT) projects have failed to fulfil their expected outcomes. A substantial amount of the literature in the field of health ICT, particularly from the social sciences, covers unsuccessful implementation projects, challenges and unforeseen consequences[23][24][25][26][27][28][29][30][31][32][33][34][35][36][37][38].

In an attempt to avoid adding to the list of health ICT developments that are either not used or used differently than expected, the eTeam-surgery project addressed the readiness for electronic communication between surgical patients and UNN prior to developing and testing the tool. In this paper, we address the readiness for electronic communication as a means to improve the quality and effectiveness of the pre-operative process.

The paper is divided into five sections. In the first section, the problem of surgical cancellations is introduced and the aim of the study is described. In the second section, the background of the study is presented. It briefly introduces the existing knowledge on e-readiness and the challenges of ICT in healthcare. As will be evident in section three, Materials and Methods, our approach to the study of e-readiness consists of a broad spectrum of qualitative methods. In this section, a brief introduction to the methodology (actor-network theory [ANT]) used to map the main actors regarding electronic patient–hospital communication in pre-operative planning is described. The diverse methods of data collection used to study e-readiness among the identified actors are also presented and explained in the third section. The results are disclosed and interpreted in the fourth section. In the last section, Discussion and Conclusions, the authors elaborate on the readiness to use electronic communication among the different actors involved in pre-operative planning, and in healthcare more generally.

II. Background

Telecare technologies are advocated by European governments and industries as innovations of great promise for improving care [39]. In the last decade, the healthcare sector has witnessed the introduction of an increased number of telemedicine applications, i.e., devices that can monitor, diagnose or treat people at a distance from the clinicians through the use of ICT [40]. Regardless of whether it is labelled 'telemedicine', 'telecare' or 'tele-monitoring', it has been reported that Norwegian patients are well prepared and able to use ICT for health purposes [41]. Patients, including the elderly and less educated [41][42], are using electronic healthcare services [43][44]. In addition to the patients' readiness for electronic communication, a tendency among healthcare workers to use personal electronic devices to support their clinical work [31][45][46] and communicate with patients [44][47][48] has also been reported.

At the same time, substantial evidence exists in the field of health ICT of less successful project implementations

[23][24][25][26][27][28][29][30][31][33][34][36][37][38]. Challenges with implementation, slow diffusion and unforeseen consequences of ICT in healthcare, particularly in hospitals, have been extensively described. This knowledge, or what we have called the 'e-readiness paradox' in the field of health informatics, shaped the research question of this paper: If patients and healthcare workers are ready for ICT in healthcare, but yet at the same time the new digital technologies are not used as expected, are Norwegian hospitals ready for electronic communication during pre-operative planning?

III. MATERIALS AND METHODS

The e-readiness paradox—which refers to the reported e-readiness among patients and healthcare workers related to studies of slow diffusion and unforeseen consequences of health ICT—illustrates the complexity in the field. It also points to the need for a broad approach to the research question, 'Are Norwegian hospitals ready for electronic communication during pre-operative planning?'

Knowledge from the field of qualitative research on health ICT, notably ANT [49] and science technology studies (STS) [50], was used as a first step in our study. The methodology employed an open, empirical approach to determine the actors in technology development, and was also used as a first step in our study to map the different actors involved in electronic communication in relation to pre-operative planning.

ANT is also called a material-semiotic method, where the concept of 'actor' is used similarly to the semiotic concept of 'actant', which means that materiality, i.e., hospitals and EHR systems, are seen as non-human actors and treated equally as human actors. An actor can be any entity that holds a position in a discourse [51]. Thus, the methodology is associated with the equal treatment of human and non-human actors, and assumes that all entities in a network can and should be described in the same terms. This is called the 'principle of generalised symmetry'.

Although it is called a 'theory', ANT does not usually explain 'why' or 'how' a network takes the form that it does [50]. ANT is a way of thoroughly exploring the relational ties within a network. The methodology is empirical and descriptive, rather than theoretical and explanatory in its approach. It consists of following the actors and mapping their actions [50]. In practical terms, this means that the first step of our study of e-readiness was to map the main actors involved in electronic communication during pre-operative planning.

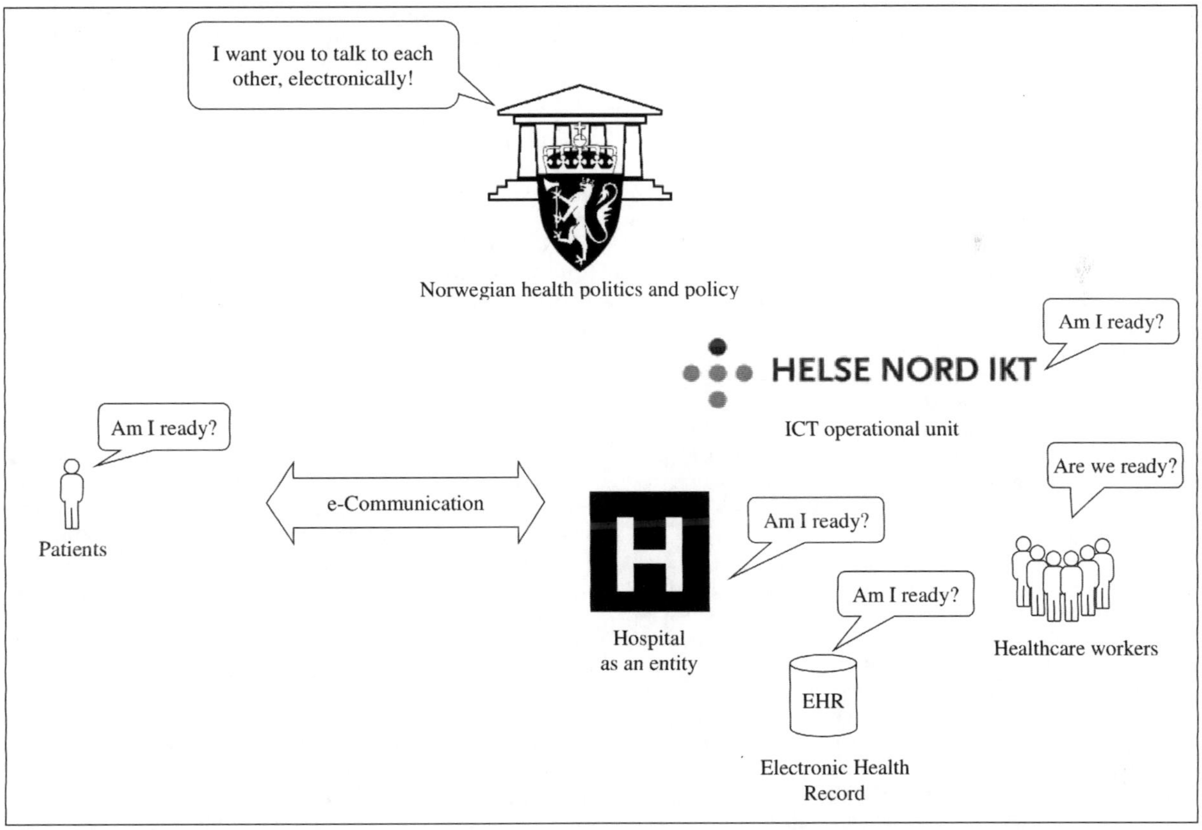

Figure 2. Actors involved in the electronic hospital–patient communication during pre-operative planning at UNN.

The process of mapping the actors identified six main human and non-human actors in need of further inquiry. The main actors, as illustrated in Figure 2, were as follows: a) Norwegian health politics and policy; b) healthcare workers; c) the hospital as an entity; d) patients; e) the EHR system and f) the ICT operational unit. Mapping actors is an extensive exercise. This paper does not reflect the entire nuanced process of mapping; however, it emphasises the need to carefully consider who the actors are in the field, as it can be much more complex than originally anticipated. For analytical reasons, the rest of the paper is organised according to the main actors.

The process of mapping the main actors created a picture of the broad network of actors involved in electronic communication in pre-operative planning. Hence, the second step of our research approach consists of five different study designs for approaching the e-readiness of the six main actors. The five study designs in step two are described below.

A. *Health politics and policy*

Norwegian health and social care are based on the classical Scandinavian welfare model, which combines financing and the provision of universally accessible services mainly within the public sector [52]. An important political goal in Norway is to provide healthcare on equal terms for all citizens, meaning that local healthcare services should be easily accessible, meet clients' needs and be provided in accordance with political priorities. The Norwegian health and social care sectors are mainly run by public institutions, which accounted for more than 95 % of the health expenses paid by governmental institutions in 1996 [53]. Despite a minor increase in private health services during the last two decades, governmental ICT strategies for the future are an important actor when studying e-readiness.

In order to gain knowledge on the governmental readiness for digitised patient–hospital communication, we conducted a study of one of the largest public sector reforms in Norway, the Coordination Reform [54]. Enhancing coordination between primary and secondary healthcare has been central in Norwegian healthcare policymaking in the last decade. In 2008, a new Minister of Health and Care worked out plans under the key term 'Coordination Reform', and the government has been implementing these administrative, structural and economic reforms since 2012 [52]. The Coordination Reform was identified as an important actor regarding e-readiness in Norwegian healthcare.

In Norway, the healthcare services are divided into four health regions, each with their own authorities aimed at planning, organising and promoting patient care [55]. One way to approach the policy implementation of the Coordination reform is to look into the regional health authority's recent strategies and investments in ICT. This study focuses on the northern region, represented by the Northern Norway Regional Health Authority (Helse-Nord RHF). In order to gain knowledge on the northern health regional ICT policy, we studied their website [56]. The ICT strategy of the northern health region was identified as one important actor in our study of e-readiness.

Since the national vision and the regional health authority's strategy for ICT in healthcare correspond, in this paper they are considered as one actor in the study of e-readiness.

B. *Healthcare workers and the Hospital as an organisation*

The UNN hospital and the healthcare workers involved in pre-operative planning are main actors in relation to e-readiness. A new communication tool must be implemented in the daily practices at the hospital and taken into use by health professionals during the pre-operative planning process. To study readiness for electronic communication between surgery patients and the hospital, we conducted an in-depth empirical observation and interview study of the pre-operative planning process at UNN. This approach consists of three phases. The empirical methodology will be described briefly; for further information, refer to [57].

Stage 1: Gather data on the hospital's representation of the elective surgery cancellation problem. Inspired by Bacchi [58], the aim was to gather knowledge on how the hospital represented the elective surgical cancellation problem. One internal report from 2008, which contained information on the use of resources involved in surgery at the hospital, was identified and studied [15]. In addition to the report, in 2012, the hospital had initiated a Lean Project (which was established by the management at UNN as an internal project to reduce the cancellation rate) to optimise the elective surgical process. Two researchers from the eTeam-surgery group followed this project. For further reference on the Lean Project, see [57].

Stage 2: Gather data on the pre-operative planning process at the hospital. In Stage 2, the pre-operative planning process in different departments at UNN was investigated. This comprised three weeks of fieldwork and interviews at the surgery and intensive care clinic while following an anaesthesiologist and an anaesthetic nurse. In addition, 13 interviews with physicians, nurses and administrative personnel were conducted in six different departments. The interviews were semi-structured, done at the workplace and lasted from 30 minutes to two hours. During the first two stages, two departments were described as more efficient. However, these departments still had a number of cancellations. One of the departments was chosen for an in-depth study in Stage 3.

Stage 3. Individual, in-depth interviews with professionals from all groups involved in pre-operative planning in a specific department. The chosen department is not revealed due to ethical reasons. In Stage 3, representatives from all the professional groups involved in the pre-operative planning process at UNN were addressed. In this specific department, extensive knowledge on the pre-operative planning process was collected. The department-specific interviews were semi-structured, done at the workplace and lasted between one to two hours.

The ANT-inspired approach helped us to map the main actors and their network of relationships. Not unexpectedly, the hospital and the involved healthcare workers were identified as main actors. Nonetheless, the approach exposed the need to treat and describe the hospital as an institution, or as an organisational entity, and the healthcare workers involved in pre-operative planning at UNN as two separate, equally important actors in the field of e-readiness. Hospital and health personnel will from now on be considered two important actors in this study of e-readiness.

C. *Patiens*

Patients are, of course, important actors in patient–hospital communication. To investigate the patients' readiness for electronic communication, we interviewed patients who had recently cancelled an elective surgery. The category was strategically chosen due to the patients' recent experience with surgical cancellations and their current incentive to communicate with the hospital on the issue. The relevant patients were identified from the hospital's EHR system. The study of the patients' readiness is part of a larger qualitative phone inquiry where a researcher from the eTeam-surgery project interviewed 16 patients who had recently cancelled elective surgery. The interviews were semi-structured, and patients were asked about their experiences with elective surgery at the hospital, patient–hospital communication and cancellations of surgery. The 16 phone interviews lasted between 15 to 60 minutes and were recorded and transcribed. In this paper, we report on one question from the phone interviews. As a follow-up to the main questions related to how the patients experienced the communication with the hospital regarding their surgical cancellations, the patients were asked, 'Do you have any thoughts on how it [the communication] could be done differently?'

Despite variations and differences among patients, the exercise of mapping actors showed that the surgery patient can be considered a main actor in our study field of e-readiness.

D. *EHR system*

The hospitals' ICT systems are obvious actors in patient–hospital communication. This paper focuses on the EHR system since it is the system that healthcare workers mainly interact with during the pre-operative planning process.

In the northern health region, the EHR systems in use are provided by DIPS ASA [51]. DIPS ASA is the leading supplier of electronic health records in Norwegian hospitals, and is a supplier of systems for the health sector with a special focus on the Scandinavian market [51]. DIPS ASA provides EHR systems to three of the four health regions in Norway. Currently, DIPS ASA, in collaboration with FIKS, an implementation project initiated by the Northern Regional Health Authority, is developing a new EHR system for the region, including a new surgery module.

In the literature, usability, integration and interoperability are identified as key topics in the development of ICT for the healthcare environment. Usability refers to the adequacy of the system for the users' needs. This is mainly reflected by the impact that the system has on the productivity of the process, e.g., efficiency, fulfilling work content, time required for data entry and interference with the physician/patient relationship. The topics, integration and interoperability, generally go hand in hand.

It was the aim of the eTeam-surgery project to combine the eTeam system with the EHR to form one system, ensuring that they function together efficiently by fulfilling the required integration. However, to do so, interoperability is required between both systems, meaning that they must be able to communicate with each other and exchange information. The use of data standards ensures that the collected patient health data can be integrated in the EHR system and shared among health entities.

In order to study the readiness for a two-way communication system between patients and the hospital integrated in the EHR system, the three aforementioned topics were investigated. Our approach was to study how the EHR was described by health personnel and by the EHR vendor. These groups were involved in separate interviews and meetings, as well as joint workshops focusing on the visions for the future, the limitations and the characteristics of the system.

E. *ICT operational unit*

The Northern Norway Regional Health Authority has established an operational unit (in Norwegian, Helse-Nord IKT) to manage, operate and develop ICT systems for the health region, which comprises a total of 11 hospitals [59]. Hence, all ICT projects involving any of the hospitals in the health region must be approved and accounted for by the ICT operational unit. The ICT operational unit was identified as one important actor regarding readiness for electronic patient–hospital communication.

In order to establish a secure online solution for electronic communication between surgical patients and UNN, a researcher from the eTeam-surgery project attended meetings and workshops over a two-year period from 2014 to 2015. During this period, the Norwegian Centre for Integrated Care and Telemedicine (NST), today called the Norwegian Centre for eHealth Research (NSE), i.e., the authors' research institution, coordinated the efforts to meet the needs and requirements set by the ICT operational unit for electronic communication. A researcher from the eTeam-surgery project attended between 13 meetings. Knowledge gained during the meetings was used to address the ICT operational unit's readiness for two-way electronic communication between the patient and UNN.

IV. RESULTS

Six main actors were identified in order to explore the readiness for two-way electronic communication between the patient and UNN within pre-operative planning. The main actors are categorised as follows:

a) *Norwegian health politics and policy:* National and regional health authorities' visions and strategies for the future of electronic communication within public healthcare;
b) *Healthcare workers*: Readiness among the professionals involved in pre-operative planning to use ICT to communicate with patients;
c) *Hospital as an entity*: Readiness within UNN as an organisation to use two-way electronic communication with patients;
d) *Patients:* Readiness among surgical patients to communicate electronically with the hospital;

e) *EHR system*: Readiness of the EHR to support two-way electronic communication with patients;
f) *ICT operational unit:* Ranking of priorities and allocation of resources regarding two-way electronic patient–hospital communication.

A. *Norwegian health politics and policy*

In the preface of the Coordination Reform, the Minister of Health and Care Services states, 'In public health spending per capita, Norway ranks among the highest of all OECD nations—but we have not achieved a correspondingly high level of health in return' [54]. However, 'With smart solutions, patients will receive proper treatment at the right place and right time. We will achieve this through the Coordination Reform'. [54]. A well-defined goal in the reform (p. 135) is that 'electronic communication should be the standard way of communicating' [54].

In line with the national ambition, an extensive ICT investment is currently being made in the northern health region, including at the UNN hospital, our site of research. The Northern Norway Regional Health Authority is investing €62.5 million in the FIKS project (from the Norwegian Felles innføring kliniske systemer) [60]. On their webpage, the regional health authority describes the FIKS project as the largest and most interesting ICT investment in northern Norway [56].

B. *Healthcare workers*

During our observations and interviews at the hospital, we did not experience any resistance from the healthcare workers towards electronic communication. On the contrary, aside from what can be described as mixed enthusiasm for 'quick IT-fixes' for complicated clinical issues, most healthcare workers expressed frustration over the current cancellation situation at the hospital. Several stressed the need for new communication tools.

A theatre nurse linked the need for new ways of communication to the current 'quick in, quick out' trend in Norwegian hospitals. The nurse emphasised that this trend requires new ways of communicating with patients prior to hospital admission in order to prepare them for surgery while they are still at home. Before the quick in, quick out movement, nurses were responsible for nursing and preparing the patient for surgery after the patient arrived at the hospital. Such preparation included, for example, cleaning, shaving and nail trimming, according to the hygienic standards required for surgery. Today, many patients are responsible for doing these tasks themselves, and they must follow the hygiene instructions provided by the hospital at home. The nurse's main concern was related to infections. In this context, an electronic communication tool between the patient and the hospital was suggested to help patients prepare for surgery.

Some of the secretaries were also very much in favour of electronic communication with patients. Secretaries are on the front line in terms of everyday communication with patients. Almost all of them expressed frustration or resignation over the current cancellation situation at the hospital, and stressed the need for better tools to book, rebook or cancel scheduled appointments.

As a professional group, with some exceptions, physicians were less troubled than nurses and secretaries by the established hospital communication procedures for patients. Physicians were deeply concerned by the cancellation rate at the hospital, but did not necessarily link it to communication issues. Some emphasised that they used the phone to contact patients if they needed additional information prior to surgery.

C. *Hospital as an entity*

During the inquiry into UNN's position on the elective surgery cancellation problem, one internal report [15] was identified and studied, and the Lean project at the surgery and intensive care ward was followed and observed. The internal report [15] acknowledges the challenges with the continuity of patient care in the region, and relates them to poor interaction between the different professional groups involved in surgical practices. The aims of the internal report [15], along with those of the Lean project, were to promote the continuity of patient care and efficient use of resources in surgery and to reduce elective surgery cancellations. Regarding e-readiness at the hospital, it is important to note that electronic collaboration as a strategy to improve the continuity of care during the pre-operative planning process is not suggested in any of the hospital initiatives.

In addition, the fieldwork at the hospital revealed internal variations between the different departments in terms of how and by whom surgeries were planned and when the planning was done. At the UNN hospital, the different departments had developed their own local practices. In some departments, senior surgeons did the pre-operative planning. In other departments, this planning involved interdisciplinary teamwork between junior and senior physicians, nurses and secretaries. Based on the empirical findings, a homogeneous structure for the pre-operative planning process at UNN could not be identified, nor was it possible to describe a standard pre-operative planning structure at the selected department. It is the authors' understanding that in order to complete the daily schedule, healthcare workers depend on personal and empirical knowledge.

The main findings from the empirical inquiry into the e-readiness at the hospital were as follows: a) the two identified hospital initiatives to reduce surgery cancellations and improve the continuity of care during the pre-operative planning process did not include two-way electronic communication; and b) heterogeneity was identified in how departments and individual professionals carry out the pre-operative planning process.

D. *Patients*

The respondents to the telephone inquiries were patients scheduled for elective surgery who had taken the initiative to call the hospital to reschedule the appointed date for surgery. Some of these patients were pleased with the existing communication with the hospital (i.e., letters and phone calls) and articulated their gratitude towards UNN. It is relevant to note that many of the grateful patients named specific health workers who had been particularly helpful during the pre-operative planning process.

A majority of the patients that participated in the telephone inquiry had experienced difficulties with patient–hospital communication and expressed readiness for new ways of communicating with the hospital. Most patients did not come up with any concrete recommendations in response to the question, 'Do you have any thoughts on how it [the communication] could be done differently?' Rather, the responses can be categorised as vague suggestions regarding the potential for online communication in modern society. One patient explicitly suggested an electronic communication system where patients could inform the hospital about specific dates when surgery was inconvenient, e.g. vacations, attending family events. It is the authors' interpretation that the patients expected some sort of interactive communication tool which they could use to participate in the planning of the surgery date. With respect to readiness for electronic communication, we did not identify any differences among patients regarding age, gender or level of education.

E. *EHR system*

During workshops and meetings with professionals from DIPS and FIKS, the EHR system was described as a central working tool for elective surgery planning. Nevertheless, the phrase 'poor functionally' was used to describe the current version of the surgery module in the EHR. This is in line with the findings from the observations and interviews at the hospital, which revealed that different individuals use this module differently in their workflow, and that different departments also use it in different ways to support the information flow. Nevertheless, the EHR system was often referred to as the future spine of the hospitals' ICT services, and huge investments were made in the development of the EHR system.

The next step in our study was to approach the EHR vendor in relation to the integration requirements and interoperability of the EHR system. The study revealed that, despite implementation and research initiatives, the lack of structured data within the EHR system hinders integration with other health ICT. Furthermore, the lack of structured data, combined with the non-use of communication standards, limits the availability of interoperability interfaces. This creates a barrier to achieve the degree of interoperability required to establish patient–hospital communication. The implementation of a communication tool is also restricted by either the inexistence or low maturity of the required platforms. However, tools for electronic communication within the EHR system are being tested at UNN, e.g., the Innsyn project [54]. Such a scenario demonstrates why the EHR system vendor, DIPS ASA, collaborates with the health authority's FIKS project to develop a new EHR system, including a new surgery module.

Professionals working for FIKS and DIPS, including nurses, physicians and engineers, as well as health personnel in general, promoted one singular system that health personnel would learn how to operate and which would provide them with relevant information and services easily and without additional log on. Quite a few professionals expressed interest and enthusiasm for the implementation of the new and future surgery module under development. Some health personnel were enthusiastic and impatient for the new version of the EHR. Still, they questioned the possibilities for integration with other ICT systems in the region, e.g., tools for electronic patient–provider interactions. The success or failure of the huge ICT investments was tied to a functional EHR system.

F. *The ICT operational unit*

In meetings and workshops with the Northern Norway Regional Health Authority's ICT operational unit, electronic communication between patients and hospitals was described as an aim for future health care. A secure system for two-way communication between patients and hospitals is a clear vision for the northern health region.

However, the ICT operational unit had several unsolved tasks on their agenda due to many large ongoing ICT projects and the lack of human resources. Furthermore, they lacked a secure platform for testing new functionalities. We also found the ambition for contributing to two-way electronic communication differed at different management levels within the ICT operation unit. The policy and willingness varied between individuals and from meeting to meeting. Neither the eTeam-surgery project nor NST were successful in reaching an agreement with the ICT operational unit on one solution for solving secure two-way electronic communication.

To this day, the ICT operational unit has not implemented a platform to support secure two-way electronic communication, nor do they have a foundation for pursuing the issue. However, the unit is supporting and testing a platform to provide patients access to their own electronic health record in an ongoing project in the health region [54].

Summing up the results, see Table I, the strong governmental vision that 'Electronic communication should be the standard way of communicating' [54] is evident in our findings from the study of e-readiness among healthcare workers at the hospital. Apart from what can be described as mixed enthusiasm for quick IT-fixes for complicated clinical issues, several healthcare workers expressed frustration with the current cancellation situation at the hospital and stressed the need for new communication tools. Similarly, and in line with the governmental vision, the patients in our study requested some sort of interactive communication tool, where they had access and could participate in the planning and decision making of the date of their forthcoming elective surgery.

TABLE I. SUMMARY OF THE TYPE OF STUDY AND RESULTS HIGHLIGHTS FOR THE IDENTIFIED ACTORS.

	Health politics and policy	**Healthcare workers**	**Hospital as an entity**	**Patients**	**EHR system**	**ICT Operational unit**
Type of study	Documentary study of the Coordination Reform [54]	Identification of UNN's representation of the elective surgery cancellation problem In-depth empirical observation of the pre-operative planning process at UNN Interviews with professionals from all groups involved in the pre-operative planning process		Phone interviews with patients that recently had an elective surgery cancelled	Interviews, meetings and workshops with vendors and users	Meetings with the ICT operational unit concerning two-way electronic communication
Results highlights	Electronic communication should be the standard way of communicating	Stressed the need for new communication tools	None of the identified hospital initiatives included two-way electronic communication Heterogeneity in how departments and individuals carry out the pre-operative planning process	Requested some sort of interactive communication tool	Lack of structured data within the EHR system hindered the integration with other health ICT solutions	Two-way electronic communication was not presented as a priority in their ICT strategy

Regarding e-readiness, electronic communication, as a strategy to improve the continuity of care during the pre-operative planning process, was not suggested in any of the hospital initiatives identified in this study. In addition, the main finding from the empirical inquiry at the hospital was heterogeneity in how departments and individual professionals carried out the pre-operative planning process. The study has also exposed that, despite implementation and research initiatives, the lack of structured data within the EHR system hinders the integration with other health ICT solutions. This creates a barrier to achieve the degree of interoperability required to establish two-way electronic communication. At this point, the regional health authority's ICT operational unit is not ready to open up for two-way electronic communication.

V. Discussion and Conclusions

Are Norwegian hospitals ready for two-way electronic communication between the patient and hospital during pre-operative planning?

Six main actors were mapped and identified as requiring further investigation in order to study readiness for two-way electronic communication within pre-operative planning at UNN. The main actors were as follows: (a) Norwegian health politics and policy; (b) healthcare workers; (c) hospital as an entity; (d) patients; (e) the EHR system and (f) the ICT operational unit.

It is the authors' conclusion that the Norwegian government states a strong wish for electronic communication in the Coordination Reform. This conclusion is in line with Tjora and Melby's [61] analysis of the reform, demonstrating the government's attention to the importance of ICT in order to succeed in healthcare coordination.

The empirical study conducted at the UNN revealed that nurses and secretaries involved in the pre-operative planning are ready for electronic communication. Even though physicians, as a professional group, reported less need for an electronic communication system than nurses and secretaries, it is the authors' interpretation that they are ready for electronic communication. The relative absence of interest is most likely linked to work tasks, such as scheduling, traveling logistics, general pre-operative information and support, which are the responsibility of nurses or secretaries. These are the ones who communicate with patients prior to surgery, not the physicians. It is also important to note that some physicians were enthusiastic about the possibility of electronic communication with patients.

The nurse's proposal of an electronic tool to help patients prepare for surgery at home illustrates how a nurse pictures the future of healthcare. The nurse's proposal includes the need for electronic communication and online care for patients while they are still at home. The shift from care towards telecare is also evident in the literature. Wyatt and Sullivan [62] state, 'In the future, health professionals may move towards spending some of their working lives as telecarers. A telecarer is a health professional who delivers responsive, high-quality information, services and support to remote patients or clients using the most appropriate communication, such as telephone, email or instant messaging'.

The study of patients' readiness identified that surgical patients expect some sort of interactive communication with the hospital. Today, hospital–patient communication is based on letters exchanged by post and/or telephone calls between patients and hospitals. This system does not allow patients to participate in the process of scheduling their forthcoming elective surgery. It is the authors' interpretation that there is a strong wish among surgery patients to coordinate minor surgery with their everyday lives, e.g., the ability to inform the hospital about the dates or weeks they are unavailable to undergo surgery. This reveals their readiness for new ways to communicate with the hospital. These findings are in line with existing literature that demonstrates how patients, including the elderly or less educated, are ready for electronic communication [41][42][43].

Regarding e-readiness at the hospital, electronic communication as a strategy to reduce surgery cancellations is not suggested in any of the identified hospital initiatives to improve planning and reduce cancellations. Further, the empirical study of the pre-operative planning process at UNN demonstrates heterogeneity in how departments and individuals carried out the planning process. It is the authors' understanding that in order to complete the daily schedule, the hospital depends on the healthcare workers' personal and empirical knowledge, proactivity and workarounds. The heterogeneity and lack of standards in the pre-operative planning processes reveal that the hospital, at this stage, lacks the organisational structure required for two-way electronic communication. Hence, the hospital is not yet prepared for two-way electronic communication with patients during pre-operative planning.

The surgery module in the EHR is an unused resource at the hospital. At present, money and resources are being invested by FIKS and DIPS ASA, the EHR vendor, to develop a new surgical module. However, two-way communication with patients is not prioritised by either of them. On the other hand, the northern health region has supported initiatives to provide patient–hospital communication within the EHR; however, to date, this is limited to one project which focuses on the patients' access to their own electronic health record. Despite this initiative, it is the authors' understanding that the current EHR is a barrier for two-way electronic communication between patients and the hospital.

The governmental vision for electronic communication has not materialised in technological solutions or in the priorities of the regional authority's ICT operational unit. Despite massive ICT investments in the region, it is the authors' understanding that the ICT operational unit, at this stage, is not ready for two-way electronic communication between patients and UNN.

This analysis is based on the recognition that in order to avoid the e-readiness paradox and to develop and implement sustainable electronic communication systems, computer scientists need to identify patterns of information and work flow. The authors conclude that Norwegian health policy strongly promotes electronic communication, and that healthcare workers and patients are ready to use new electronic tools, while the hospital as an entity, the current EHR system and the ICT operational unit are not yet ready for two-way electronic communication between patients and UNN.

While the analysis, and particularly the conclusions, might be debated, they are still relevant on multiple levels. For the eTeam-surgery project, the study of e-readiness have a great impact on future work. Should resources be spent on the development and implementation of new technology or should they be spent primarily on gathering knowledge on the organisation and how to prepare the hospital for electronic communication? A relevant question for future work is as follows: Is heterogeneity in pre-operative planning processes exclusive to our site of research, or are local practices and differences among individuals and between departments common in Norwegian hospitals? What about the regional ICT operational units? Are they powerful actors in all Norwegian hospitals, or are our findings related to local issues, such as lack of resources and bad timing?

In an applied context, the analysis has relevance for policy makers, managers and stakeholder in the healthcare sector, e.g., health authorities, vendors and large ICT projects. Are the findings applicable to other hospitals in Norway? What if today's hospitals, as organisational entities, are not ready for electronic communication with patients? Will a new EHR, additional resources and different priorities in the ICT operational unit solve the barriers for two-way electronic communication, or do the reported findings illustrate how work is organised and done in healthcare today?

In the scientific field of health informatics, the debate on e-readiness requires that the concepts of 'user involvement' and 'user centred', as well as the users' role, need to be revisited. As demonstrated in this paper, the health policy, the hospital and the EHR system are important non-human actors that need to be studied, analysed and accounted for in relation to the question of e-readiness. A methodology that practices the equal treatment of human and non-human actors, and assumes that all entities in a network can and should be described in the same terms, is useful to map important actors. As we have demonstrated, readiness for electronic communication is not exclusively about the interaction between patients and health personnel. This multi-method approach revealed that some actors are ready for electronic communication while others are not.

We argue that in order to avoid the e-readiness paradox and develop and implement sustainable electronic communication systems, it is not only human actors—patients and healthcare professionals—who need to be ready for electronic communication.

ACKNOWLEDGMENT

The authors would like to thank the Northern Norway Regional Health Authority, Helse-Nord, for funding the research project HST 1119-13 and HST 1125-13.

REFERENCES

[1] K. Dyb, C. Granja, S. R. Bolle, and G. Hartvigsen, "Online Patients in an Offline Health Care Sector: Are Hospitals Ready for Electronic Communication with Patients?," The Seventh International Conference on eHealth, Telemedicine, and Social Medicine, 2015, pp. 26-30.

[2] M. Johnston, "Anxiety in Surgical Patients," Psychological medicine, vol. 10, pp. 145-152, 1980.

[3] M. J. Pritchard, "Identifying and Assessing Anxiety in Pre-Operative Patients," Nursing standard, vol. 23, p. 35, 2009.

[4] B. Ivarsson, P. O. Kimblad, T. Sjoberg, and S. Larsson, "Patient Reactions to Cancelled or Postponed Heart Operations," J Nurs Manag, vol. 10, pp. 75-81, 2002.

[5] M. J. Lacqua and J. T. Evans, "Cancelled Elective Surgery: An Evaluation," The American Surgeon, vol. 60, pp. 809-811, 1994.

[6] B. Denton, J. Viapiano, and A. Vogl, "Optimization of Surgery Sequencing and Scheduling Decisions under Uncertainty," Health Care Manag Sci, vol. 10, pp. 13-24, 2007.

[7] W. N. Schofield, G. L. Rubin, M. Piza, Y. Y. Lai, D. Sindhusake, M. R. Fearnside, and P. L. Klineberg, "Cancellation of Operations on the Day of Intended Surgery at a Major Australian Referral Hospital," Medical Journal of Australia, vol. 182, pp. 612-615, 2005.

[8] P. Sanjay, A. Dodds, E. Miller, P. J. Arumugam, and A. Woodward, "Cancelled Elective Operations: An Observational Study from a District General Hospital," J Health Organ Manag, vol. 21, pp. 54-58, 2007.

[9] T. L. Trentman, J. T. Mueller, S. L. Fassett, C. L. Dormer, and K. P. Weinmeister, "Day of Surgery Cancellations in a Tertiary Care Hospital: A One Year Review," Journal of Anesthesia & Clinical Research, vol. 1, 2010.

[10] A. C. f. L. Authorities, t. N. H. S. i. England, and Wales, Operating Theatres: Review of National Findings: Audit Commission, 2003.

[11] S. Yoon, S. Lee, H. Lee, H. Lim, S. Yoon, and S. Chang, "The Effect of Increasing Operating Room Capacity on Day-of-Surgery Cancellation," Anaesthesia and intensive care, vol. 37, p. 261, 2009.

[12] A. González-Arévalo, J. I. Gómez-Arnau, F. J. DelaCruz, J. M. Marzal, S. Ramírez, E. M. Corral, and S. García-del-Valle, "Causes for Cancellation of Elective Surgical Procedures in a Spanish General Hospital," Anaesthesia, vol. 64, pp. 487-493, 2009.

[13] A. González - Arévalo, J. Gómez - Arnau, F. DelaCruz, J. Marzal, S. Ramírez, E. Corral, and S. García - del - Valle, "Causes for Cancellation of Elective Surgical Procedures in a Spanish General Hospital," Anaesthesia, vol. 64, pp. 487-493, 2009.

[14] M. Knox, E. Myers, I. Wilson, and M. Hurley, "The Impact of Pre-Operative Assessment Clinics on Elective Surgical Case Cancellations," Surgeon-Journal of the Royal Colleges of Surgeons of Edinburgh and Ireland, vol. 7, pp. 76-78, 2009.

[15] R. Busund, "Rapport Fra Prosjekt: Optimal Ressursutnyttelse Av Opperasjonskapasiteten I Unn," Universitetssykehuset Nord-Norge, Norway 2008 (in Norwegian).

[16] M. B. Ferschl, A. Tung, B. Sweitzer, D. Huo, and D. B. Glick, "Preoperative Clinic Visits Reduce Operating Room Cancellations and Delays," Anesthesiology, vol. 103, pp. 855-859, 2005.

[17] M. R. Rai and J. J. Pandit, "Day of Surgery Cancellations after Nurse-Led Pre-Assessment in an Elective Surgical Centre: The First 2 Years," Anaesthesia, vol. 58, pp. 692-699, 2003.

[18] A. R. Seim, T. Fagerhaug, S. M. Ryen, P. Curran, O. D. Sæther, H. O. Myhre, and W. S. Sandberg, "Causes of Cancellations on the Day of Surgery at Two Major University Hospitals," Surgical Innovation, vol. 16, pp. 173-180, 2009.

[19] W. A. van Klei, C. L. Rutten, K. G. Moons, B. Lo, J. T. Knape, and D. E. Grobbee, "Limited Effect of Health Council Guideline on Outpatient Preoperative Evaluation Clinics in the Netherlands: An Inventory," Ned Tijdschr Geneeskd, vol. 145, pp. 25-9, Jan 6 2001 (in dut).

[20] C. Granja, K. Dyb, S. R. Bolle, and G. Hartvingsen, "Reduced Elective Surgery Cancellations through Patient Involvement in Pre-Operative Planning in Norway," eTELEMED 2014, The Sixth International Conference on eHealth, Telemedicine, and Social Medicine, 2014, pp. 164-169.

[21] C. Granja, K. Dyb, S. R. Bolle, and G. Hartvigsen, "On the Development of a Hospital-Patient Web-Based Communication Tool: A Case Study from Norway," MEDINFO 2015: EHealth-enabled Health: Proceedings of the 15th World Congress on Health and Biomedical Informatics, IOS Press, 2015, pp. 438-442.

[22] N. Gale and H. Sultan, "Telehealth as 'Peace of Mind': Embodiment, Emotions and the Home as the Primary Health Space for People with Chronic Obstructive Pulmonary Disorder," Health & place, vol. 21, pp. 140-147, 2013.

[23] J. Starling and S. Foley, "From Pilot to Permanent Service: Ten Years of Paediatric Telepsychiatry," Journal of Telemedicine and Telecare, vol. 12, pp. 80-82, 2006.

[24] P. Whitten, B. Holtz, and L. Nguyen, "Keys to a Successful and Sustainable Telemedicine Program," International journal of technology assessment in health care, vol. 26, pp. 211-216, 2010.

[25] P. Zanaboni and R. Wootton, "Adoption of Telemedicine: From Pilot Stage to Routine Delivery," BMC medical informatics and decision making, vol. 12, p. 1, 2012.

[26] M. Berg, "Implementing Information Systems in Health Care Organizations: Myths and Challenges," International journal of medical informatics, vol. 64, pp. 143-156, 2001.

[27] R. Heeks, "Health Information Systems: Failure, Success and Improvisation," International journal of medical informatics, vol. 75, pp. 125-137, 2006.

[28] C. May, M. Mort, F. S. Mair, and T. Finch, Telemedicine and the Future Patient: Risk, Governance and Innovation: Economic and Social Research Council, 2005.

[29] C. May and N. T. Ellis, "When Protocols Fail: Technical Evaluation, Biomedical Knowledge, and the Social Production of 'Facts' About a Telemedicine Clinic," Soc Sci Med, vol. 53, pp. 989-1002, Oct 2001.

[30] M. Mort and A. Smith, "Beyond Information: Intimate Relations in Sociotechnical Practice," Sociology, vol. 43, pp. 215-231, 2009.

[31] S. Dünnebeil, A. Sunyaev, I. Blohm, J. M. Leimeister, and H. Krcmar, "Determinants of Physicians' Technology Acceptance for E-Health in Ambulatory Care," International Journal of Medical Informatics, vol. 81, pp. 746-760, 2012.

[32] KS, "Ikt I Helse- Og Omsorg 2008-2012 - Strategi- Og Handlingsplan," Oslo 2008 (in Norwegian).

[33] H. K. Andreassen, "What Does an E-Mail Address Add?-Doing Health and Technology at Home," Social Science & Medicine, vol. 72, pp. 521-528, 2011.

[34] N. Schreurs. (2012) Fiasko Eller Fremtid? *Computerworld.* (in Norwegian)

[35] J. C. Wyatt and F. Sullivan, "Ehealth and the Future: Promise or Peril?," BMJ, vol. 331, pp. 1391-1393, 2005.

[36] K. Dyb, T. Solvoll, E. Rygh, and T. Sørensen, "Analysing the Use of a Telestroke Service," International Journal On Advances in Life Sciences, vol. 5, pp. 179-187, 2013.

[37] K. Dyb and S. Halford, "Placing Globalizing Technologies: Telemedicine and the Making of Difference," Sociology, vol. 43, pp. 232-249, 2009.

[38] A. G. Sandaunet, "The Challenge of Fitting In: Non - Participation and Withdrawal from an Online Self - Help Group for Breast Cancer Patients," Sociology of health & illness, vol. 30, pp. 131-144, 2008.

[39] J. Pols and D. Willems, "Innovation and Evaluation: Taming and Unleashing Telecare Technology," Sociology of health & illness, vol. 33, pp. 484-498, 2011.

[40] N. Oudshoorn, "Diagnosis at a Distance: The Invisible Work of Patients and Healthcare Professionals in Cardiac Telemonitoring Technology," Sociology of health & illness, vol. 30, pp. 272-288, 2008.

[41] P. E. Kummervold, D. Gammon, S. Bergvik, J.-A. K. Johnsen, T. Hasvold, and J. H. Rosenvinge, "Social Support in a Wired World: Use of Online Mental Health Forums in Norway," Nordic journal of psychiatry, vol. 56, pp. 59-65, 2002.

[42] L. Millsopp, S. Frackleton, D. Lowe, and S. Rogers, "A Feasibility Study of Computer-Assisted Health-Related Quality of Life Data Collection in Patients with Oral and Oropharyngeal Cancer," International journal of oral and maxillofacial surgery, vol. 35, pp. 761-764, 2006.

[43] P. M. Webb, G. D. Zimet, J. D. Fortenberry, and M. J. Blythe, "Comparability of a Computer-Assisted Versus Written Method for Collecting Health Behavior Information from Adolescent Patients," Journal of Adolescent Health, vol. 24, pp. 383-388, 1999.

[44] S. Santana, B. Lausen, M. Bujnowska-Fedak, C. Chronaki, P. E. Kummervold, J. Rasmussen, and T. Sorensen, "Online Communication between Doctors and Patients in Europe: Status and Perspectives," Journal of medical Internet research, vol. 12, 2010.

[45] M. N. Boulos, S. Wheeler, C. Tavares, and R. Jones, "How Smartphones Are Changing the Face of Mobile and Participatory Healthcare: An Overview, with Example from Ecaalyx," Biomedical engineering online, vol. 10, p. 24, 2011.

[46] B. Dolan. 72 Percent of Us Physicians Use Smartphones. Available: http://mobihealthnews.com/7505/72-percent-of-us-physicians-use-smartphones/ October 2014.

[47] H. Atherton, P. Sawmynaden, A. Sheikh, A. Majeed, and J. Car, "Email for Clinical Communication between Patients/Caregivers and Healthcare Professionals," The Cochrane Library, 2012.

[48] J. Car and A. Sheikh, "Email Consultations in Health Care: 2—Acceptability and Safe Application," Bmj, vol. 329, pp. 439-442, 2004.

[49] J. a. J. H. Law, Ed., *Actor Network Theory and After*. Oxford: Blackwell Publishing 2005, p.^pp. Pages.

[50] B. Latour, Science in Action: How to Follow Scientists and Engineers through Society: Harvard university press, 1987.

[51] K. B. Asdal, Brita; Moser Ingunn, Teknovitenskapelige Kulturer. Oslo, Norway: Spartacus, 2001.

[52] T. I. Romøren, D. O. Torjesen, and B. Landmark, "Promoting Coordination in Norwegian Health Care," International Journal of Integrated Care, vol. 11, p. e127, 10/07 2011.

[53] T. Sæter and V. Heimly, "En Kort Beskrivelse Av Helsevesenet I Norge Tjenester, Enheter Og Organisering, Trender Og Tall," KITH R 13/96, Oslo 1996.

[54] Helse- og omsorgsdepartement, "Samhandlingsreformen," Det Kongelige helse- og omsorgsdepartement 2009 (in Norwegian).

[55] Helse-NordRHF. Om Helse Nord Rhf [Internet]. Available: http://www.helse-nord.no/om-helseforetaket/category23156.html 1th of april.

[56] Helse-Nord. Følg Med På Nord-Norges Største Og Mest Spennende Ikt-Satsning! Available: http://www.helse-nord.no/fiks/category31090.html (in Norwegian).

[57] C. Granja, K. Dyb, E. Larsen, S. Bolle, and G. Hartvigsen, "Methodology for Health Care Process Modelling: Bringing the Health Care Complexity into Health It System Development," Scandinavian Conference on Health Informatics, Grimstad, 2014, pp. 17-21.

[58] C. Bacchi Lee, Analysing Policy: Whats the Problem Represented to Be?, Frenchs Forest: Pearson Australia, 2009.

[59] Helse-Nord. Helse Nord Ikt. Available: http://www.helse-nord.no/helse-nord-ikt/category40976.html February 2016.

[60] Helse-Nord. Dette Er De Nye Systemene. Available: http://www.helse-nord.no/helse-nord-nytt/dette-er-de-nye-systemene-article90645-1526.html 4 November 2013 (in Norwegian).

[61] A. T. Hagen and L. Melby, Samhandling for Helse: Kunnskap, Kommunikasjon Og Teknologi I Helsetjenesten. Oslo: Gyldendal Akademisk, 2013.

[62] J. C. Wyatt and F. Sullivan, "Abc of Health Informatics: Ehealth and the Future: Promise or Peril?," BMJ: British Medical Journal, vol. 331, p. 1391, 2005.

International Journal on Advances in Life Sciences, vol 8 no 1 & 2, year 2016, http://www.iariajournals.org/life_sciences/

Assessing Visitor Engagement in Science Centres and Museums

Wolfgang Leister, Ingvar Tjøstheim, Trenton Schulz
Norsk Regnesentral
Oslo, Norway
{wolfgang.leister, ingvar.tjostheim, trenton.schulz}@nr.no

Göran Joryd, Andreas Larssen, Michel de Brisis
Expology
Oslo, Norway
{goran, andreas, michel}@expology.no

***Abstract*—Science centres and museums struggle to measure how engaging specific installations are for visitors. We present a framework for assessing visitor engagement by using non-intrusive technologies. We present a profile for mapping out an installation over eight dimensions and how we created it. We also present techniques for performing assessments using the facial expressions of visitors and asking short, targeted questions to visitors. Combining these together results in a fast assessment that happens as a visitor interacts with an installation. We have performed evaluations of three different installations in three science centres: one looked at the role of competition in exhibits, one at ways to automate the assessment, and another at how altering components like narrative in an installation affects the assessment's result. The assessment framework and classification method work in multiple installations and form the basis for a new tool for measuring engagement in a visitor centres and museums.**

Keywords*—*assessment; installations; science centres; museums; visitor engagement.

I. INTRODUCTION

Science centres and museums present exhibitions, installations, and educational programmes that should engage visitors for self-education on a subject and to inspire the visitors to learn more. There is little data showing how well these installations perform in transferring knowledge to the visitors. Similarly, there is little data to determine whether modifying an installation increases a visitor's engagement. Previously, we proposed a concept for a system that can give evidence to these questions in real-time [1]. This concept was supported by a further study [2], and the current paper builds on this work.

Our objective is to measure the performance of installations, but we assume we cannot measure this directly. Instead, we assess the engagement of visitors while they use the installation and retrieve parameters and objective data from the installation and its context. We intend to avoid time-consuming observations by the museum staff and keep intrusive methodologies, such as questionnaires, to a minimum.

We argue that we can assess dimensions of engagement in an installation using subjective assessment and automated observations of technical data from the installations, physiological data of the visitor, camera data, behaviour, etc. These data are used to estimate the performance of the installation, and whether adjusting these installations contribute to a better engagement and experience.

Observations by museum personnel tend to focus on the visitor instead of the installation. Common methods for collecting data from visitors include interviews and questionnaires. But long interviews or questionnaires might be intrusive for the visitor, and the answers are given in retrospect, i.e., not *in situ*. Our approach is to use observations from sensors to retrieve data about a visitor's engagement. Electronic questionnaires will be tailored so that only relevant questions will appear. Thus, the visitor will not be bothered more than absolutely necessary.

First, we present an overview of related work, showing the installation-centric and visitor-centric view of studies (Section II). Then, we show the approach of our proposed framework for assessing engagement (Section III). We present the Visitor Engagement Installation (VEI) profile to characterise installations using eight dimensions (Section IV). An assessment of selected installations follows (Section V). Finally, we present our conclusion (Section VI).

II. RELATED WORK

It is well-documented that science centres and museums perform visitor studies. While demographic data about visitors and information on their enjoyment is assessed, the impact of these data is more difficult to grasp [3, p.169]. Many science centres have performed visitor studies to varying degrees. These studies include statistical data from ticket sales, questionnaires, feedback from visitors, observations, and larger studies [4] [5] [6]. Several studies focus on the learning aspect. Other studies evaluated whether the visitors enjoyed their visit, whether an installation works as intended, and how installations could be changed for a better presentation of the contents.

Science centres are informal learning environments [7] that are distinct from classrooms because they offer free-choice learning [8][9], i.e., visitors can choose which activities to participate in and they can leave at any time. Visitor studies have been performed since the late nineteenth century. In 1884, Higgins [10] mentions that observations of visitors and asking them for remarks might lead to valuable information.

Lindauer [11] presents a historical perspective of methodologies and philosophies of exhibit evaluations. Lindauer lists only a few methods that perform measurements using simple metrics of counting or measuring time. In the literature, the majority of evaluations in science centres deal with the assessment of learning, often using a longitudinal approach [12], i.e., observing a subject or installation over time. Šuldová and Cimler [13] suggest that engagement can be assessed more instantaneously and be used as a part of learning assessment, supporting Sanford's [14] claim that "some compelling evidence links visitor engagement to learning".

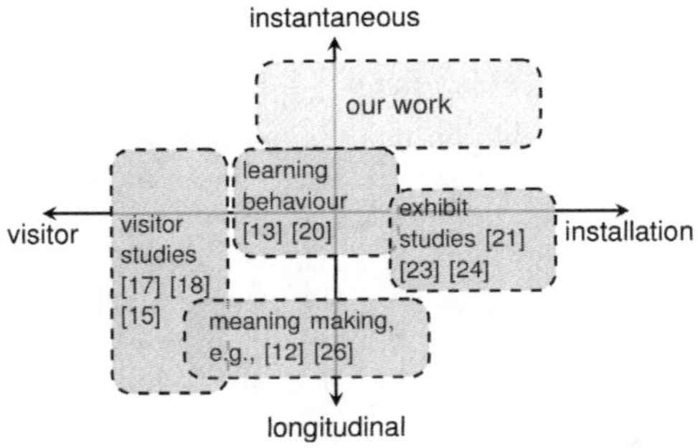

Figure 1: Classification of selected work in visitor studies

We align the literature along two axes, as illustrated in Figure 1: the vertical axis denotes the span between *longitudinal* and *instantaneous* assessment; the horizontal axis denotes whether the assessment is *visitor* or *installation*-centric. In general, assessing an installation also needs to take an assessment of the visitor into account.

A. Visitor-Centric View

According to McManus [15], the visitor instead of the artefact has been the focus in visitor studies since the 1980s. These visitor studies include demographic characteristics and segmentation, behavioural and knowledge gain studies, and visitor focused studies. Yalowitz and Bronnenkant [27] give a review of methodologies for timing and tracking visitors in exhibitions, also giving advice on how to perform assessments of visitor behaviour. Various methodologies have been developed to examine the behaviour of visitors in museums in detail [28] [29] [30] [31].

Dierking and Falk [16] present the Interactive Experience Model, which is a visitor-centric model. They define the interactive experience influenced by three contexts: *1*) the personal context, *2*) the physical context, and *3*) the social context. Falk and Storksdieck [17] use the principle of identity-related motivation to place visitors into five identity types: *1*) the explorer; *2*) the facilitator; *3*) the professional and hobbyist; *4*) the experience seeker; and *5*) the spiritual pilgrim. Variables, such as prior knowledge, experience, interest, visitor agenda, and social group are encapsulated in these identity types. This line of research has been further studied [18][19].

Barriault and Pearson [20] present frameworks that analyse the learning experience near instantaneously by identifying learning-specific behaviour observed by cameras and microphones installed within an installation. Šuldová and Cimler [13] refine these methods, but still depend on manual analysis.

Recently, Pierroux and Steier [32] presented the Visitracker, a tablet-based system for registering the visitors' behaviours. The graphical interface replaces the manual note-making, but a human observer is still needed to register the events.

In longitudinal visitor studies, observations and sense-making [26] are often used. In sense-making, qualitative mental models, understanding events, and an iterative approach for interpretation of situations (e.g., the data/frame theory of sense-making [33][34]) are in the foreground. But we are interested in concrete measurements and quantitative and descriptive data based on machine-retrievable data and questionnaires that allow us to get an instant result.

B. Installation-Centric View

In the installation-centric view, the science centre assesses installations rather than the visitors. The developers of installations need to consider the aspects of attractiveness, usability, being educational, etc.

Shettel et al. [21] present a more installation-centric approach where they evaluate exhibits by means of visitor observations and questionnaires using the technology available at the time, such as video tape recordings. They observe how visitors behave toward installations to determine how effective an exhibit is.

Alt and Shaw [22] present a study where visitors characterise installations using a list of phrases, both positively and negatively loaded. The phrases mentioned most often are then compared with the goals of the museum to identify where the installations can be improved.

Spegel [23] presents the *Expogon*, a graphical classification used as a mind map for exhibit planners when going through a museum. Note that the purpose of the Expogon is to stimulate and inspire on a subjective (qualitative) basis rather than to measure. The Expogon breaks down the exhibition medium into six elements: *1*) narrative, *2*) space, *3*) visitor, *4*) objects, *5*) time, and *6*) sender. Each element consists of fifteen hexagons representing categories, ten pre-filled and five empty for additional categories. The researcher wanders through an exhibition and notes observations on the Expogon. Thus, it is a qualitative tool that allows brainstorming when evaluating an exhibition. The Expogon gives hints to an evaluator on what to improve in an exhibition. However, it does not reflect to what degree the six elements are fulfilled. To rectify that, we developed a different approach with the VEI profile [1] that we extended in the current paper.

Young [24] suggests that developers need to advocate for the visitors and think as a visitor; Young recommends a cyclical development process. Allen [25] presents a study of three different versions of an exhibit for the purpose of studying dimensions of interactivity.

C. Observation Methodology

Traditionally, visitor studies use assessments where observers are placed near the installations. These observers make notes of the visitors' actions related to their use of the installations. Methods include counting and making notes. The visitors are often asked to fill out questionnaires related to their visit. However, such methods are often perceived as being intrusive and, thus, can reduce the visitor's experience.

Tröndle et al. [35] show an innovative possibility of combining movement tracking, physiological data (heart rate, skin conductivity, etc.), and psychological data. Any single senor for measuring has weaknesses and limitations when used in visitor studies [14]. However, when using multiple sensors concurrently the result will become better, provided that the method and the weight factors in an estimation model are calibrated correctly. Oppermann [36, p.145] posits that "a multi-method approach allows researchers to be more confident about their results".

We found few examples where physiological data for studies in science centres were used [35, 37]. Other data sources are mimics from video- and image data, prosody in sound data (e.g., intonation, pitch, strength), gesture recognition, or other bio-physiological sensor data.

For semi-automatic data assessment of visitor engagement, some components, such as face recognition and emotion recognition, are already available [38, 39]. Others have studied how to assess visitors' physical reaction using galvanic skin response in art exhibitions [37]. It is known that observations, e.g., visual or auditive recordings, and physiologic data measured with sensors are correlated with visitor experience; however, there is no unambiguous correlation.

Our assessment methodology is inspired by the model described by Russell [40], which shows a relation between psycho-physiological reactions and emotions. O'Brien [41] posits that engagement has been defined as *to involve the user emotionally when interacting with a system.* Engagement can be quantified by *focused attention* [42], measured, e.g., by using questionnaires after a museum visit or after the use of an installation. A modified version of the *Menorah Park Engagement Scale* together with the observation tool by Šuldová and Cimler [13] can be used for observing and classifying into categories [42]. See also the work by Barriault and Pearson [20], Griffin [43], and Griffin and Paroissien [44].

Wilhelm and Grossmann [45] and Nacke and Lindley [46] have shown the connection between emotions and psycho-physiological reactions, such as skin conductance, breath (strength, frequency), ECG, and EEG. These reactions has been used systematically in *quality assessment* studies [47].

Picard [48] coined the term *affective computing* to describe using computers and sensors to interpret emotions. Hoque et al. [49] show examples where facial expressions in camera images are used to interpret emotions. Ben Ammar et al. [50] have shown adaptive systems, e.g., within learning. Emotion recognition is a field where expressions can be interpreted (e.g., facial expressions, gestures, movements, voice) or physiological reactions (e.g., skin conductance or changes in the face colour [51]). A recent research challenge is the interpretation of multi-modal expressions.

Witchel et al. [52, 53] posit that non-instrumental movement inhibition can be used as a manifestation and proxy for engagement. According to them, cognitive engagement is an embodied phenomenon that can attenuate certain types of non-instrumental movements, such as larger postural movements and self-adaptors. They also found that non-instrumental movement disinhibition can be an indicator of engagement, e.g., during breaks, or disengagement, e.g., when occurring during active parts of a presentation. Their experiments were performed for screen-based visual stimuli. Research is needed to evaluate whether their arguments apply to visitors in science centres and museums interacting with installations.

D. Engagement and Gamification

Gamification is the application of game-design elements and game principles in non-game contexts [54, 55] to improve user engagement, productivity, learning, flow, etc. Kapp [56] presents gamification in the context of learning. We argue there are similar opportunities for using installations and gamification elements when it comes to learning.

Dixon [57] gives a brief history of the concept of player types, starting with Bartle's [58] concept of four player types. Marczewski [59] presents user types for game players, comparable to the classification by Falk and Storksdieck [17] for installations. Marczewski classifies users into *philanthropists*, *achievers*, *free spirits*, *socialisers*, *disruptors*, and *players*; he explains their roles and preferences.

Legault [60] presents twelve gamification elements that apply to e-learning. These elements are: *1*) narrative (story, protagonist, antagonist, plot), *2*) rules, *3*) player control, *4*) discovery and exploration possibilities, *5*) interactivity, *6*) feedback (provide a cue to player about progress), *7*) time constraints (create a sense of urgency), *8*) loss aversion (humans prefer avoiding losses; loss is twice as powerful as a gain), *9*) continuous play (after interruption), *10*) reward, *11*) levels (achieving different levels, goals, or challenges), and *12*) competition.

Hamari et al. [61] presents a literature review of empirical studies on gamification. In this context, they classify literature that refers to motivational affordances into ten categories: *1*) points, *2*) leaderboards, *3*) achievements and badges, *4*) levels, *5*) story and theme, *6*) clear goals, *7*) feedback, *8*) rewards, *9*) progress, and *10*) challenge. Weiser et al. [62] present a taxonomy of motivational affordances for meaningful gamified and persuasive technologies. Their taxonomy comprises of elements (including assignments, achievements, leaderboards, reminders, points, virtual goods, friends), mechanics (including feedback, rewards, education, competition, challenges, cooperation), and general design principles (including personalise experiences, offer meaningful suggestions, support user choice, respect stages of behaviour change, and provide user guidance). In our current paper, we will consider suitable gamification elements to extend our previous work.

III. Approach

Museums and science centres are places where visitors learn and gain knowledge through encountering and engaging with installations. These installations are complex systems that need to perform in their context together with the visitors. We take an installation-centric approach over a visitor-centric approach since we are interested in how the installations and potential changes of installations will perform. Also in the installation-centric view, it is important to observe visitors, study what they do, and determine whether the installations work as intended.

Our methodology is illustrated in Figure 2. We start by selecting the characteristics of installations. Currently, we use the VEI profile (Section IV) to identify potential characteristics

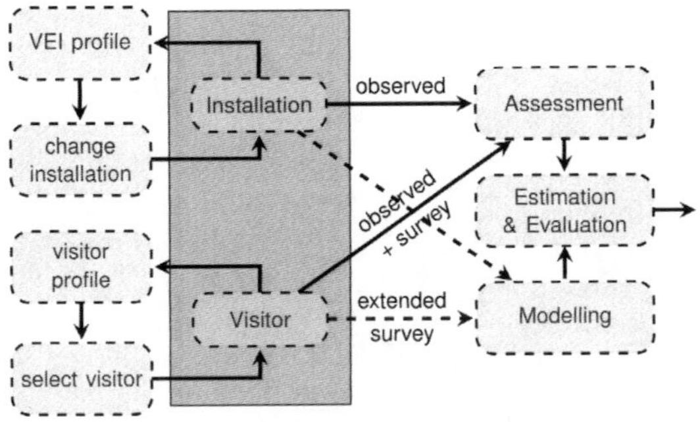

Figure 2: Method applied in our research to evaluate properties of installations

along eight dimensions that could be altered to achieve a design goal for the installation. The success of these changes can be evaluated, and the results can be compared.

Whether a goal is achieved can be assessed using the assessment framework presented in Section III-A. Additionally, the visitor characteristics, e.g., the characteristics described by Falk and Storksdieck [17], are used for the selection of visitors that act as respondents in the assessment.

The evaluation process for an installation, shown on the right side of Figure 2, consists of a *modelling phase* and an *assessment phase*. In the modelling phase, we establish an estimation model that allows us to estimate values and parameters from only few available data. If we succeed with this, it will be sufficient in the assessment phase to retrieve few data and use only a few questions to the visitor to extract a rich body of information about the installation.

We use all available data about the installation and the visitor in the modelling phase to establish the estimation model (stippled lines in Figure 2). In current work, we use statistics and regression analysis to establish correlations between data, but we intend to use a machine learning approach [63] to establish more complex models later.

In the assessment phase, we use the previously established estimation model. We gather information about visitors' interaction with the installation using diverse data, observation, and responses from selected questions to estimate a goal (e.g., how satisfied or engaged a visitor is during a visit).

When performing the assessment we need to find the minimum number of sources to provide a valid score. The goal is to make the assessment as non-intrusive as possible. Questionnaires may be used, but the questions that are asked are targeted and will only be about the installation or an aspect of the installation. Using this methodology, museums can then change the installation, run a new assessment and see the effects of the change in the installation's VEI profile.

A. Assessment Framework

We propose an assessment framework that uses objective assessment, physiological responses, and estimation models to derive evidence of how a visit is perceived for individuals and groups of subjects. A generalised version of this assessment framework is presented by Leister [64].

An important requirement is that the assessment methods are not perceived as being intrusive. Intrusive assessment methods are usually only applicable in a lab setting, as they reduce the quality of experience (QoE) and, thus, impact the result of an assessment negatively.

Engagement and visitor experience cannot be measured directly. They are latent constructs. From measurable data and an estimation model trained by our machine learning approach we intend to derive a measure of experience of the visitors using an installation. It is similar to a satisfaction index and can be used to evaluate an installation.

Our assessment framework (Figure 3) consists of four layers: *Layer I*: the *Scenario Layer* presents the artefact, the subject, the action or interaction of the subject, other subjects, and, to some extent, observers; *Layer II*: the *Data Collection and Observer Layer* describes which data are collected from the elements of the scenario. *Layer III*: the *Assessment Layer* describes the types of assessment performed; and *Layer IV*: the *Assessment Process Layer* describes how the assessed data are processed further for the evaluated properties.

B. The Data Collection and Observer Layer

From a technical perspective, we classify whether these data in the Data Collection and Observer Layer (Layer II) as *1*) data automatically retrieved and processed, e.g., log files, technical parameters, event lists, sensor data, or physiological data; *2*) data from surveys and questionnaires; these data are often coded and analysed after the visitors have left the site, and the answering process might be intrusive; *3*) data from observations by an external observer; or *4*) static data stored, available, or known, e.g., from databases, or historical data.

C. The Assessment Layer

For defining the categories used in the Assessment Layer (Layer III), we adapt the assessment categories presented by Leister and Tjøstheim [65] into the following components: *a*) subjective assessment based on questionnaires and ratings, *b*) objective assessment based on measurements of the artefact, *c*) physiological assessment based on sensor data from a subject, *d*) behaviour and interaction assessment based on observations of the subject and the subject's behaviour and interaction with both the object and other subjects, *e*) observation of the subject and interaction with other visitors, and *f*) objective and subjective context information – including visitor type.

D. The Assessment Process Layer

The Assessment Process Layer (Layer IV) describes how the data from the Assessment Layer are processed. In Figure 3, the impact of these data is shown with bold arrows. Additionally, values with dashed lines could be taken into consideration. Data that are visualised with dotted lines are used in the calibration process when creating the estimation model or for evaluation purposes. Most of these data cannot be automatically processed and need human intervention of some kind.

Layer IV contains the following elements:

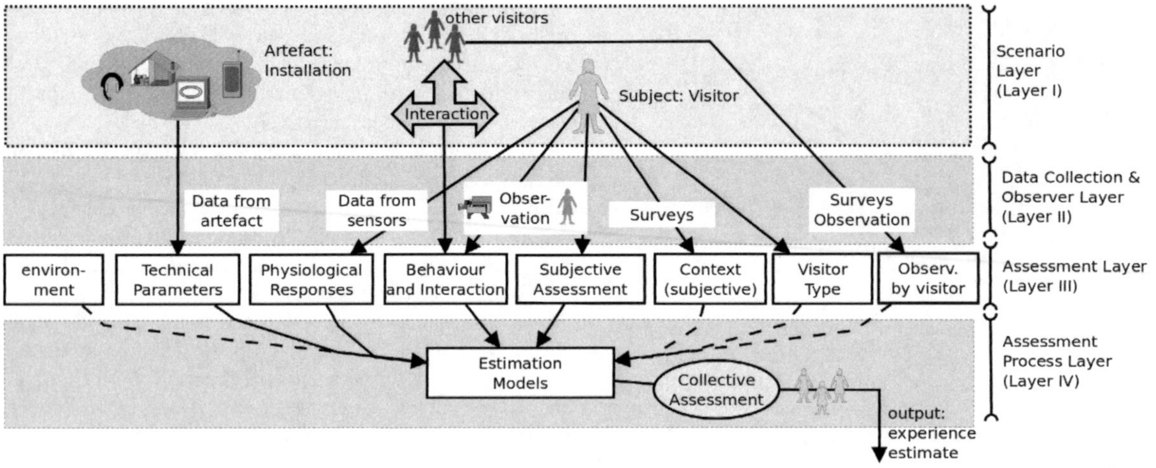

Figure 3: Four-layer assessment framework for engagement of visitors using installations in science centres and museums.

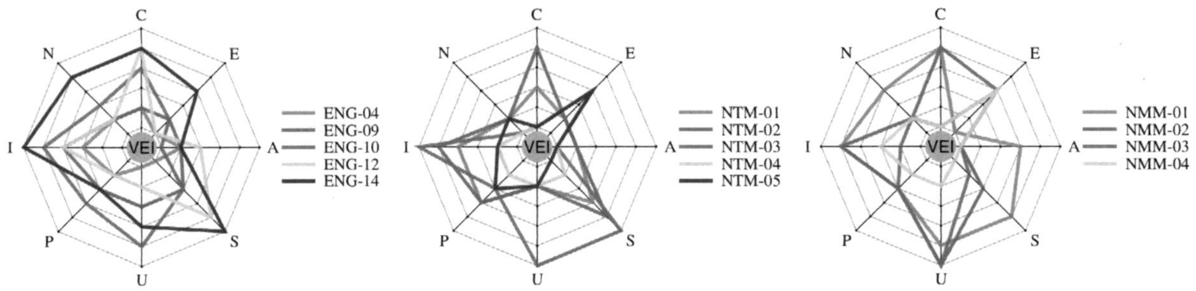

Figure 4: The VEI profile for selected installations in three science centres.

1) Estimation Model: The estimation model is a mathematical model that takes measurable assessment data as input and returns estimated values expressed in suitable metrics. The estimation model usually returns an estimated value for one subject at a time since personal data specific to the subject are involved in the calculation. Machine learning approaches [63] can be used to implement the estimation model.

2) Collective Assessment: Collective assessment presents the rating for one installation based on the individual assessments by many subjects.

3) Measures for evaluated properties: The result of the assessment process consists of measures for the evaluated properties. This can be a vector of values that will be used in the process that requires such assessment data.

IV. Characterising Installations

Installations can have a variety of qualities and characteristics. The design of these installations is important for the engagement and experience outcome for visitors. However, the assessment of the installation design is often unstructured. To develop a more structured way of quantifying the characteristics in an installation, we developed the Visitor Engagement Installation (VEI) profile that can assess properties of installations and give hints for the designers on how to improve the experience.

To characterise installations, we developed the original VEI profile [1] in an iterative process with three sciences centres: the Engineerium (ENG), the Norwegian Museum of Science and Technology (NTM), and the Norwegian Maritime Museum (NMM).

The VEI profile was developed from a set of requirements for a well-working installation given by the participating science centres. From these requirements, we selected a set of dimensions that we considered sufficiently orthogonal and tried these on a set of fourteen selected installations (see Figure 4). These installations range from simple vitrine exhibits to complex games or simulations where several visitors compete.

We performed several iterations of the finding process for the set of dimensions until the requirements for common science centre installations were covered.

Most studies that evaluate installations in science centres evaluate the impact of one dimension, such as interactivity, on the visitor. For this, observations of visitors are performed with various degrees of the dimension in question. However, we did not find a profile that characterises installations in multiple dimensions directly from an objective perspective, i.e., from only evaluating the installation.

A. The current Visitor Engagement Installation (VEI) Profile

The VEI profile characterises installations. The original defined by Leister et al. [1] had six dimensions, but we saw a need to extend the profile to make it more relevant for installations. We made adjustments to some of the definitions, e.g., the narrative has been adapted from its previous definition to findings from the literature [66]. We retrieved candidate dimensions from the gamification literature [56] [60] [61] [62]. This list is shown in TABLE I. We excluded dimensions that are related to already existing dimensions. This results in our current version with eight dimensions:

1) *Competition* (C): the degree of competition in an installation. It ranges from no competitive elements, to competition as a single player with high scores, to competing concurrently with other players.
2) *Narrative* (N): the degree the installation's narrative impacts the visitor. It ranges from no narrative (simply observing an object) to a fully developed, dramatic narrative with a story arc and characters.
3) *Interaction* (I): the degree of interaction between the visitor and the installation. It ranges from no interaction, to making choices that have consequences, to visitors creating their own content.
4) *Physical* (P): the degree of physical activity the visitor must perform when using the installation. It ranges from observing, i.e., no significant physical activity, to full body motion over time in realistic settings.
5) *Visitor (user) control* (U): the degree a visitor can control the use of the installation. It is also characterised by the size of the possibility space of user interactions. It ranges from no control over the installation (e.g., you can only read things in one order) to the ability to control the flow of the installation and add to its possibility space.
6) *Social* (S): the degree of social interaction between visitors. It ranges from a design for one visitor only, to groups of visitors interacting by themselves, to visitors that need to cooperate together to use the installation.
7) *Achievements* (A): the degree a visitor needs to be aware of achievements when using an installation. It is also characterised by the degree of feedback from the system. It ranges from no achievements to having a visitors achievements made concrete and the choices and their consequences displayed.
8) *Explore* (E): the degree of exploration or discovery for visitors in the installation. Exploration often can be done by trying out things with the possibility of failing, and thus learning from the failures. It ranges from predefined experiences to to exploring timeliness and possibilities spaces without penalty.

The new dimensions are *achievements* and *explore*. We also considered the dimensions such as *time constraints*, *cooperation*, *reminders*, and *challenges*, but did not find these most relevant for installations. The VEI profile is flexible and dimensions could be exchanged or more could be added depending on the purpose of a study.

TABLE I: Gamification elements and motivational affordances used to extend the VEI profile

#	Element	VEI	References	Comment
1	competition	C	[1] [60] [62]	
2	narrative	N	[1] [60] [61]	adjusted
3	interactivity	I	[1] [60]	
4	physical	P	[1]	
5	user control	U	[1] [60]	adjusted
6	social	S	[1]	
7	achievements	A	[67] [61] [62]	new
8	explore, discover	E	[60]	new
9	time constraints	T	[60]	considered
10	assignments, goals	(N)	[62] [61]	
11	challenges	(N)	[61] [62]	
12	feedback	(A)	[60] [62]	
13	rules	(U)	[60]	
14	loss aversion	(C,S)	[60]	
15	continuous play	–	[60]	
16	rewards	(A)	[60] [62]	
17	levels	(A)	[60] [61] [62]	
18	chance	(U)	[56]	
19	points	(A,C)	[61] [62]	
20	leaderboards	(A,C)	[61] [62]	
21	feedback	(A)	[60] [61]	
22	rewards	(A)	[61]	
23	progress	(A)	[61]	
24	reminders	(N)	[62]	
25	virtual goods	(A)	[62]	
26	friends, teams	S	[62]	
27	cooperation	(S)	[62]	

External influences are not taken into account in the VEI profile since these are not properties of the installation. Thus, physical factors, such as noise, light or smell need to be handled separately. We also exclude properties that belong to the context, such as social factors, institutional factors, or recent incidents personally or globally.

Each dimension has a value from 0 to 5; the higher the value, the more that dimension is present in an installation. TABLE II presents the description of the values for each dimension.

We posit that increasing each of these dimensions will potentially increase the visitor's engagement up to a point. At some point, the installation becomes too demanding or complex and the engagement will likely drop. There are implicit dependencies between dimensions for any installation, i.e., a change in one dimension may affect others. Where these points and dependencies are will depend on the installation.

B. Using the VEI profile to measure changes in engagement

We applied the VEI profile to installations from the three science centres: five at ENG, five at NTM, and four at NMM. The VEI profiles of these installations are shown in Figure 4.

We assessed installations with visitors. We wanted to determine whether a change in one dimension of the VEI profile would result in a change of the visitor's engagement. For example, the assumption that a change in an installation with a C-factor (competition) of 3 to 4 would increase the visitor engagement could be tested by measuring the visitor engagement with the originally designed installation, make changes in the installation to increase the C-factor (e.g., making the competition with other visitors happen in real-

TABLE II: EXPLANATION OF THE VALUES USED IN THE VEI PROFILE.

	0	1	2	3	4	5	
C	visitor observes only; no competition element.	inst. has several components; result must be achieved to proceed or succeed.	visitor receives a score; competition with the installation (machine).	competition with other visitors asynchroneously.	competition with other visitors in real-time.	challenge in team; influence on other players' result.	C
N	no narrative elements added; object can only be observed.	non-dramatised story; explaining text only.	narrative structure with limited use of narrative or scenographic elements.	narrative structure with rich use of narrative elements in a scenographic setting.	a developed dramatised story with a narrative universe in a scenographic setting.	a visual, immersive environment with a strong dramatised narrative story.	N
I	no interaction with object; observe only.	primarily no interaction; visitor can do something with the installation.	some interaction, such as "continue", "stop", "yes/no"; installation reacts.	moderate degree of interaction; choices influence outcome.	high degree of interaction; choices have consequences; content is stored.	visitor creates some of the content or develops narrative.	I
P	no physical activity; observation only.	push buttons; touch screen; hold or touch object.	visitor moves betw. parts of installation; enter installation; guided tour.	some activity, e.g., operating pumps; throwing balls.	full body-motion; longer physical activity.	full body motion over time; performing physical task in real setting.	P
U	controlled; visitor is observer; linear structure.	controlled; linear structure or chronological succession of events.	installation is built up in sequences; conditions must be met to proceed to next phase.	visitor can make choices, but the choices have no effect on the flow of the installation.	visitor controls flow, but installation limits its choices; multiple parallel narratives.	visitor has high degree of control; creative process.	U
S	single visitor.	single visitor, others observe.	several installations used independently from each other.	single visitor while others observe and engage and cheer.	installation intended for several simultaneous visitors.	multi-visitor installation; visitors must cooperate.	S
A	no specific achievements possible with installation.	immediate feedback on failure or success; countdown timer.	achievement collected, only shown at the end.	current status represents achievements; progress bar; graphical visualisation.	achievements are shown; points, lists, gadgets are displayed.	achievements are shown; choices and their consequences are displayed.	A
E	installation allows pre-defined views only.	view from several perspectives.	explore while progress is stopped.	explore while progress is ongoing.	can dissect installation with recover-functionality.	can follow timelines or branches in possibility space.	E
	0	1	2	3	4	5	

time), and then measure the visitor engagement for the altered installation. We are interested in the relative changes of the assessed engagement-related values when testing installations with modified versions that have a different VEI profile. For some installations, a change in one dimension of the VEI profile might be too small to result in a significant change in engagement.

C. Characterising Exhibitions Using the VEI profile

Besides single installations, the VEI profile can be used to characterise exhibitions or groups of installations. For example, the graphical representation of the VEI profile for selected installations in Figure 4 suggests that physical activity is characterised as low for these installations, as is the A-dimension. Also, the N-dimension seems to be low, with the exception of two recently developed installations that are based on longer narratives. We also observe differences between the three sites regarding their overall profile characterised by mean values and variance of the respective VEI profiles.

Assuming that installations in an evaluation form one ensemble, we can visualise this ensemble's characteristics as a whole. In the above example, we recognise that the physical (P) dimension has rather low values for these installations. An exhibition designer could consider to increase the P-value by making changes to the installation for the sake of giving visitors a better experience or to decrease values, e.g., the U-dimension in the above example. In other cases, a good mix of characteristics could be the objective of an exhibition.

When an installation designer decides to make changes on the basis of the VEI profile, both the original and the modified installation need to be assessed with regard to how engaging these installations are. In the field of visitor studies several approaches are possible, such as observations, questionnaires, or assessment using diverse sensors. In the ideal case, the assessment is minimally intrusive, does not bother the visitor, and can be performed in a short time.

V. ASSESSMENT OF SELECTED INSTALLATIONS

We performed assessments to analyse the correlations between the various data and layers in our assessment framework. We did not aim at creating the complete estimation function, but to find evidence that it is feasible to create a function using properties and correlations.

We performed three assessments. In the first assessment, we assume that competition, i.e., the C-dimension of the VEI profile, has an impact on the visitor's engagement. Using a quiz game, we compared subjective data of winners, losers, and single players of a quiz game. When analysing the data, we received unexpected results with the interpretation of observed smiles: in a competition situation, visitors that answer wrong to an quiz question tended to smile more often than when answering correctly.

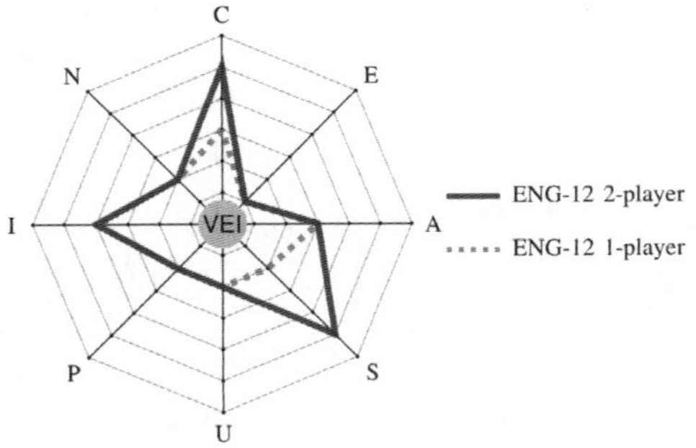

Figure 5: VEI profiles of the ENG-12 installation when two players compete (solid line) and a single player version (dashed line); the single player version has lower values for C and S.

In the second, still ongoing assessment, we evaluated the impact of the C-dimension in a more complex game that lasts fifteen to twenty minutes. From the experiences of the first assessment, we tried to automate the assessment using events from the installation and using a Kinect to retrieve the players' emotions.

The third assessment looks at the influence of the narrative. For this, we used a type of installations where visitors toss balls on a wall without many explanations. After changing the narrative, we make similar evaluations with a different group of visitors and compare the results.

A. The Influence of Competition to Experience

The installation *Footprint eQuiz* at the Engineerium, here denoted as ENG-12, shall challenge the visitors with questions about different environmental perspectives, show how the oil and gas industry takes responsibility, and how they work to minimise the negative impact on the environment. The installation provides an understanding of different ways we can lower our energy consumption to reduce the environmental impact.

ENG-12 is a game where up to two players compete by answering questions related to energy and the environment. There are two levels available, beginner and expert. The installation consists of two stations with two large buttons each, an orange one and a blue one. ENG-12 starts with a short introduction before ten questions are shown on the screen in sequence. As a question is shown, a timer starts counting down to zero. Players answer by pressing either button before the timer reaches zero. Players receive points for a correct answer and bonus points based on how quickly they answered. Players lose points when answering incorrectly but the score cannot go below zero. After the ten questions, a summary with the number of points scored for each player is presented.

1) Experiment Setup: Figure 5 shows the VEI profile of ENG-12 with the solid line. We also show a version where only one player answers questions with the dotted line. This change lowers the values of both the C-dimension and the S-dimension.

Figure 6: The installation ENG-12 at the Engineerium during the assessment.

Figure 6 shows the installation ENG-12 during the assessment. In addition to the installation, we have installed two cameras that observe each of the players, one camera that observes the scene from behind, and, for each player, a human observer makes notes. The video footage is used both for manual analysis and automated analysis of facial expressions using the Face Reader software by Noldus [68][69]. We also made changes to the installation's software to record all events (e.g., which button is pressed, points awarded, and player scores).

The observers note visitor's mood using a simplified valence tracker [70], i.e., whether the visitor is excited-positive, excited-negative, or calm-neutral for each quiz question. These values are compared with the outcome of the Face Reader software. The self-reported data by the visitors consist of a self-developed questionnaire for ENG-12 and a 20-item PANAS scale [71]. Since we are interested in the the positive affect (i.e., the PA of the PANAS), we omitted factors that express negative emotions (e.g., *guilty* or *scared*) that hardly can be an impact from the use of the installation.

We performed tests to ensure that the preliminary technical setup is in place and working. This includes logging the events from the installation (objective data), interpretation of the video footage and light conditions, usefulness of the questionnaires and valence tracker, and conformance with the Norwegian privacy laws. Still, challenges needed to be addressed, such as lighting problems or adjustments in the questionnaires (some items of the PANAS adjectives seem not to be understood by the target group; as a consequence, we did not use these items).

2) Results: We asked students from school classes that visit the Engineerium to use ENG-12 with our assessment equipment and observed them as described above. In five sessions between October 2014 and March 2015 we assessed data from 33 winners, 34 losers, and 20 single players. All participants were between the ages of 14 to 16. The data from

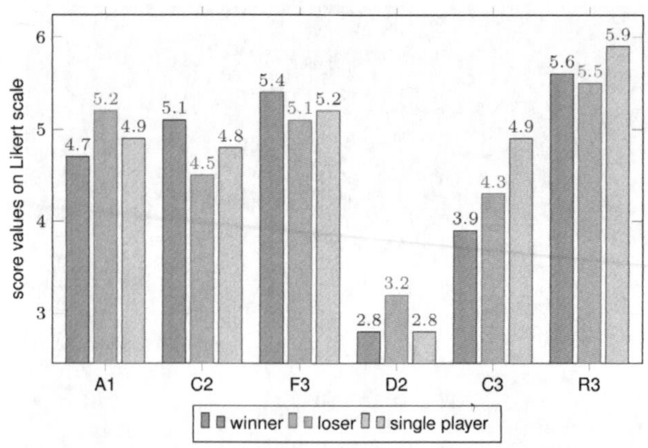

Figure 7: Response scores on a Likert scale for winners ($n = 33$, mean game score: 2009), losers ($n = 34$, mean game score: 1391), and single players ($n = 20$, mean game score: 1790) for the subjective constructs A_1, C_2, F_3, D_2, C_3, and R_3.

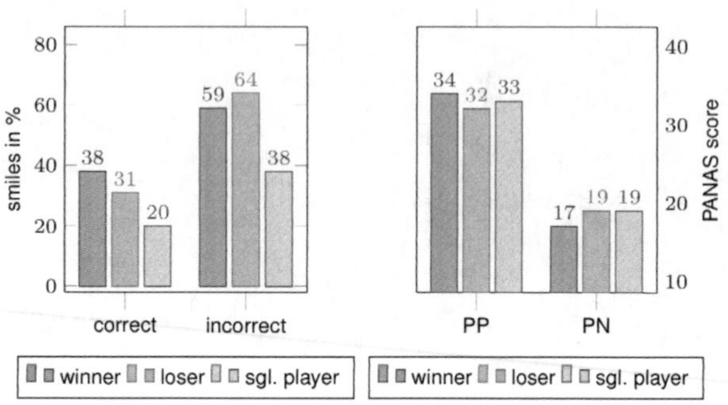

Figure 8: Smiles for correct answers and incorrect answers (left) and PANAS Positive and Negative values (right) for the three player categories winner, loser, and single player. While the differences for PANAS are small, there are significant differences for the number of smiles: visitors smile more often when answering incorrect.

TABLE III: PANAS SCORES FROM THE EXPERIMENT.

PANAS	Pos.	Neg.
winners ($n = 33$)	34.0	16.5
losers ($n = 34$)	31.5	18.5
single players ($n = 20$)	33.0	18.6
all ($n = 87$)	32.3	17.4
std. dev. ($n = 87$)	6.1	6.1

one of the winners was discarded due to an irregularity (he played the game twice). We are aware that the number of single players is too low to give a significant result, and one of the single player responses is an outlier. So, we refrain from interpretations of the single player data.

We show results from the subjective answers the players gave after having played ENG-12 with six selected questions in Figure 7. In TABLE III, we show the mean values of the positive and negative PANAS scores for the three groups and the mean value. We note that the standard deviation is in a similar range as published by Watson et al. [71] for assessments in the moment.

Figure 8 shows that we registered significantly more smiles when an incorrect answer was given than when a correct answer was given, independent of whether they ended up the winner or loser of the game after 10 questions. These smiles occur *before* the players know they win or lose the competition. We also observe that the number of smiles is significantly reduced for the single player games. We interpret this as a smile does not necessarily expresses happiness about answering correctly, as we first assumed. Instead, we need to re-interpret the smile to have a different function, e.g., a social function; this fact is supported from the field of psychology [72][73]. Yet the high number of smiles, specifically when answering incorrectly, show that the visitors are engaged and show emotions; they are not indifferent. This also shows that it is feasible to register engagement automatically.

The questions from the questionnaires is given in TABLE IV. The interpretation of the assessed data from these

TABLE IV: FORMULATION OF THE QUESTIONS FOR THE VARIABLES F_i, C_i, A_i, R_i, L_i, D_i, AND E, TAILORED FOR THE INSTALLATION ENG-12. ALL QUESTIONS ARE ON A SCALE FROM 1 (LOW) TO 7 (HIGH).

F_1	Did you have fun using the installation and answering the questions?
F_2	How much did you like the installation?
F_3	The installation was entertaining.
C_1	I concentrated so that I could answer as fast as possible.
C_2	I read the question. If I did not remember the text, I read it once more.
C_3	I was focused and carefully read the question. Then I made my judgement before I answered.
A_1	Do you want to use the installation one more time?
A_2	Do you want to use the installation a second time to improve your score?
A_3	At your next visit to the science centre, do you think you will use this installation?
R_1	Would you recommend the installation to others who are with you today?
R_2	I would like to recommend the installation to someone I know.
R_3	I will recommend the installation to other visitors to the science centre.
L_1	After reading a question, but unsure about the answer, was topic something that I would like learn more about.
L_2	After you had answered, did your interest in the subject increase or decrease?
L_3	The installation triggered my interest to learn more about energy and the environment.
D_1	How easy or difficult was it to answer the questions?
D_2	I thought the questions were too difficult.
E	How engaging was the installation?

questionnaires show small differences between winners and losers. However, a trend is visible: losers find the quiz questions somewhat more difficult (D_2). While they show lower engagement (R_3), their intention to answer again (A_1) and to learn more (C_3) is higher. They also report less fun (F_3) and less concentration (C_2). The PANAS scores show a similar trend, i.e., winners have a higher positive score while losers have a higher negative score. Note, however, that the differences are rather small. We also note that the trends in these responses are as expected between winners and losers. The data for the single players are not as expected, but due to low data quality we refrain from an interpretation.

We cannot say for certain that VEI profile's C-dimension has an impact on the QoE because we do not have sufficient data yet. We need more data for single player games. Yet the fact that winners and losers have different values in the questionnaire and PANAS in the expected manner (that is winners have higher values than losers), shows that the C-dimension has an impact on two-player games. If it did not, the data from these two groups would be the same.

3) Automating the Value Tracker: In our experiments, we used both a valence tracker operated by observers and the FaceReader software by Noldus [68][69]. Both assessment methods have advantages and disadvantages. The valence tracker is a manual method performed by an observer where the opinion of the observer plays a role. But in the experiment settings, it is quite easy to make mistakes when registering emotions, e.g., having to focus on both players can make it difficult to capture the emotion from both before the start of a new question since the players normally are indifferent as they read a question. Thus, two observers were used in the experiments for ENG-12. The video footage can be used in the case of doubts, but this is time consuming. A human observer is visible for the visitors. Thus, for the experiment unwanted communication between visitor and observer can occur that might influence the result, also referred to as the Hawthorne effect [74]. In our studies, we have not taken this effect into account.

In our experiments, the automated face expression recognition fails in about half of the cases. The reasons for these failures include lighting problems (the light settings in science centres are often problematic for such analysis) and positioning of the cameras (these should be installed so that they do not obstruct essential parts of the installation). Other problems occur when visitors temporarily turn their heads away or make hand movements that partially obstruct their face.

We discussed some of these issues with the developers of the FaceReader software. We found out that we ideally should have the camera at the same height as the head of the visitor rather than filming from a lower position. However, having the observation cameras at the ideal height might obstruct important parts of the installation. When placing cameras in future installations, the camera placement needs to be planned carefully; we also intend to evaluate whether competing technologies suffer from similar impacts.

4) Optimising Questionnaires: When assessing engagement and other properties, questionnaires are still necessary. However, to reduce intrusiveness of such questionnaires, we integrate the questionnaire into the natural flow of the visit and reduce the number of questions. To find out which questions are representative, we combined the positive-negative affect-instrument (PANAS) with a survey-instrument that captures the subjective experience of the player [2].

For this study, the evaluation used a within subject design, i.e., all students used eQuiz as a competition between two players. For the analysis, we used partial least squares (PLS), which is a structural equation modelling technique [75] that can simultaneously estimate measurement components and structural components (i.e., the relationships among these constructs). The PLS algorithm is a sequence of regressions

TABLE V: Summary of measurement scales for Research Model 1.

Construct	Measurement[a]	Factor Loading
Concentration composite reliability: 0.74	C_1	0.81
	C_2	0.42
	C_3	0.83
Enjoyment and engaging composite reliability: 0.88	F_1	0.68
	F_2	0.86
	F_3	0.87
	E	0.81
Intention to use again composite reliability: 0.88	A_1	0.85
	A_2	0.77
	A_3	0.89
Positive Affect composite reliability: 0.84	Active	0.72
	Excited	0.64
	Enthusiastic	0.77
	Inspired	0.67
	Attentive	0.78
Negative Affect composite reliability: 0.83	Nervous	0.77
	Afraid	0.77
	Frustrated	0.60
	Scared	0.69
	Upset	0.67

[a]The formulation of the measures for the questionnaires are presented in TABLE IV

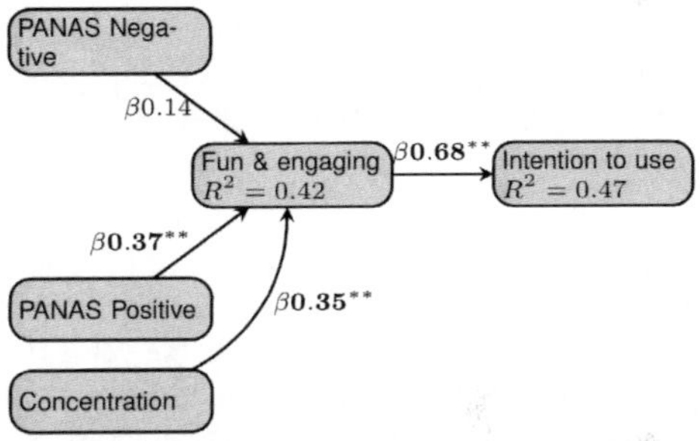

Figure 9: Research Model 1: PLS to predict the intention of use for all visitors. Independent variables: PANAS Negative (PN) and PANAS Positive (PP) are emotional factors; concentration is a cognitive factor; fun & engaging is a hedonic factor. ** = significant; $n = 67$

in terms of weight vectors.

PLS does not require a large sample size [76], and it is not a pre-requisite that the research models are based on comprehensive theories [77][78]. Still, a research model should have a theoretical foundation, although it might contain exploratory aspects.

We used statistical modelling with *smartPLS* version 2.0 M3 [79] to analyse the data and to compare the three models. TABLE V shows the composite reliability of the constructs and the factor loading for each item that need to meet the evaluation criteria for partial least square modelling [75]. We updated these results from Tjøstheim et al. [2] by increasing the number of respondents ($n = 67$). In TABLE V, note that C_2 has low factor loading. This is caused by winners answering differently from losers (winners tend not to read questions more than once).

We created a dependent variable *intention to use the eQuiz again*. The R^2 variable is a measure of the proportion a dependent variable is explained by the independent variables

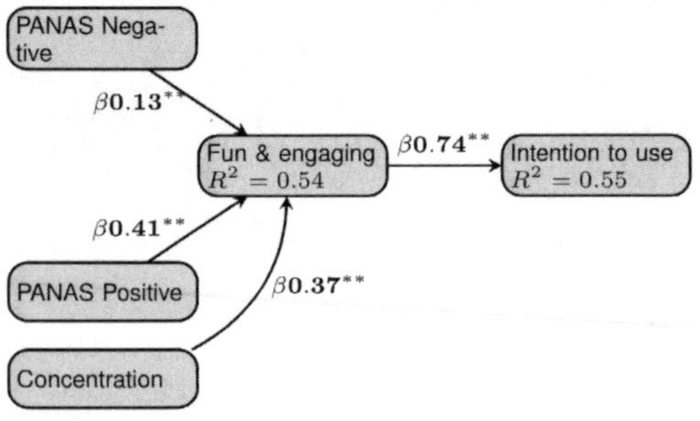

Figure 10: Research Model 2: PLS to predict the intention of use for losers. ** = significant; $n = 34$

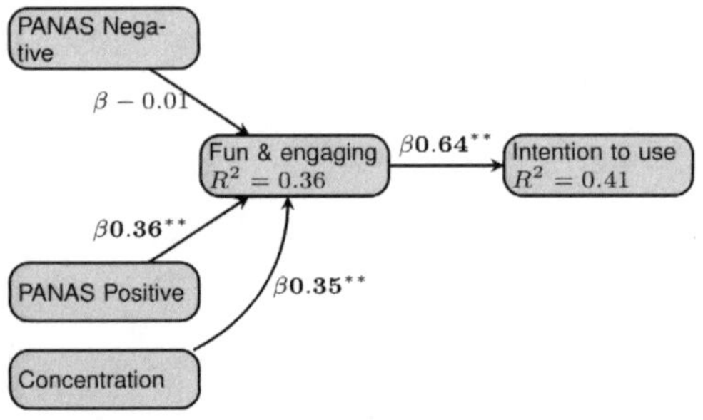

Figure 11: Research Model 3: PLS to predict the intention of use for winners. ** = significant; $n = 33$

in the model. For the winners and the losers of the game, the factors *fun* and *engaging* explain the new variable at 41% and 55%, respectively; combining them together the value is 47%. We updated the graphs to include data from more participants. Figures 9, 10, and 11 show graphical representations of these dependencies for all participants, losers, and winners.

For someone managing a science centre, it seems like a good choice to ask visitors if they enjoyed playing eQuiz. It is valuable to know the answer to this question, and it might give the science centre an indication of whether the visitors are interested in using the installation again.

B. Automating the Assessment Process

The installation *The Motorway of the Ocean* in the exhibition *At Sea* at the Norwegian Maritime Museum (NMM), here denoted as NMM-01, is a game that teaches players the roles of people employed in shipping and tasks related to shipping from the perspective of a ship owner.

In NMM-01, up to four players compete against each as ship owners. Each player controls one vessel. Through the course of the game, players make informed decisions as the ships travel across the ocean to its destination. These decisions can be the speed of the vessel, whether or not to take on extra cargo, bunker oil, deal with weather or pirates, and so on. Each player is placed behind a console where the player can control the game while the current progress for all ships is shown on a projection wall, visible for all.

For NMM-01, we used a similar setup as in the previous case, i.e., we used camera observation, observation by human observers using a simplified valence tracker, and questionnaires including PANAS, a questionnaire about emotions, and a questionnaire containing knowledge questions. The visitors answered this questionnaire both before and after the game was played. This questionnaire was developed to assess the impact of the game on visitor's knowledge.

We used Kinect II devices to create the video footage for the analysis. Additionally, we analysed whether emotions can be assessed using the Kinect-API that allows to retrieve the parameters *smile* and *engaged*. Currently, the software for this type of assessment is under development. The first tests show that smiles in the faces of the visitors can be recognised, but *engaged* currently only means that the visitor is facing the Kinect. Some technical issues with the API and a low number of test subjects has resulted in insufficient data so far.

The idea is to compare the results from the valence tracker, the face reader using the video-footage from the Kinect II, and the results from the Kinect-API. We noted that the nature of NMM-01 is not making people smile much; thus, the retrieved data are not sufficient to come to a conclusive answer.

Results from the questionnaires show that NMM-01 shows the highest values in terms of engagement, as can be seen for the value E in Figure 12.

C. The Influence of the Narrative to Experience

The installation *Solar Cell*, here denoted as NTM-01, is part of the exhibition *Energy Tivoli* at the Norwegian Museum for Science and Technology. The installation presents a wall where an atom and its electrons are drawn. The goal is to visualise how energy is created from colliding protons and election. The visitor throws balls representing photons at the four outer electrons on the wall during a given time. The original installation only instructed visitors to hit the outer electrons. For our experiment, the first half of the participants ($n = 39$) use the installation unchanged. The second half of the participants ($n = 36$) received more information. We extended the installation with an additional board explaining the role of electrons and photons before participants started throwing balls. We compute this as a change in VEI profile for the N-dimension from Level 2 to Level 3.

In our evaluation, we asked visitors to use NTM-01, either the original or modified version, and fill out the questionnaires as described for the other evaluations. The evaluation shows that the engagement increases slightly for the modified version, but this increase is not statistically significant with the number of participants that took part in the study. This result is expected since the changes to the narrative are small. We also noted that the participants were focused on throwing as many balls as possible within the time-limit of twenty seconds.

We observed that the unmodified NTM-01 shows a low score for the *intention to learn* (L_i); after the modification, this score increases by about one point on the Likert scale. We refer to Figure 12 explained in Section V-D for more details.

NTM-01 was included in our study because it is an example of a common type of installations in science centres that

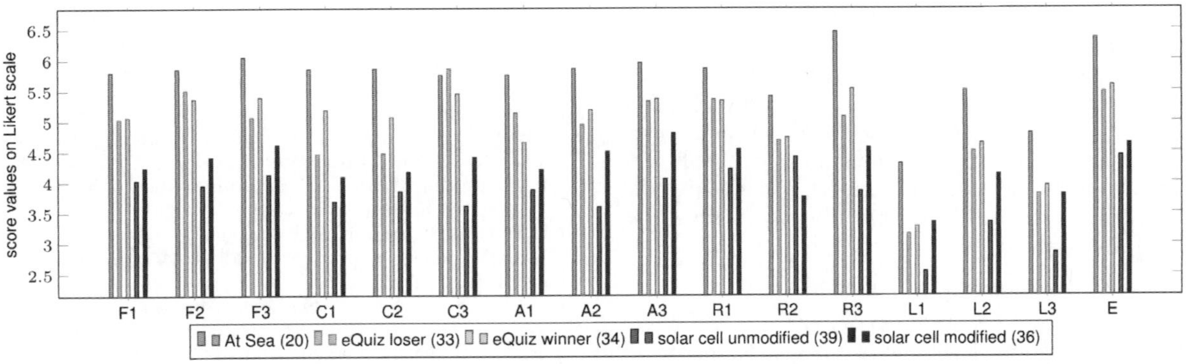

Figure 12: Response scores for five tests on a Likert scale for the independent variables for *fun* F_i and *concentrate* C_i and the dependent variables for *play again* A_i, *recommend* R_i, and *intention to learn* L_i. Each variable occurs in three different questions. We also show the results for the variable *engaging* E.

emphasise user's physical activity, the time use is rather short, and little written information is presented.

D. Comparing the installations

In Figure 12, we show the results from the questionnaires for the installations NMM-01, ENG-12 (separate results for winners and losers), and NTM-01 (modified and unmodified) in a graph for several dimensions of engagement. The graphs show the results for the independent variables *Fun* (F) and *Concentrate* (C), and the dependent variables *Play Again* (A), *Recommend* (R), *Intention to Learn* (L), and *Engaging* (E). The trend in these data is clear: NMM-01 has the highest engagement factor, followed by ENG-12, while both versions of NTM-01 score significantly lower.

For each of the variables F, C, A, R, and L, three different questions were asked; this was done to combat possible effects from the way the questions were presented. One question was asked for variable E. The formulation of the questions is shown in TABLE IV for ENG-12. For the installations NMM-01 and NTM-01, similar questions were asked with some modifications due to the nature of the installation.

As Figure 12 shows, there are some differences between the responses in one category, but most of these follow the same pattern; with the exception of R_2 score.

Our goal was to find a suitable proxy question for engagement. The analysis shows that the values for F_i are most structurally similar to the variable E. This result suggests that studies in science centres can use one of the questions about fun as a proxy for engagement instead of asking several questions to the visitors. It is straightforward to implement these shorter questionnaires in an application on a mobile device or another part of a system that facilitates studies in science centres and museums.

VI. Conclusion and Future Work

Science centres and museums are interested in having engaging exhibits to attract visitors. Our methodology to assess and analyse visitors' engagement can be a new instrument in finding these exhibits. Our assessments at three science centres used installations with different properties to find evidence that we can use measured values from installations, sensors, and cameras to estimate visitor engagement. We can use this evidence to reduce the size of questionnaires down to a few questions. Using the evidence and the short questionnaires gives us a good way to find and assess engagement.

This shows that our methodology can work in different kinds of science centres with different subject matter and ways of interacting with an exhibit. We have explored different methods for gathering data. We focused on emotions visible on the visitor's face combined with data from the installation, and short questionnaires, but other methods could also work such as skin conductivity or heart rate. There are few limits to data sources, but each source could add to complexity.

The science centres and those creating exhibits benefit from this research. They can use our tools in the design phase and in maintenance to find what engages visitors and how a change affects engagement. The tools described in the article, like the VEI profile, is actively used in the design phase of exhibitions. The results of the assessments have been integrated into an exhibition management system to give science centres a better means to perform visitor studies.

There are several research paths forward. Further refinement and validation of our work so far needs to be performed. This includes a comparison with results from researchers who use qualitative analysis methods, such as sense-making.

When extending our methodology to use more sensors and collecting all available data using the Internet of Things (IoT) [80], several challenges occur. The sensors of the IoT can produce large amounts of data that need to be analysed. This amount of data is much larger than being processed in traditional visitor studies. Methods described in the research field of *big data* [81] need to be applied, including statistical methods and machine learning. Collecting large amounts of data will result in challenges for the visitor's privacy that are beyond the usual privacy consent agreements in museums when visitor studies are performed. Analysing ample sensor data using big data technology could potentially identify visitors through their behaviour in situations when they should be able to experiment unobserved. Both legal and technological measures must be considered to find a balance between all-encompassing visitor studies and privacy.

Finally, our current work does not take into account the visitor's identity type, as defined by Falk and Storksdieck [17], nor the visitor's expectation to a science centre visit. Our studies were performed with students who are a homogeneous group. But we suspect that the visitor's engagement will vary for the different identity types. Thus, an integration of the VEI profile with visitor type and the subjective context of a visitor needs to be considered.

VII. Acknowledgments

The work presented here has been carried out in the project VisitorEngagement funded by the Research Council of Norway in the BIA programme, grant number 228737.

References

[1] W. Leister, I. Tjøstheim, G. Joryd, and T. Schulz, "Towards assessing visitor engagement in science centres and museums," in Proc. PESARO 2015, The Fifth Intl. Conf. on Performance, Safety, and Robustness in Complex Systems and Applications. IARIA, 2015, pp. 21–27.

[2] I. Tjøstheim, W. Leister, A. Larssen, and T. Schulz, "The role of emotion and enjoyment for QoE – a case study of a science centre installation," in Proc. QoMeX 2015, The 7th Intl. Workshop on Quality of Multimedia Experience, A. Skodras, Ed. IEEE, 2015, pp. 1–6.

[3] L. Rennie and D. Johnson, "Visitors perceptions of changes in their thinking about science and technology following a visit to science center," Visitor Studies, vol. 10, 2007, pp. 168–177.

[4] S. Nordal, "Science centres and museums as rooms of revelation (RoR) in science education," in Abstracts for seminar on Science Centre Research. skolelab.no, Sep. 2012. [Online]. Available: https://skolelab.no/innsendt/skolelab.no/kurs/711/2012-10-08_135131-Nordal_Abstract_DeWitt_seminar.pdf [Accessed: 1. June 2015].

[5] M. I. Dahl and D. Stuedahl, "Transforming children's participation and learning in museums: From singular dialogues to a multilayered explorative experience," in Proc. The Transformative Museum. Roskilde University, May 23-25 2012.

[6] P. M. F. Sandsmark, "Mellom opplevelse og læring – en undersøkelse av publikums bevegelse og interaksjon i utstillingen Skipet," bachelor's thesis, Høgskolen i Telemark, 2011, in Norwegian.

[7] A. Hofstein and S. Rosenfeld, "Bridging the gap between formal and informal science learning," Studies in Science Education, vol. 28, no. 1, 1996, pp. 87–112.

[8] J. H. Falk and L. D. Dierking, "Learning from the outside in," in Lessons without limit: How free-choice learning is transforming education. New York: Altamira press, 2002, pp. 47–62.

[9] J. Wang and A. M. Agogino, "Cross-community design and implementation of engineering tinkering activities at a science center," in Proc. FabLearn 2013: Digital Fabrication in Education, Stanford, 27-28 October 2013, pp. 1–4. [Online]. Available: http://fablearn.stanford.edu/2013/papers/ [Accessed: 2 Feb 2015].

[10] H. H. Higgins, "Museums of Natural History : Part I. Museum Visitors". D. Marples, 1884.

[11] M. Lindauer, "What to ask and how to answer: a comparative analysis of methodologies and philosophies of summative exhibit evaluation," museum and society, vol. 3, no. 3, November 2005, pp. 137–152.

[12] P. Pierroux and A. Kluge, "Bridging the extended classroom: Social, technological and institutional challenges," Nordic Journal of Digital Literacy, vol. 6, no. 3, 2011, pp. 115–120.

[13] A. Šuldová and P. Cimler, "How to assess expirience – the new trend in research technique, use in nonprofit sector of entertainment and educational industries," Marketing a obchod, vol. 4, 2011, pp. 115–124.

[14] C. W. Sanford, "Evaluating Family Interactions to Inform Exhibit Design: Comparing Three Different Learning Behaviors in a Museum Setting," Visitor Studies, vol. 13, 2010, pp. 67–89.

[15] P. M. McManus, "Museum and visitor studies today," Visitor Studies, vol. 8, no. 1, 1996, pp. 1–12.

[16] L. D. Dierking and J. H. Falk, "Redefining the museum experience: the interactive experience model," Visitor Studies, vol. 4, no. 1, 1992, pp. 173–176.

[17] J. Falk and M. Storksdieck, "Using the contextual model of learning to understand visitor learning from a science center exhibition," Science Education, vol. 89, 2005, pp. 744–778.

[18] J. H. Falk, "The impact of visit motivation on learning: Using identity as a construct to understand the visitor experience," Curator, vol. 49, no. 2, 2006, pp. 151–166.

[19] J. H. Falk, J. Heimlich, and K. Bronnenkant, "Using identity-related visit motivations as a tool for understanding adult zoo and aquarium visitors' meaning-making," Curator: The Museum Journal, vol. 51, no. 1, 2008, pp. 55–79.

[20] C. Barriault and D. Pearson, "Assessing Exhibits for Learning in Science Centers: A Practical Tool," Visitor Studies, vol. 13, 2010, pp. 90–106.

[21] H. H. Shettel, M. Butcher, T. S. Cotton, J. Northrup, and D. C. Slough, "Strategies for determining exhibit effectiveness," American Institutes for Research, Washington, DC, Tech. Rep. Report No. AIR E-95-4/68-FR, April 1968.

[22] M. B. Alt and K. M. Shaw, "Characteristics of ideal museum exhibits," British Journal of Psychology, vol. 75, no. 1, 1984, pp. 25–36.

[23] D. Spegel, "Expogon," 2014. [Online]. Available: www.expogon.org [Accessed: 1. June 2015].

[24] D. L. Young, "A phenomenological investigation of science center exhibition developers' expertise development," Ph.D. dissertation, The University of North Carolina at Chapel Hill, 2012, 174 pages.

[25] S. Allen, "Designs for learning: Studying science museum exhibits that do more than entertain," Sci. Ed., vol. 88, no. S1, 2004, pp. S17–S33.

[26] D. M. Russell, M. J. Stefik, P. Pirolli, and S. K. Card, "The cost structure of sensemaking," in Proc. of the INTERACT '93 and CHI '93 Conference on Human Factors in Computing Systems. New York: ACM, 1993, pp. 269–276.

[27] S. S. Yalowitz and K. Bronnenkant, "Timing and tracking: Unlocking visitor behavior," Visitor Studies, vol. 12, no. 1, 2009, pp. 47–64.

[28] J. Baur, Ed., "Museumsanalyse – Methoden und Konturen eines neuen Forschungsfeldes", ser. Kultur- und Museumsmanagement. Bielefeld: Transcript, 2010, in German.

[29] V. Kirchberg, "Besucherforschung in Museen: Evaluation von Ausstellungen," in Museumsanalyse: Methoden und Konturen eines neuen Forschungsfeldes, ser. Kultur- und Museumsmanagement, J. Baur, Ed. Bielefeld: Transcript, 2010, pp. 171–184, in German.

[30] S. Macdonald, Ed., "A Companion to Museum Studies", 1st ed. Wiley-Blackwell, Aug. 2010.

[31] C. Martindale, "Recent trends in the psychological study of aesthetics, creativity, and the arts," Empirical Studies of the Arts, vol. 25, no. 2, 2007, pp. 121–141.

[32] P. Pierroux and R. Steier, "Making it real: Transforming a university and museum research collaboration into a design product," in Design as Scholarship in the Learning Sciences, V. Svihla and R. Reeve, Eds. London: Routledge, 2016, ch. 9, pp. 115–126.

[33] G. Klein, B. Moon, and R. R. Hoffman, "Making sense of sensemaking 1: Alternative perspectives," IEEE Intelligent Systems, vol. 21, no. 4, Jul. 2006, pp. 70–73.
[34] ——, "Making sense of sensemaking 2: A macrocognitive model," IEEE Intelligent Systems, vol. 21, no. 5, Sep. 2006, pp. 88–92.
[35] M. Tröndle, S. Greenwood, V. Kirchberg, and W. Tschacher, "An integrative and comprehensive methodology for studying aesthetic experience in the field: Merging movement tracking, physiology, and psychiological data," Environment and Behaviour, August 9 2012, pp. 1–34.
[36] M. Oppermann, "Triangulation – a methodological discussion," International Journal of Tourism Research, vol. 2, no. 2, 2000, pp. 141–146.
[37] W. Tschacher, S. Greenwood, V. Kirchberg, S. Wintzerith, K. Van Den Berg, and M. Tröndle, "Physiological correlates of aesthetic perception of artworks in a museum." Psychology of Aesthetics Creativity and the Arts, vol. 6, no. 1, 2012, pp. 96–103.
[38] M. Ghijsen, "Facial expression analysis for human computer interaction," Ph.D. dissertation, University of Twente, 2004.
[39] V. Terzis, C. N. Moridis, and A. A. Economides, "Measuring instant emotions based on facial expressions during computer-based assessment," Personal and Ubiquitous Computing, 2011, pp. 1–10.
[40] J. A. Russell, "A circumplex model of affect," Journal of Personality and Social Psychology, vol. 39, 1980, pp. 1161–1178.
[41] H. L. O'Brien, "The influence of hedonic and utilitarian motivations on user engagement: The case of online shopping experiences," Interact. Comput., vol. 22, no. 5, Sep. 2010, pp. 344–352.
[42] K. S. Judge, C. J. Camp, and S. Orsulic-Jeras, "Use of montessori-based activities for clients with dementia in adult day care: Effects on engagement," American Journal of Alzheimers Disease and Other Dementias, vol. 15, no. 1, 2000, pp. 42–46.
[43] J. Griffin, "Learning science through practical experiences in museums," International Journal of Science Education, vol. 20, no. 6, 1998, pp. 655–663.
[44] D. Griffin and L. Paroissien, Eds., "Understanding Museums – Museums, education and visitor experience". National Museum of Australia, 2011. [Online]. Available: http://nma.gov.au/research/understanding-museums/Museums_education.html [Accessed: 1. June 2015].
[45] F. H. Wilhelm and P. Grossmann, "Emotions beyond the laboratory: Theoretical fundaments, study design, and analytic strategies for advanced ambulatory assessment," Biological Psychology, vol. 84, 2010, pp. 552–569.
[46] L. E. Nacke and C. A. Lindley, "Affective ludology, flow and immersion in a first-person shooter: Measurement of player experience," CoRR, 2010. [Online]. Available: http://arxiv.org/abs/1004.0248 [Accessed: 1. June 2015].
[47] A. G. Money and H. Agius, "Analysing user physiological responses for affective video summarisation," Displays, vol. 30, no. 2, Apr. 2009, pp. 59–70.
[48] R. Picard, "Affective computing," MIT Media Laboratory, Perceptual Computing Section, Tech. Rep. 321, Nov. 1997.
[49] M. E. Hoque, J. Hernandez, W. Drevo, and R. W. Picard, "Mood meter: Counting smiles in the wild," in Proc. UbiComp'12. New York, NY, USA: ACM, 2012.
[50] M. Ben Ammar, M. Neji, A. Alimi, and G. Gouardères, "The Affective Tutoring System," Expert Systems with Applications, vol. 37, no. 4, 2010, pp. 3013–3023.
[51] M. Brand, F. Klompmaker, P. Schleining, and F. Weiß, "Automatische Emotionserkennung – Technologien, Deutung und Anwendungen." Informatik Spektrum, vol. 35, no. 6, 2012, pp. 424–432, in German.
[52] H. J. Witchel, C. E. I. Westling, J. Tee, A. Healy, R. Needham, and N. Chockalingam, "What does not happen: quantifying embodied engagement using nimi and self-adaptors," Participations, vol. 11, no. 1, May 2014, pp. 304–331.
[53] H. J. Witchel, C. E. I. Westling, J. Tee, R. Needham, A. Healy, and N. Chockalingam, "A time series feature of variability to detect two types of boredom from motion capture of the head and shoulders," in Proceedings of the 2014 European Conference on Cognitive Ergonomics, ser. ECCE '14. New York, NY, USA: ACM, 2014, pp. 37:1–37:5.
[54] K. Huotari and J. Hamari, "Defining gamification: a service marketing perspective," in International Conference on Media of the Future, Academic MindTrek '12, Tampere, Finland, October 3-5, 2012, A. Lugmayr, Ed. ACM, 2012, pp. 17–22.
[55] S. Deterding, D. Dixon, R. Khaled, and L. Nacke, "From game design elements to gamefulness: defining "gamification"," in Proceedings of the 15th International Academic MindTrek Conference: Envisioning Future Media Environments, ser. MindTrek '11. New York, NY, USA: ACM, 2011, pp. 9–15.
[56] K. M. Kapp, "The Gamification of Learning and Instruction: Game-based Methods and Strategies for Training and Education", 1st ed. Pfeiffer & Company, 2012.
[57] D. Dixon, "Player types and gamification," Proceedings of the CHI 2011 Workshop on Gamification, 2011.
[58] R. A. Bartle, "Hearts, Clubs, Diamonds, Spades: Players Who Suit MUDs," Journal of MUD Research, 1996.
[59] A. Marczewski, "User types," in Even Ninja Monkeys Like to Play: Gamification, Game Thinking and Motivational Design. CreateSpace Indep. Publ. Platform, 2015, pp. 65–80.
[60] N. Legault, "Gamification techniques: How to apply them to e-learning," E-Learning Heroes, March 2015. [Online]. Available: https://community.articulate.com/articles/gamification-techniques-how-to-apply-them-to-e-learning [Accessed: 1. June 2015].
[61] J. Hamari, J. Koivisto, and H. Sarsa, "Does gamification work? – a literature review of empirical studies on gamification," in System Sciences (HICSS), 2014 47th Hawaii International Conference on, Jan 2014, pp. 3025–3034.
[62] P. Weiser, D. Bucher, F. Cellina, and V. De Luca, "A Taxonomy of Motivational Affordances for Meaningful Gamified and Persuasive Technologies," in Proc. 3rd International Conference on ICT for Sustainability (ICT4S), ser. Advances in Computer Science Research, vol. 22. Paris: Atlantis Press, 2015, pp. 271–280.
[63] C. M. Bishop, "Pattern Recognition and Machine Learning (Information Science and Statistics)". Secaucus, NJ, USA: Springer-Verlag New York, Inc., 2006.
[64] W. Leister, "Towards a generic assessment framework – ideas and cases," Norsk Regnesentral, NR Notat DART/02/2015, 2015.
[65] W. Leister and I. Tjøstheim, "Concepts for User Experience Research," Norsk Regnesentral, NR Note DART/14/2012, 2012.
[66] M. Cavazza and D. Pizzi, "Narratology for interactive storytelling: A critical introduction," in Technologies for Interactive Digital Storytelling and Entertainment: Proc. Third International Conference, TIDSE 2006, Darmstadt, Germany, December 4-6, 2006, S. Göbel, R. Malkewitz, and I. Iurgel, Eds. Berlin, Heidelberg: Springer, 2006, pp. 72–83.
[67] J. Hamari and V. Eranti, "Framework for designing and evaluating game achievements," in Proc. 2011 DiGRA International Conference: Think Design Play. DiGRA/Utrecht School of the Arts, January 2011.
[68] V. Terzis, C. N. Moridis, and A. A. Economides, "Measuring instant emotions during a self-assessment test: The use of facereader," in Proc. Measuring Behavior 2010, A. J. Spink, F. Grieco et al., Eds., Eindhoven, The Netherlands, 2010, pp.

192–195.

[69] B. Zaman and T. Shrimpton-Smith, "The facereader: Measuring instant fun of use," in Proceedings of the 4th Nordic Conference on Human-computer Interaction: Changing Roles, ser. NordiCHI '06. New York, NY, USA: ACM, 2006, pp. 457–460.

[70] G. Chanel, K. Ansari-Asl, and T. Pun, "Valence-arousal evaluation using physiological signals in an emotion recall paradigm," in Systems, Man and Cybernetics, 2007. ISIC. IEEE International Conference on, oct 2007, pp. 2662–2667.

[71] D. Watson, L. A. Clark, and A. Tellegen, "Development and validation of brief measures of positive and negative affect: the PANAS scales," Journal of Personality and Social Psychology, vol. 54, 1988, pp. 1063–1070.

[72] E. Jaffe, "The psychological study of smiling," APS Observer, vol. 23, no. 10, 2010.

[73] M. G. Frank and P. Ekman, "Physiologic effects of the smile," Directions in Psychiatry, vol. 16, no. 25, December 1996, pp. 1–8.

[74] H. A. Landsberger, "Hawthorne revisited: management and the worker : its critics, and developments in human relations in industry". Cornell University, 1958.

[75] W. W. Chin, "The partial least squares approach to structural equation modeling," in Modern Methods for Business Research, G. A. Marcoulides, Ed. Hillsdale, NJ: Lawrence Erlbaum Associates, 1998, pp. 294–336.

[76] C. Fornell and D. F. Larcker, "Evaluating structural equation models with unobservable variables and measurement error: Algebra and statistics," Journal of marketing research, vol. 18, no. 3, 1981, pp. 382–388.

[77] D. Barclay, C. Higgins, and R. Thompson, "The partial least square (PLS) approach to casual modeling: Personal computer adoption and use as an illustration," Technology, vol. 2, no. 2, 1995, pp. 285–309.

[78] W. W. Chin, B. L. Marcolin, and P. R. Newsted, "A partial least squares latent variable modeling approach for measuring interaction effects: Results from a monte carlo simulation study and an electronic-mail emotion/adoption study," Info. Sys. Research, vol. 14, no. 2, Jun. 2003, pp. 189–217.

[79] C. M. Ringle, S. Wende, and A. Will, "SmartPLS 2.0 (M3)," 2005. [Online]. Available: www.smartpls.de [Accessed: 24. Feb 2015].

[80] K. Ashton, "That 'Internet of Things' Thing," RFID Journal, 22 July 2009.

[81] J. S. Ward and A. Barker, "Undefined by data: A survey of big data definitions," CoRR, vol. abs/1309.5821, 2013.

Multiscale Cancer Modelling in Terms of Copyright

Iryna Lishchuk, Marc Stauch
Institut für Rechtsinformatik
Leibniz Universität Hannover
Hannover, Germany
e-mail: lishchuk@iri.uni-hannover.de, stauch@iri.uni-hannover.de

Abstract—**Multiscale cancer modelling is a complex process, which requires interdisciplinary effort. Simulation of tumor progression across multiple scales by computer models is a challenge for scientists, and determining applicable legal protection is a challenge for lawyers. Insofar as a computer model is defined as a computer program, software copyright comes into play. At this stage several questions arise: What elements in computer modelling are copyrightable? Is the modelling work protected? What about copyright in a hyper-model design? The intellectual effort and investment deployed into the modelling have been tested against the requirements of copyright. In fact, various elements in cancer modelling, which express original creative input, qualify as copyright works and are protected as such.**

Keywords-computer models; software copyright; copyright in compilations; non-literal expression; program expression.

I. Introduction

Computational modelling is one of the IT tools applied in support of the oncology of today. Simulation of cancer progression by computer models allows predicting variations more efficiently, saving time, money, and materials. The potential of computer modelling is that the models can be used as clinical decision support systems in future. For this, a decision based on prediction of a model must prove better than a decision based on the clinical standards of today. Apart from scientific and technological issues raised by computer modelling, a number of legal issues need to be solved before the cancer models are released into practice. One of the legal issues relates to protectability of cancer models by copyright [1]. Who is entitled to do what with existing models and data sets? What elements are copyrightable and what elements are void of copyright? How to deal with the risk that cancer models could be regarded as mere ideas and not protected as such? Before we approach and suggest answers to these questions, let us explore the substance of cancer modelling first.

Cancer is a complex disease. Heterogeneous types of tumor cells, uncontrolled behavior and invasion of tumor into the healthy tissue, interplay among the cells themselves and the microenvironment make cancer a challenging object for research and treatment [2]. The oncology of today requires an interdisciplinary approach and is increasingly supported by specialized software solutions. *In silico* cancer modelling is one of the IT solutions in this field. *In silico* oncology aims to improve cancer knowledge and treatment by creating reliable computer predictions.

Simulation of cancer progression in space and time requires the use of multiscale cancer modelling. Multiscaling is realized *in silico* by constructing elementary models (the ones, which correspond to elementary biological processes) and relation models (the ones, which reflect relations across them) into multiscale hyper-models [3]. "*A model is considered to be "multiscale" if it spans two or more different spatial scales and/or includes processes that occur at two or more temporal scales.*" [2]. The four main biological scales, which are being modelled are the atomic, molecular, microscopic and macroscopic scales [2]. Processes, which occur at the atomic level, are linked to the processes at a higher level. The composite multiscale constructs of models (hyper-models or integrative models) are then able to synthesize and imitate the biological processes at several temporal and spatial levels (molecular, cellular, etc.) at once.

Research on multiscale cancer modelling is ongoing. In particular, the ICT research project CHIC, full title "*Computational Horizons In Cancer (CHIC): Developing Meta- and Hyper-Multiscale Models and Repositories for In Silico Oncology*", conducts research on multiscale cancer modelling [3]. The CHIC project "*proposes the development of clinical trial driven tools, services and secure infrastructure that will support the creation of multiscale cancer hyper-models (integrative models).*" [3]. The research groups from the different project partner institutions contribute single-scale models (from molecular to compartment models), which are then combined into integrated multiscale hyper-models. Linking and interplay between the models in CHIC is shown in Figure 1[4].

In general, the process of multiscale cancer modelling can be divided into three main stages:

(a) Scientific modelling: at this stage, modelers study the tumor types and biological processes, selected for simulation, investigate the types of cells and interactions among them, define algorithms and modelling techniques, which are capable to capture such processes best.

(b) Coding: at this stage, the tumor models are transformed into executable form. In this process, either the already developed tumor models are broken down into simpler models or computer codes of elementary biological processes (biomechanics) are developed anew.

(c) Hyper-modelling: in this step, the elementary models, each of which represents a biological process at a single scale, are combined with the other models into multiscale hyper-models. In this course, spatiotemporal simulation of tumor types, addressed by the CHIC project,

i.e., Wilms tumor, glioblastoma multiforme (GBM) and non small cell lung cancer (NSCLC), is achieved [3].

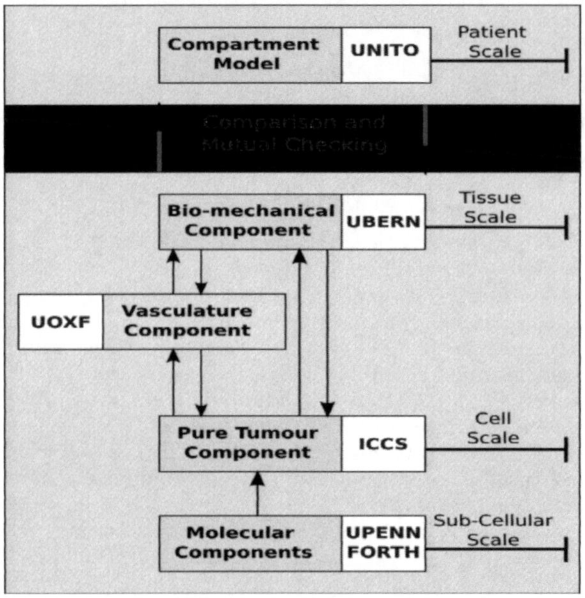

Figure 1. Schematic of the planned modelling framework for the CHIC project with the angiogenesis/vascular component highlighted.

The research on cancer modelling is of clinical relevance and is motivated by the perspective of using multiscale cancer models as a clinical decision support tool [4]. Meanwhile, the legal research in the project is concerned with the legal implications surrounding amalgamation of models. In terms of Intellectual Property (IP) law, this includes identifying the type of protection applicable to cancer models and hyper-models, the limits of such protection, and the conditions that protectable elements of the models must fulfill in order to be protected.

The cancer models, encoded into computer programs, constitute subject matter protectable by copyright [5]. However, these are not only the codes of computer models themselves, which may enjoy such protection. Also, the scientific modelling work as well as designing models into hyper-models may be copyrightable.

In this paper, we explore copyright, as a type of protection applicable to the cancer modelling, investigate in how far the scope of copyright reaches to protect the intellectual input and investment, deployed in the cancer modelling. We start with the overview of general principles of copyright in Section II, look at how it may apply in turn to three main aspects of modelling: coding models (Section III), structuring of the hyper-models (Section IV), and scientific modelling (Section V). Conclusions finalize the paper in Section VI.

II. What Copyright Protects

The law of copyright has its requirements, which all copyrighted works must fulfill in order to be protected. These criteria are examined below.

A. Protection of Computer Programs by Copyright

Copyright is a traditional type of protection, which both European and International law grant to computer programs. Article 4 WIPO Copyright Treaty [6] and Article 10 TRIPS Agreement [5] protect computer programs as literary works within the meaning of the Berne Convention (1886) [7]. The same principle is followed by European copyright law. Article 1 of Directive 2009/24/EC on the legal protection of computer programs (Software Directive) [8] recognizes computer programs as an object of copyright protection in the EU.

B. What Copyright Protects

The Software Directive protects programs, which are "*original in the sense that it is the author's own intellectual creation.*" [8]. Original intellectual creation is a basic requirement for copyright protection. It is equally applicable to the other copyright works, be it software, writings, photographic works, or other works protectable by copyright [7]. No other criteria, such as whether a program is functional, or how many lines of code it has, etc. are relevant for copyright. However, the requirement of originality has its own interpretation in terms of copyright.

As interpreted by the Court of Justice of the European Union (CJEU), which dealt with copyrightability of text extracts from articles in the case *Infopaq International A/S*, "*it is only through the choice, sequence and combination of those words that the author may express his creativity in an original manner and achieve a result which is an intellectual creation.*" [9].

In the case *SAS Institute Inc.* [10], which concerned copyright in a manual to a computer program, the court applied the same interpretative criteria in relation to copyrightability of elements, which appear in the manual and/or are implemented by a computer program. In doing so, the court reached the conclusion that "*the keywords, syntax, commands and combinations of commands, options, defaults and iterations ..., figures or mathematical concepts ..., considered in isolation*" are not an intellectual creation of the author of a computer program and are not protected by copyright. In this context, the keywords, syntax, commands, and other items in a computer program, considered in isolation, refer to the elements of a programming language, in which computer programs are normally written. By contrast, where the programmer "*by the choice, sequence and combination of those words, figures or mathematical concepts*" in his program succeeds in expressing his creativity in an original manner, this intellectual creation justifies the protection by copyright [10].

C. Program Expression for the Purposes of Copyright

Another important issue for copyright protection concerns the need for 'expression'. Thus, a characteristic feature of copyright is that the scope of protection is limited to the expression of a work, and not the underlying idea. (In contrast, "*a valid patent does not protect the expression of an idea but the underlying substance of it*" [11]; patent protection for cancer models is outside the scope of this paper).

As noted earlier, it has been established by EU and International copyright law, that the program object code and the source code constitute the object of protection of software by copyright. In this regard, the Software Directive in Article 1 (2) grants copyright protection to "*the expression in any form of a computer program*" [8]. What counts as a program expression for the purposes of the Directive has been established by the CJEU in its case law.

In particular, the CJEU considered this issue in the case *Bezpečnostní softwarová asociace – Svaz softwarové ochrany (BSA)* [12]. The main question raised was whether the graphic user interface (GUI) of a computer program counts as a form of a program expression within the meaning of Article 1(2) of the Directive and is thus protected. The court, in consideration of the international copyright law, and Article 10 TRIPS Agreement, held that the source code and the object code of a computer program constitute forms of expression entitled to be protected by copyright. Indeed, the court took the view that this applies to any expression of a program, which permits reproduction in different computer languages. As it stated, "the *object of the protection conferred by that directive is the expression in any form of a computer program which permits reproduction in different computer languages, such as the source code and the object code.*" [12]. The source code of a program usually constitutes a script, written in a human readable form. The source code, compiled into a binary executable, constitutes the object code, which gives the final instructions to the computer [13].

What can be drawn from this decision is that only such expression of a program, which permits reproduction or re-creation of a program into other computer languages, counts as a program expression for the purposes of software copyright in the EU.

Usually, such reproduction or translation of a program into other programming languages is possible from the source code. The source code for a program is normally written in one or another programming language. The programming languages, which are mostly used in computer programming now, are Java, C^{++}, Python, etc. The choice of a language is mostly dictated by a system, upon which a program is intended to run. The reason is that some computers operate and can read only certain languages [13].

However, in contrast to the code itself, which because of its creative substance, is considered as a copyright work, the programming languages, which, in principle, "*comprise ideas and principles*", on which the programs operate and which do not expose creative substance, are not protected by copyright.

A programming language itself, be it C or Java, is normally composed of keywords and other symbols and may include a set of pre-written programs to carry out various operations, such as displaying something on the screen or retrieving the cosine of an angle [14]. According to the CJEU *SAS Institute* decision [10], the programming language, which comprises ideas and principles, both as isolated symbols, figures, keywords, mathematical concepts, etc., which constitute the material of that programming language, are not copyrightable.

Subsequently, in its decision *SAS Institute* [10], the CJEU considered the copyrightability of various other elements, which appear or are implemented by a computer program. The following elements were concerned:

"(a) the selection of statistical operations which have been implemented in the First Program;

(b) the mathematical formulae used in the Manual to describe those operations;

(c) the particular commands or combinations of commands by which those operations may be invoked;

(d) the options which the author of the First Program has provided in respect of various commands;

(e) the keywords and syntax recognised by the First Program;

(f) the defaults which the author of the First Program has chosen to implement in the event that a particular command or option is not specified by the user;

(g) the number of iterations which the First Program will perform in certain circumstances?" [10].

On the basis of the previous case law, and rules established in the Infopaq and BSA decisions, the court concluded that "*neither the functionality of a computer program nor the programming language and the format of data files used in a computer program in order to exploit certain of its functions constitute a form of expression of that program for the purposes of Article 1(2) of Directive 91/250.*" [10]. The CJEU supported its decision by the argument that "*... to accept that the functionality of a computer program can be protected by copyright would amount to making it possible to monopolise ideas, to the detriment of technological progress and industrial development.*" [10]. Such decision may also be interpreted the following way: that the elements, which express rather general principles than original creativity, belong to the domain of science and should not be monopolized, even if included into a copyright work. On the other hand, computer programs, which implement functions of a particular programming language, both as original structuring of such functions for being accessible via language application programming interfaces (API), may qualify as copyrightable expression and be protected as such.

Apart from the program source code and the object code, which are recognized as a literal expression of a program, there are also other hidden elements, which define perception of a program by a user. Such elements constitute a non-literal expression. And the question on protection of non-literal expression of a program by software copyright has been raised for consideration and, indeed, answered by the courts.

D. Non-Literal Program Expression

The courts in common law countries, such as the UK and US, tend to extend the scope of software copyright beyond the literal code to a program non-literal expression. The UK courts approach copyrightability of non-literal expression as follows: "*Consideration is not restricted to the text of the code... That must be right: most literal copyright works involve both literal matter (the exact words of a novel or computer program) and varying levels of abstraction (plot, more or less detailed of a novel, general structure of a computer program)."* [15].

The elements of a non-literal expression may include a program structure, sequence and organization, a program "*look and feel*", input and output routines [16]. It is often the case that structure of one program is imitated or reproduced by another program. This is a typical case when copyright protection for a program structure is sought. The legal issue raised before the courts here is: "*whether the structure (or sequence and organization) of a computer program is protectable by copyright, or whether the protection of the copyright law extends only as far as the literal computer code."* [14].

The reason why protection of the program structure by copyright is important is that structuring a program often takes more time and intellectual effort than writing the code. And the UK copyright law gives much importance to protecting the skill and labor, which the authors deployed in their works, be it programmers or writers [16]. Let us consider why such protection is justified.

The code of a program itself consists of a set of instructions to the computer and is an end product of a complex software development process [17]. The latter process often occurs in several steps. First, the problem that needs to be solved by a computer is identified. In the next step, the outline for the solution follows. In the outline, the programmer breaks down the solution into smaller units called 'subroutines' or 'modules', each of which handles elements of a problem. The outline can be laid down in the form of a flowchart [13]. The next step is organizing the modules and subroutines into a program structure. A program structure is dictated by "*the functions of the modules in a program together with each module's relationships to other modules"* [14]. Usually, modules are arranged in such a way that a problem is solved in a more efficient way. As interpreted by the court in the case *Whelan Associates Inc. v. Jaslow Dental Laboratory* [13], "*Although two programs could produce the same result, one might be more efficient because of different internal arrangements of modules and subroutines. Because efficiency is a prime concern in computer programs (an efficient program being obviously more valuable than a comparatively inefficient one), the arrangement of modules and subroutines is a critical factor for any programmer.*" [13].

After defining the structure, a programmer decides about what data is needed, where and how the data should be introduced and how it should be combined with the other data. It is when the data is arranged into data files [13]. Once the program design is ready, the coding begins. The coding is a comparatively small part of programming. "*By far the larger portion of the expense and difficulty in creating computer programs is attributable to the development of the structure and logic of the program, and to debugging, documentation and maintenance, rather than to the coding."* [13].

Against this technical background, it becomes clear why the structuring of a program and arranging its modules in an efficient way takes a large amount of skill and work, is often copied and deserves protection on its own.

E. Idea-Expression Dichotomy

One of the tricky legal questions, which needs to be solved when copyright in non-literal expression of a program is sought, is separation of copyrightable expression from principles and ideas, which are non-copyrightable as such.

One of the general principles of copyright is that "*Copyright protection shall extend to expressions and not to ideas"* [5]. "*Procedures, methods of operation*" are not copyrightable as such [5]. This principle applies to copyright in computer programs as well. The Software Directive says: "*For the avoidance of doubt, it has to be made clear that only the expression of a computer program is protected and that ideas and principles which underlie any element of a program, including those which underlie its interfaces, are not protected by copyright under this Directive. In accordance with this principle of copyright, to the extent that logic, algorithms and programming languages comprise ideas and principles, those ideas and principles are not protected under this Directive. In accordance with the legislation and case-law of the Member States and the international copyright conventions, the expression of those ideas and principles is to be protected by copyright."* [8].

Therefore, in order to grant copyright to non-literal program expression and, what is more important, to justify such decision, courts need to separate copyrightable expression from non-protectable ideas. Different criteria to address copyrightability of non-literal program expression have been elaborated.

F. Abstraction-Filtration-Comparison Test

One of the tests in this regard, which is widely applied both in the US [14] and in the UK [15], is the "*abstract-filtration-comparison*" test. The test has been established in the US case *Computer Associates International, Inc. v. Altai, Inc.* [18]. The idea behind the test is to reward programmers with copyright protection for creating "*innovative utilitarian works containing expressions*" and to leave non-protectable technical expressions in the public domain for further use [16].

The test comprises three steps. First, an original copyright program is broken down into its structural components according to the levels of abstraction. The second step extracts certain non-copyrightable structures (discussed below) until the copyrightable substance remains. In the third step, the portion of copying is compared with the copyrightable expression, left in the structure of the original program. Finally, it is estimated in how far such copying is substantial to justify infringement of copyright in a software program [18].

The three types of structures, identified as precluded from copyright, comprise: (a) elements dictated by efficiency, (b) elements required by external factors, (c) elements taken from the public domain.

1) Elements Dictated by Efficiency

First, copyright does not extend to structures dictated by efficiency (the doctrine of merger). Accordingly, copyright will not subsist in the expression, which is "*necessarily incidental to the idea being expressed*." In this step, it is determined "*whether the use of this particular set of modules is necessary efficiently to implement that part of the program's process" being implemented*." [18]. If efficiency dictates that the choice of modules is limited to just a few workable solutions (such as one or two options), such selection of modules may not be protectable as such.

2) Elements Dictated by External Factors

Secondly, copyright does not extend to structures dictated by external factors. External factors in software programming may include: compatibility requirements, mechanical specifications, computer manufacturer design standards, industry demands, and common programming practices. In US copyright law, this is known as the 'scenes a faire' doctrine [18]. In consequence, a particular set of modules, which need to be present as an integral part in all programs of the same category are non-copyrightable.

3) Elements Taken from the Public Domain

Thirdly, copyright does not protect structures that are found in the public domain. The rationale here is that such material, which is included as an element in a copyrightable work, may itself not be appropriated as it should remain free for use by the community [18].

G. Abstraction-Filtration-Comparison Test in Practice

The application of the test in practice may be illustrated by reference to the case Oracle America, Inc. v. Google Inc., C 10-03561 WHA, which dealt with copyrightability and copyright infringement in interfaces of the programming language Java. The central question related to the "*extent to which if at all, certain replicated elements of the structure, sequence and organization of the Java application programming interface are protected by copyright*" [14].

Java is a powerful object oriented programming language, developed by Sun Microsystems, first released in 1996, and acquired by Oracle in 2010. Java has a number of pre-written programs, called "methods", which invoke different functions, such as retrieving the cosine of an angle. These methods are grouped into "classes" and organised into "packages". Software developers get access to those classes through the Java APIs [19]. In 2008 Java APIs had 166 "packages", split into more than six hundred "classes", all divided into six thousand "methods".

Google built its Android platform for the smartphones using the Java language and, according to Oracle, "*utilized the same 37 sets of functionalities in the new Android system callable by the same names as used in Java*" [14]. By doing that, Google wrote its own implementations of the methods and classes, which it needed. The only one substantial element, which Google copied from Java into Android was the names and headers of 37 API packages in question. Such copying of the headers amounted to replication of the structure, sequence and organisation of Java APIs. Oracle claimed copyright infringement, and Google defended with fair use, arguing that Java is an open solution (which Oracle did not dispute) and there was no literal copying of the Java code.

The court of first instance trying the case qualified the headers and method names in Java APIs as non-copyrightable, referring to the interpretation criteria of the US Copyright Office: "*Even if a name, title, or short phrase is novel or distinctive or lends itself to a play on words, it cannot be protected by copyright*." [20]. This lends support to non-protectability of isolated code items by copyright, as similarly recognized by the CJEU [10].

As regards copying of the declarations and duplicating the command structure of Java APIs, the judge found that the command structure of Java APIs amounts to a method of operation – a material not subject to copyright in the US [20].

In Java programming, the specific declarations in the Java APIs designate a method. A method can be implemented in different ways, but is invoked by that specific declaration only. The command format, used to call the methods in Java, reads:

"java.package.Class.method()."

Here, a formula "*a = java.package.Class.method()*" sets the field "a", which is equal to the return of the method called. For example, the following call would call the method from Java:

"int a = java.lang.Math.max (2, 3)"

This command line would instruct the computer to fetch "*the max method under the Math class in the java.lang package, input "2" and "3" as arguments, and then return a "3," which would then be set as the value of "a."* [14].

As interpreted by the judge, in Java, each symbol in a command structure is more than a simple name - each symbol carries a task to invoke a pre-assigned function.

Considering that for using Java class methods software developers need to replicate the Java declarations, the judge qualified the command structure of Java APIs as a method of operation – a functional element essential for interoperability, not subject to the US Copyright Act. This position was based on the merger doctrine and non-copyrightability of structures dictated by efficiency: "*... When there is only one way to express an idea or function, then everyone is free to do so and no one can monopolize that expression.*" [14]. However, on appeal, the Federal Circuit Court reversed that ruling [21]. The appellate court found the declaring code and the structure, sequence and organization of packages in Java APIs were entitled to be protected by copyright.

The appellate court supported its decision by the argument that Java programmers were not limited in a way how they could arrange the 37 Java API packages at issue and had a choice to organize these API packages in other ways. For instance, instead of using the command format "*java.package.Class.method()*": language – package – class – method, the same method could be called by the format: method – class – package – language. By making a decision to arrange the declarations in Java in this way and by having also other choices, the programmers were not prevented by the factor of efficiency, which would preclude copyright. By that, the programmers had a scope to exercise their creation, which they, in view of the court, exercised, indeed. This creation, realized in sequencing the Java APIs, amounted to a copyrightable expression. Against these considerations, the court concluded that, "*the structure, sequence, and organization of the 37 Java API packages at issue are entitled to copyright protection.*" [21].

Google argued fair use and petitioned the US Supreme Court to hear the case. The US Supreme Court, referring to the opinion of the US Solicitor General, denied the petition. In the result, a new district court trial began. On 26 of May 2016 the district court jury found that Google´s Android does not infringe Oracle copyrights, because Google´s re-implementation of 37 Java APIs in question amounted and was protected by fair use. According to a Google spokesperson, *"Today's verdict that Android makes fair use of Java APIs represents a win for the Android ecosystem, for the Java programming community, and for software developers who rely on open and free programming languages to build innovative consumer products."* [22].

These cases deal with challenging questions of copyright law: free use of programming language APIs, copyrightability of APIs and an attempt "*to control APIs with copyright law*" and counter-balance between copyrights and *"fair use"* [22]. As established, the APIs, although elements necessary for interaction between computer programs, but which amount to intellectual creation, can be protected by copyright, at least in the opinion of one court of appeals. At the same time, as the jury found, the APIs, although protected by copyright, may be reused in other software systems, if such re-use is covered by fair use of open and free programming languages, like Java.

III. CANCER MODELS IN TERMS OF COPYRIGHT

In this section, we return to the cancer modeling. We explore the substance of cancer models and look into transformation of scientific models into computer models.

First of all, the two types of models need to be differentiated here: scientific models, which represent the biological processes, and computer models, which implement these biological processes *in silico*.

In the context of the CHIC project, scientific models are defined as: "*finalized cognitive constructs of finite complexity that idealize an infinitely complex portion of reality through idealizations that contribute to the achievement of knowledge on that portion of reality that is objective, shareable, reliable and verifiable.*" [23]. These scientific models correspond to the biological processes, being simulated. The scientific models are then implemented *in silico* via computer models. Computer models represent an executable form of scientific models, these being encoded into computer programs.

A computer model is defined as: "*a computer program that implements a scientific model, so that when executed according to a given set of control instructions (control inputs) computes certain quantities (data outputs) on the basis of a set of initial quantities (data inputs), and a set of execution logs (control outputs).*" [24].

As noted above, computer programs constitute subject matter protectable by copyright [5]. Cancer models, written in computer programs, may also enjoy such protection, if they stand the requirements for protection. Such requirements would be fulfilled, if a model amounts to a copyrightable expression and is expressed in a form, which counts as a program expression for the purposes of software copyright.

According to the case law of the CJEU, a model would constitute a copyrightable expression, when the modeller "*through the choice, sequence and combination*" of commands in the model code succeeds to "*express his creativity in an original manner and achieve a result which is an intellectual creation*" [9]. On the other hand, if the code is generated automatically, for example by automatic translation into another programming language, or compiled from bits of code collected from the public domain, copyright protection in such code is most likely to be denied.

As regards the form of expression recognized by the software copyright, so it is the "*source or object code*", which constitute the object of copyright protection [5]. In relation to cancer models it means that models, which are exposed either in the source code or provided as binary executables shall be protected by copyright.

Models, as computer programs, are written in one or another programming language. In the context of the CHIC project, the models, embodied in source code, are mostly written for interpreted languages, such as Python, Perl or MATLAB; models provided in object code are usually binary C or C++ compiled executables [25]. Hence, it is the code, in which a model is embodied, be it the source code written in Python or the code compiled in C++, which is protected by copyright. The programming language itself, the biological process implemented by the model, the general process of its implementation, both as formats of data files, used by exchange of data between the models remain outside the scope of copyright subsisting in the model.

Once we established that computer models are protectable by copyright as computer programs and defined the scope of copyright protection, we can proceed to the next step, namely copyright issues in the hyper-modelling. At this stage it is necessary to state that copyright in the models, arranged into a hyper-model, remains by the models and does not extend to protecting the hyper-model as a copyright work on its own [5].

IV. COPYRIGHT IN HYPER-MODELS

In this section, we continue the analysis on the scope of copyright in relation to cancer modeling. We proceed to a higher stage, namely coupling singe models into multiscale integrative hyper-models and explore the protection of hyper-models by copyright.

A. *Substance of Hyper-Models*

As mentioned at the beginning of this paper, multiscale cancer modelling is achieved by combining simple models into hyper-models.

In the context of the CHIC project, hyper-models comprise "*choreographies of component models, each one describing a biological process at a characteristic spatiotemporal scale, and of relation models/metamodels defining the relations across scales.*" [3]. The aim thereby is that in a given instance (e.g., when populated with the data for a given patient) the hyper-model is able to reproduce biological processes of tumor progression, involving multiple phenomena, which are respectively captured in single models. A hyper-model "*emerges from the composition and orchestration of multiple hypomodels, each one of which is capable of simulating a specific entity or phenomenon... and can simulate an entity or phenomenon that may be more complex than the ones simulated by each separate simpler model.*" [24].

The term hyper-model first appeared in 2008 in relation to Virtual Physiological Human (VPH) [26]. In 2011, the notion of a hyper-model was used in the context of computer science. Here, it was defined as "*a concrete instance of an integrative model, built as the orchestration of multiple computer models that might run on different computers at different locations using different simulation stacks.*" [27]. The first implementation, based on web services, was tested on biochemical models [28].

An inter-play between single-scale models in a multiscale hyper-model is illustrated in Figure 1 [4]. Processes at lower levels occur faster, and processes at higher levels last longer. Above all it is the accurate correlation of these multiscale processes that makes multiscaling such a tricky task and requires the greatest intellectual effort. The scales are correlated to one another by the exchange of data where the output of one model serves as an input into another. The relevant interaction may be demonstrated by the example of a model of tumor growth where a cellular bio-model is coupled to biomechanical simulations [29].

Here the model of tumor growth is composed from a macroscopic mechanical model, which provides directions of least pressure in the tissue, and the cellular level model, which simulates concentration of cells inside the element. The microscopic cell-level model requires the direction of cells proliferation by tumor growth as an input. This information is provided by the mechanical model. In its turn, the biomechanical model requires the number of cells inside the element. This information is calculated by the biological cell-level model and fed to the biomechanical model. This model of tumor growth is structured via interplay and exchange of information between the models. The flowchart of the diffusion/mass effect coupling simulation used to simulate tumor growth is represented in Figure 2 [29].

The hyper-modelling execution process, such as the one just illustrated, is to a large extent facilitated and semi-automated by the underlying technical infrastructure. For example, in CHIC, the process of hyper-modelling is managed by VPH-Hypermodelling Framework [25].

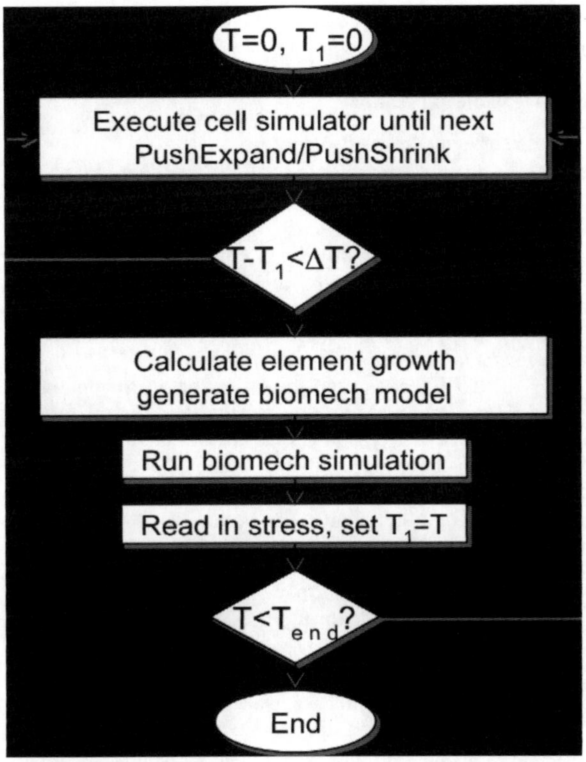

Figure 2. Flowchart of the diffusion/mass effect coupling simulation used to simulate tumor growth. Since the cellular effects occur at a different time scale than the mechanical reaction, the biomechanical calculations occur only after a larger number of cycles of the cell simulator biomodel.

However, at the earlier stage of hyper-model construction, human input into the hyper-model design is indispensable. Particularly where the intention is to use hyper-models for decision support in the clinical setting, the automatic linking of hypo-models is unjustifiably risky and thus not acceptable [30]. The hypermodelling strategy, the hypomodel linking, the hypomodel integration are instead normally designed by the modelling party, who has substantial expertise in the field of bioinformatics. Moreover, it is this intellectual input, deployed in designing a hyper-model, which may render a hyper-model copyrightable.

B. Hyper-Models in Terms of Copyright

The appropriate concept for qualifying hyper-models in terms of copyright is arguably that of a "compilation". Compilations, as works protectable by copyright, are introduced by Article 5 WIPO Copyright Treaty [6], Article 10 (2) TRIPS Agreement [5]. The TRIPS Agreement provides: "*Compilations, whether in machine readable or other form, which by reason of the selection or arrangement of their contents constitute intellectual creations shall be protected as such.*" [5].

The crucial aspect in a compilation, which attracts copyright, is the intellectual creation represented by "*the selection or arrangement*" of its contents. In the case of

CHIC hyper-models, such intellectual creation may reside in the innovative "*composition and orchestration of multiple hypomodels*", thereby making a hyper-model a copyrightable expression.

C. Requirements for Protection

By applying the criteria of the *Infopaq* decision [9], which defined the meaning of the "*author's own intellectual creation*" in the context of copyright, we can conclude that a hyper-model would constitute a copyrightable expression if the modeller "*through the choice, sequence and combination*" of models in a hyper-model "*may express his creativity in an original manner and achieve a result which is an intellectual creation.*" [9].

In the CJEU case law, the cases on infringement of copyright in compilations are mostly tried in relation to database rights. The reason is that the Directive 96/9/EC on the legal protection of database (Database Directive) [31] also uses the term "compilation", referring to collections of works, data or other materials. In this case, copyright is a reward for the intellectual effort, which the author deployed in the selection or arrangement of the database contents [31].

The criteria for measuring originality in a database structure have been established by the CJEU in the case *Football Dataco Ltd. and Others v Yahoo! UK Ltd. and Others* [32]. Accordingly, the originality is measured by the creativity, which the database maker expressed in making free and creative choices, thereby stamping his work with a personal touch. On the other hand, if the technical constraints leave no space for creativity, the element of originality in structuring a database will be lacking and copyright protection unjustified [32].

If we compare the criteria for measuring creativity in a program structure, considered above, with the criteria of originality in the compilation, we can see that these criteria inter-relate with each other. Thus, if the author (either a programmer or a compiler of a database) had some scope to express his creativity in arranging the modules in a program (or contents into a database) by making free and creative choices, such a structure may show originality and qualify as an intellectual expression in terms of copyright. On the other hand, if creativity may not be realized because of some technical considerations, then further analysis is needed to decide if and how far copyright in the structure may still arise. In particular, the abstraction-filtration-comparison test and criteria, defined by the courts, when they dealt with protection of a program structure by copyright, may be a helpful pointer.

D. Abstraction-Filtration-Comparison Test in Relation to Hyper-Models

By way of providing a concrete illustration, let us return to the hyper-model of tumor growth that we described above. This hyper-model consists of two component models: a microscopic cellular bio-model, which shows the concentration of cells in the invaded tissue, and a macroscopic biomechanical model, which provides directions of least pressure in the tissue. The overall function of the hyper-model is to simulate the "*geometrical evolution of the tumor predicted at the cellular level.*" [29].

First, according to the test, we dissect a hyper-model into its structural components, i.e., component models. The hyper-model structure is dictated by the two component models and a relationship, in which they stand to each other.

1) Elements Dictated by Efficiency

In this step we look whether the choice and combination of component models is "*necessary efficiently to implement*" the function of this hyper-model [18].

If a bio-model of cell concentration and biomechanical model of tissue are the only models and the combination of these models by exchange of output/input data is the only way, in which the geometrical evolution of the tumor predicted at the cellular level can be simulated, then it may mean that this hyper-model is designed in this way for reasons of efficiency. In such a case, the modeler's choice of component models and their combination has merged with the underlying idea and according to the merger doctrine may not be appropriated by copyright.

Otherwise, where the "*evolution of the tumor predicted at the cellular level*" may be represented by the selection of other models and/or the combination of these or other models in another way, then the hyper-model structure may qualify as copyrightable expression.

2) Elements Dictated by External Factors

According to the "*scenes a faire*" doctrine, copyright does not extend to structures dictated by external factors. In the context of programming, such external factors may be the compatibility requirements, computational constraints, demands of the industry being served, widely accepted programming practices, etc.

For example, elements or methods, which are "indispensable or at least standard" in the computer modelling community, would be non-copyrightable. Such elements may include: computational demands (especially for models, which work on higher spatial and temporal resolutions), or requirements of Digital Model Repository (an innovative platform for exchange of cancer models), or criteria of Heterogenous Multiscale Method (HMM) etc. [2].

3) Elements Taken from the Public Domain

The other group of elements, which may not be copyrighted, are elements taken from the public domain. Copyright may not be claimed in "*an expression which, if not standard, then commonplace in the computer software industry.*" [18].

For example, if the component models, selected for a hyper-model in question, and/or the method of combining them have been borrowed from the public domain, the same models and/or a method of coupling them should remain outside the scope of copyright, residing in the hyper-model.

Admittedly, the above analysis is somewhat schematic, and rather hypothetical. Its main purpose is to outline some general guidance and describe some general cases when certain elements and structures may be protectable by copyright or not.

V. Copyright in Preparatory Design Material

Another major element in cancer modelling, which also consumes much of intellectual effort and may well deserve copyright protection, is the scientific modelling. In fact, the modelling work, which precedes the development of computer codes for the cancer models, may qualify as a preparatory design material to a program and fall under the scope of software copyright [8].

Scientific modelling comprises the works, performed in the course of transforming scientific models, which represent biological processes, into computer models, which are encoded into computer programs. Such modelling work comprises various stages. These include identification of elementary processes for simulation (e.g., cell cycling, the angiogenesis process, declination of a cell to apoptosis after a particular treatment, etc.), the definition of modelling techniques - discrete, continuum, or hybrid, and development of computer codes for the cancer models, which would simulate those biological processes *in silico* [2]. These modelling steps follow in many aspects the stages of software development. First, the problem to be solved by a computer is analyzed, then methods of solving the problem are adopted and stages of running the program are identified. Subsequently, detailed instructions for a computer to perform operations necessary for the execution of those stages are developed [17].

As noted earlier, both as a computer program stands at the end of a long development process [17], the model code, embodied in the computer program, results from the foregoing modelling work. A code of a program may be compared with the top of the iceberg, which reaches out above the water, where 90% percent is hidden under the surface. And the same may be said in relation to the earlier scientific modelling (which precedes the computer modelling).

In view of the skill, time and labour, which a programmer invests in studying a problem, elaborating a solution for it and making the solution executable by a computer, a justifiable legal question arises whether this pre-programming work is protectable by software copyright. In fact, in contrast to some of the other issues considered in this paper, here the Software Directive gives a reasonably clear affirmative answer. It does so by extending the scope of software copyright to the preparatory design work that preceded the creation of a computer program.

In particular, Article 1 (1) of the Directive, when defining the object of protection, provides: "*the term 'computer programs' shall include their preparatory design material.*" [8]. What is meant by the "*preparatory design material*" is explained in Recital 7. A preparatory design work must be of a kind "*such that a computer program can result from it at a later stage.*" [8]. As further interpreted by the CJEU, the preparatory design work counts as protectable by the directive along with a program, if it is "*capable of leading, respectively, to the reproduction or the subsequent creation of such a program*" [12].

In contrast to the form of program expression, protectable by copyright, there are no specific requirements to the form or mode, in which the preparatory design work must be expressed. According to one decision of the Federal Court of Justice of Germany, which dealt with protection of preparatory design work by copyright, it will be sufficient that the development documentation was recorded in writing, such as: data flow plans, designs of commands and information cycles, exhibits of scientific or technical art, expressed in any form, including mathematical, or technical or graphic symbols [33].

It follows, in relation to copyright in the kind of work occurring in CHIC, that modelling work (a) documented in writing, such as laid down in flow charts, exhibits, etc., and (b) attributable to a specific model, which if necessary may be rewritten from these materials, has a good chance of being protected by software copyright along with the model code.

VI. Conclusions

From the above observations it follows that various elements in cancer modelling, which express original creative input, may be protectable by copyright. If protection is not achievable by software copyright, then protection under the ordinary law of copyright may be another option.

First, the models may enjoy copyright protection as computer programs. Second, copyright in the model may extend to the modelling work, underlying the model code. Third, hyper-models, structured in an innovative and original way, may count as compilations in terms of copyright and be protected as such.

Dealing with the computer models first. Where the models are encoded into computer programs and exposed either in source code or as compiled executables, they shall be protectable by software copyright. Apart from the code, also the modelling documentation can be copyrightable along with the model code, provided it records the steps, conducted in the course of developing a model and may be used as a basis for reproducing the model concerned.

Whereas, the model code, the outline of a model, the flowcharts and exhibits, produced by the modelling, are protectable by copyright, the programming language, in which models are written, the processes and functionality, which cancer models implement, the data files and algorithms, utilized by the models in performing their functions, may not be covered by copyright subsisting in the model.

Hyper-models, which are composed of single models, arranged in specific relations to each other, may also be protectable by copyright as compilations. For this, a hyper-model must show an original and creative structure. Copyrightability of a hyper-model as a compilation is to a large extent dictated and can be measured by the degree of originality and creativity, deployed in a hyper-model structure. Thus, in cases where a hyper-model is structured in a particular way because there have been no other ways of arranging models into it (such as when the model arrangement is pre-determined by technical requirements), it will not attract copyright. The same goes for cases where the structure relies on elements taken from the public domain. In contrast, if the modeler was not so restricted and expressed creativity stamping a hyper-model structure "*with his personal touch*", such investment shall not remain unrewarded and copyright in his work will likely be recognized.

ACKNOWLEDGMENT

The research leading to these results has received funding from the European Union Seventh Framework Programme FP7/2007-2013 under grant agreement No 600841.

REFERENCES

[1] I. Lishchuk and M. Stauch, "Copyright in Multiscale Cancer Modelling," The Fifth International Conference on Advanced Communications and Computation (INFOCOMP 2015) IARIA, June 2015, pp. 117-122, ISSN: 2308-3484, ISBN: 978-1-61208-416-9.

[2] T. Deisboeck, Z. Wang, P. Macklin, and V. Cristini, "Multiscale Cancer Modelling," Annual Review of Biomedical Engineering, vol. 13, pp. 127-155, August 2011, doi: 10.1146/annurev-bioeng-071910-124729.

[3] "Computational Horizons In Cancer (CHIC): Developing Meta- and Hyper-Multiscale Models and Repositories for In Silico Oncology", CHIC, Project, Objectives, <http://chic-vph.eu/project/objectives> 2016.05.30.

[4] J. A. Grogan, P. K. Maini, J. Pitt-Francis, and H. M. Byrne, "Simulating Tumour Vasculature at Multiple Scales," In G. Stamatakos and D. Dionysiou (Eds): Proc. 2014 6th Int. Adv. Res. Workshop on In Silico Oncology and Cancer Investigation – The CHIC Project Workshop (IARWISOCI), Athens, Greece, Nov.3-4, 2014 (www.6thiarwisoci.iccs.ntua.gr), pp. 23-26, (open-access version), ISBN: 978-618-80348-1-5.

[5] Agreement on Trade-Related Aspects of Intellectual Property Rights, the TRIPS Agreement, Annex 1C of the Marrakesh Agreement Establishing the World Trade Organization, Marrakesh, Morocco, 15 April 1994.

[6] WIPO Copyright Treaty (WCT), Geneva, 20 December 1996.

[7] Berne Convention for the Protection of Literary and Artistic Works of September 9, 1886.

[8] Directive 2009/24/EC of the European Parliament and of the Council of 23 April 2009 on the legal protection of computer programs, Official Journal of the European Union (OJEU), L 111/16 – 111/22, 5 May 2009.

[9] CJEU, Judgment of 16 July 2009, Case C 5/08, Infopaq International A/S v Danske Dagblades Forening, ECLI:EU:C:2009:465.

[10] CJEU, Judgment of 2 May 2012, Case C 406/10, SAS Institute Inc v World Programming Ltd, ECLI:EU:C:2012:259.

[11] A. M. St. Laurent, "Understanding open source & free software licensing," O'Reilly, 1. Edition, 2004, p. 2.

[12] CJEU, Judgment of 22 December 2010, Case C 393/09, Bezpečnostní softwarová asociace (BSA) – Svaz softwarové ochrany v Ministerstvo kultury, ECLI:EU:C:2010:816.

[13] U.S. Court of Appeals, Third Circuit, Ruling of August 4, 1986, Whelan Associates Inc. v. Jaslow Dental Laboratory, Inc., et al, 797 F.2d 1222, 230 USPQ 481.

[14] U.S. District Court for the Northern District of California, Ruling of 31 May 2012, Case C 10-03561 WHA, Oracle America, Inc., v. Google Inc.

[15] Ibcos Computers Ltd v Barclays Mercantile Highland Finance Ltd [1994] FSR 275.

[16] S. Stokes, "Digital Copyright, Law and Practice," 4. Edition, Hart Publishing, Oxford and Portland, Oregon, 2014, p. 121.

[17] Report by the International Bureau, "Measures to enhance International Cooperation in the field of Legal Protection of Computer Software," WIPO expert group on the legal protection of computer software, First Session, Geneva, 27-30 November 1979, LPCS/I/2, 30 September 1979.

[18] U.S. Court of Appeals for the Second Circuit, Ruling of 17 December 1992, Computer Associates International, Inc. v. Altai, Inc., 982 F.2d 693, 2d Cir.1992.

[19] Oracle, Java Platform, Standard Edition 6, API Specification, Overview, <http://docs.oracle.com/javase/6/docs/api/> 2016.05.30.

[20] U.S. Copyright Office, Circular 34; "Copyright Protection 'Not Available for Names, Titles or Short Phrases," rev. January 2012.

[21] U.S. Court of Appeals for the Federal Circuit, Ruling of 09 May 2014, Oracle America, Inc., v. Google Inc., Appeals from the United States District Court for the Northern District of California in No. 10-CV-3561.

[22] J. Mullin, "Google beats Oracle—Android makes "fair use" of Java APIs," Ars Technica, 26 May 2016, <http://arstechnica.com/tech-policy/2016/05/google-wins-trial-against-oracle-as-jury-finds-android-is-fair-use/> 2016.05.31.

[23] M. Viceconti, "A tentative taxonomy for predictive models in relation to their falsifiability," Philos. Transact A Math Phys Eng Sci, vol. 369, pp. 4149-61, 1954.

[24] CHIC, Deliverable No. 7.1, "Hypermodelling specifications," CHIC, Downloads, Deliverables, Public Deliverables, <http://chic-vph.eu/uploads/media/D7-1_Hypermodelling_specifications.pdf> 2016.05.30.

[25] D. Tartarini, K. Duan, N. Gruel, D. Testi, D. Walker, and M. Viceconti, "The VPH Hypermodelling Framework for Cancer Multiscale Models in the Clinical Practice," In G. Stamatakos and D. Dionysiou (Eds): Proc. 2014 6th Int. Adv. Res. Workshop on In Silico Oncology and Cancer Investigation – The CHIC Project Workshop (IARWISOCI), Athens, Greece, Nov.3-4, 2014 (www.6thiarwisoci.iccs.ntua.gr), pp. 61-64, (open-access version), ISBN: 978-618-80348-1-5.

[26] M. Viceconti, "A tentative taxonomy for predictive models in relation to their falsifiability," Philos. Transact A Math Phys Eng Sci, vol. 369, pp. 4149-61, 1954.

[27] P. A. Fishwick, "Hypermodelling: an intergrated approach to dynamic system modelling," Journal of simulation, vol. 6, pp. 2-8, February 2012.

[28] R. Guha, "Flexible Web Service Infrastructure,for the Development and Deployment of Predictive Models," Journal of Chemical Information and Modellling, vol. 48 (2), pp. 456-464, 2008.

[29] F. Rikhtegar, E. Kolokotroni, G. Stamatakos, and P. Büchler, "A Model of Tumor Growth Coupling a Cellular Biomodel with Biomechanical Simulations," In G. Stamatakos and D. Dionysiou (Eds): Proc. 2014 6th Int. Adv. Res. Workshop on In Silico Oncology and Cancer Investigation – The CHIC Project Workshop (IARWISOCI), Athens, Greece, Nov.3-4, 2014 (www.6thiarwisoci.iccs.ntua.gr), pp.43-46 (open-access version), ISBN: 978-618-80348-1-5.

[30] G. Stamatakos, et al., "The Technologically Integrated Oncosimulator: Combining Multiscale Cancer Modelling with Information Technology in the In Silico Oncology Context," IEEE J Biomedical and Health Informatics, vol. 18, no. 3, pp. 840-854, May 2014.DOI:10.1109/JBHI.2013.2284276.

[31] Directive 96/9/EC of the European Parliament and of the Council of 11 March 1996 on the legal protection of databases, OJEU, Nr. L 77/20-77/28, 27 March 1996.

[32] CJEU, Judgment of 1 March 2012, Case C 604/10 Football Dataco Ltd. and Others v Yahoo! UK Ltd. and Others, ECLI:EU:C:2012:115.

[33] Federal Court of Justice of Germany, Judgment of 09 May 1985, Case I ZR 52/83, BGHZ 94, pp. 276 – 292.

Optimization of Lead Design and Electrode Configuration in Deep Brain Stimulation

Ruben Cubo
Department of Information Technology
Uppsala University
Box 337
751 05 Uppsala, Sweden
Email: ruben.cubo@it.uu.se

Mattias Åström
Department of Biomedical Engineering
Linköping University
581 85 Linköping, Sweden
Email: matas@imt.liu.se

Alexander Medvedev
Department of Information Technology
Uppsala University
Box 337
751 05 Uppsala, Sweden
Email: am@it.uu.se

***Abstract*—Deep Brain Stimulation (DBS) is a medical treatment whose exact underlying biological mechanism is unknown. Yet, DBS is an established therapy in a number of neurological and mental disorders. Mathematical models aiming at a better understanding of DBS through the simulation of the electrical field in the brain have been developed in the past years. This study covers *in silico* optimization of the electrical stimuli delivered to the brain by means of a Finite Element model individualized through medical imaging data. The goal is to cover a given target volume with stimulation for full therapeutic effect while limiting the spread of the stimuli beyond the target border, to avoid undesirable side effects. The fraction of the activated tissue volume within the target and the fraction of the stimulation field that spreads beyond it are computed in order to quantify the performance of the stimuli. Two readily available leads are treated: a state-of-the-art lead using single active contact and a field-steering one in multiple active contact stimulation. Further, in order to obtain insights into lead design, hypothetical leads with different geometric characteristics are as well considered. The obtained results suggest that simplified models give a reasonably good approximation to optimal contact selection when compared to clinical data. Configurations with multiple active contacts might improve stimulation in some cases, although there is no general tendency. The lead design study suggests that row segmentation with three or four contacts per row is a good option. In addition, the stimulation performance was generally better for the designs where the contacts were closer to each other. This study thus confirms the importance of mathematical modeling in DBS as an inexpensive way of obtaining optimal stimulation settings and lead designs.**

Keywords*–*Deep Brain Stimulation; Optimization; Lead Design; Convex optimization; Field Steering; Parkinson Disease.

I. Introduction

Deep Brain Stimulation (DBS) is a surgical procedure that consists of delivering electrical stimuli, usually rectangular biphasic pulses, to a certain target in the brain by using one or several surgically implanted leads. Initially DBS was considered a purely medical problem. However, nowadays there is a lot of interest in the engineering community, e.g., by suggesting alternative set ups based on modeling [1]. The goal of the therapy is the alleviation of symptoms of various neurological diseases, such as Parkinson's Disease (PD) [2], epilepsy [3], dystonia [4], and others. DBS has mostly replaced surgical lesioning and ablation procedures because of its reversibility, flexibility, and individualization potential [5]. The interest in DBS has spread to other areas of medicine and applied to treatment of psychiatric diseases, such as e.g., schizophrenia [6] or Tourette Syndrome [7]. In the case of PD, since the implantation and programming procedure is quite complicated and costly compared to pharmacotherapy [8], [9], physicians usually choose advanced patients for this procedure, when drugs such as levodopa have lost effectiveness or have severe side effects [2] such as fluctuations, dyskinesias or toxicity [10]. Some studies suggest in fact that an earlier implantation could be beneficial [11].

The principle of DBS is in delivering mild electrical pulses via a chronically implanted lead, whose active contacts are positioned in the subcortical area, where a stimulation target is usually defined. Prior to the operation, patients undergo an extensive clinical examination as well as medical imaging. Based on the images, the physician pinpoints a target area, which is in PD usually located in the basal ganglia, with the subthalamic nucleus (STN) being of particular interest. A few weeks after the surgery, the patients undergo a lengthy trial-and-error programming period to properly tune the stimuli.

Since the underlying physiological mechanism of DBS and its long-term effects on the brain still remain unknown, the therapeutical outcome is difficult to predict. Furthermore, because of uncertainties in the position of the lead or suboptimal stimulation settings, the stimulated volume might go beyond the target causing undesirable side effects [12]. Shaping the stimuli so that the stimulated volume covers the intended target and does not spill outside of it is thus important for maximization of the therapeutical benefits and minimization of the side effects.

Currently used lead designs (see Fig. 1(a)) were originally adopted from cardiac pacing technology [13] and have not evolved much since then. Meanwhile, further insights into neuromodulation obtained in recent years by using computer modeling based in Finite Element Method (FEM), multiphysics simulation and neuron models, along with the exponential improvement of computational capabilities, open up for more sophisticated and individualized solutions. The aim of these is to shorten the programming time and to better understand the underlying mechanisms [14].

Addressing the shortcomings of the currently used designs, novel DBS electrodes have been developed by such companies as Boston Scientific (USA), St. Jude (USA) and Sapiens (The Netherlands, now part of Medtronic). These leads could be configured in more versatile spacial settings and take advantage of field steering techniques to shape the stimuli. As seen from the geometry of the lead contacts in Fig. 1,

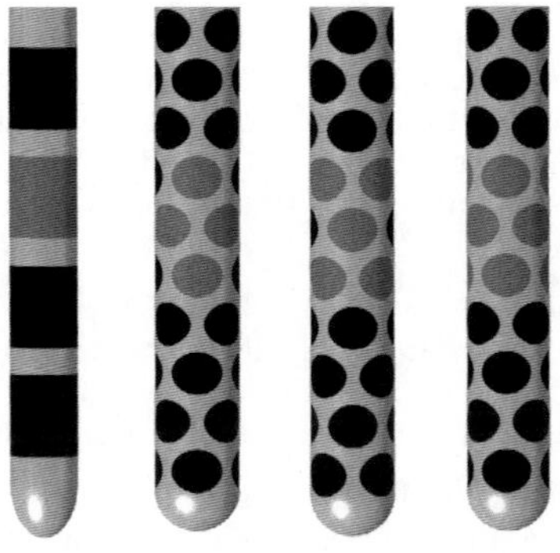

Figure 1. Lead configurations for the conventional lead (a), field-steering Diamond-4 (b), X-5 (c) and X-8 (d). Active contacts are marked in red.

while the conventional state-of-the-art lead delivers a radially symmetric stimulation over the whole cylindrical contact, the field steering one is capable of asymmetrical stimulation that can be tailored to the target area anatomy [15].

Apart from the existing electrodes, hypothetical ones can be tested inexpensively by using realistic computer models. Analysis of design degrees of freedom in novel leads can be performed as well, going beyond the ones already in clinical usage or in development. Segmentation schemes were analyzed in [16] while row separation as a design degree of freedom was treated in [17] and [18]. This study compares the performance of different lead schemes with respect to row segmentation and the separation between the rows, see Fig. 2. At the moment, it is a topic receiving close review, with some of the companies mentioned above already developing schemes with optimized geometry.

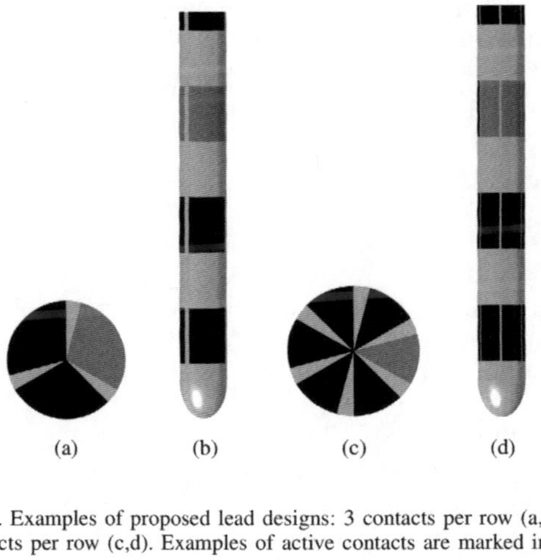

Figure 2. Examples of proposed lead designs: 3 contacts per row (a,b) and 6 contacts per row (c,d). Examples of active contacts are marked in red.

Another possibility for electrode performance improvement is offered by a multicontact stimulation approach, i.e., manipulating the stimuli simultaneously at two or more active contacts. It allows further shaping of the activated tissue volume, thus providing more flexibility.

This study focuses on stimuli optimization using the aforementioned leads, i.e., how to choose a configuration such that a target volume, given by the STN or other volumes, is stimulated. Due to the possible side effects, the stimulation should be kept low outside the target. Within this framework, the objectives of the present study are:

- To compare the optimization results for a state-of-the-art lead to the active contact and to the stimulation amplitude in clinical data, which are assumed to provide good outcome. Both single contact and multicontact approaches are evaluated.
- To optimize a field steering lead over a set of contact configurations and compare to the state-of-the-art lead, in order to assess possible advantages of the former.
- To suggest design guidelines for several proposed leads. Apart from the overspill, stimulation amplitudes are also analyzed to address potential issues with safety and battery life.

Results obtained in this paper suggest that a simplified DBS model can reasonably well predict which contact or contacts are used for stimulation by medical personnel according to clinical data, at least for the analyzed lead population. Further, the described optimization algorithm obtains a significant improvement in the overspill with field steering asymmetrical stimulation compared to symmetrical stimulation provided by a state-of-the-art lead. In addition, the proposed approach makes a useful design tool for new leads, although given the variability of clinical data drawing reliable conclusions is difficult.

The rest of the paper is composed as follows. In Section II, an overview of the FEM mathematical model is given, along with different neuronal stimulation quantification schemes. Afterwards, the core optimization technique used is presented. The state-of-the-art lead is analyzed in Section III with one or two contacts used for stimulation. The field steering lead is analyzed and compared with the state-of-the-art one in Section IV. In Section V, hypothetical electrode designs are analyzed and the results obtained summarized. Conclusions and limitations are discussed in Section VI.

II. Models and Methods

In this section, the FEM model, the neuronal stimulation quantification scheme and the optimization procedure used for this study are described in detail.

A. Electric Field Model

The first step to compute optimized stimuli is to obtain the electric field distribution given the lead geometry. The electric potential is evaluated by solving the equation of steady currents in the brain tissue:

$$\nabla \cdot (\sigma \nabla u) = 0, \tag{1}$$

where u is the electric potential, σ the electric conductivity, and ∇ is the gradient operator. The electric field $\mathbf{E}$ is obtained by taking the negative gradient of u:

$$\mathbf{E} = -\nabla u. \tag{2}$$

An analytical solution to the model given by (1) does not exist in most cases, but it can be integrated numerically using, e.g., a FEM solver. The model considered in this study consists of three main components: the bulk brain tissue, the lead, and an encapsulation layer surrounding the lead.

The bulk tissue is represented as a cube with a side of 0.4 m centered at the tip of the lead that is grounded on the outer surfaces to simulate the ground in the implanted pulse generator. Although the brain tissue is heterogeneous and anisotropic in reality, these effects are beyond the scope of this paper, see [14] and [19] for details. Although the brain tissue is made of several components, e.g., white matter, gray matter, cerebrospinal fluid and blood vessels, its conductivity is approximated for this study as homogeneous with $\sigma = 0.1$ S/m [20].

In addition to the former two components, an encapsulation layer is formed around a lead implanted in the brain due to the reaction of the body to foreign objects [21]. However, the thickness and conductivity of it are still open to debate and might be patient specific. Following [14], a 0.5 mm thick layer with a conductivity of 0.18 S/m is introduced.

Several lead designs are considered in this study:

- A widely used state-of-the-art lead (Fig. 1(a)) with cylindrical contacts, a height of 1.5 mm, and a separation between contacts of 0.5 mm. Its diameter is 1.27 mm.
- A field-steering lead with elliptical contacts. To facilitate field steering, the rows are rotated 45° to each other with respect to the lead axis, as shown in Fig. 1(b),1(c), and 1(d). Its diameter is 1.27 mm as well.
- Hypothetical leads derived from the state-of-the-art one. Two design degrees of freedom are considered:
  - Row segmentation: As seen in Fig. 2, one possibility is to split each row in a number of contacts. For this study, rows with 2, 3, 4 and 6 contacts are considered.
  - Row separation: Different separation schemes are simulated to investigate the influence of other contacts' proximity on the shape of the electric field. In this study, contact row separations are in the range 0.25-1.5 mm. For comparison, depending on the lead model, the state-of-the-art leads have separations of 0.5 or 1.5 mm between contacts.

The stimulation signal is modeled as a Dirichlet boundary condition at the active contacts surface while the non-active contacts are left floating. It should be noted that model (1) is a linear partial differential equation, and thus, it is enough to compute the field distribution for a unit stimulus and then scale it accordingly, which transformation will simplify the computations.

The model has been implemented in COMSOL 4.3b (Comsol AB, Sweden). The solutions obtained by the FEM solver were then equidistantly gridded on a $70 \times 70 \times 60$ grid centered at the lead tip and expanding 16 mm in the axes perpendicular to the lead and 20 mm in the lead axis, in order to be exported for further processing.

Several field distributions were computed:

- State-of-the-art lead: Distributions with one active contact and the rest floating were evaluated at first. In addition to that, field distributions with the grounded inactive contacts were computed. This was done to enable summing up them for the multicontact approach, since the effect of one active contact on the others when they are left floating can be computed.
- Field steering lead: Distributions for the considered configurations (Diamond-4, X-5, X-8, shown in Fig. 1) were obtained for each row of contacts.
- Hypothetical leads: Similarly to the computations for the state-of-the-art leads, field distributions for one active contact with the rest floating were computed at first and, in addition, distributions with grounded inactive contacts were computed for the multicontact approach. It is useful in particular when considering two rows stimulating with different amplitudes.

B. *Quantification of activated volumes*

Volumes of activated tissue can be quantified by using axon models [22]. While axon models yield precise results, the procedure is computationally expensive and the topology and connectivity of the neuron network must be known to some degree. Other approaches involve functions that approximate the activated volume without taking into account the anatomy of the neurons, such as Rattay's activation function [23] or the electric field [24]. These have the advantage of requiring less computations and only stationary analysis. However, using second derivatives might result in numerical issues, in particular in the area near the lead. Furthermore, it was shown that the electric field provides more robust means of quantifying neuronal stimulation [24]. Thus, the electric field will be used in this study. The activated neurons are distinguished from the rest by applying a threshold to the electric field, with the threshold value depending on the neuron anatomy and the characteristics of the stimulation pulse itself [24].

To place the electric field pre-computed by the FEM solver at the proper position, conventional translation-rotation algebra is utilized. Assuming that the tip of the lead is at the origin, the set of operations is given by:

$$\mathbf{E}_{\mathrm{eval}} = R_{\mathrm{rot}} R_z \mathbf{E} + \mathbf{x}_{\mathrm{lead}}, \tag{3}$$

where $\mathbf{E}$ and $\mathbf{E}_{\mathrm{eval}}$ are the original and positioned electric field vectors respectively, R_{rot} is a rotation matrix that aligns the field with the given lead vector, R_z is a rotation matrix with respect to the Z axis (used for field steering), and $\mathbf{x}_{\mathrm{lead}}$ is the lead position.

Once the field is properly positioned and filtered with the aforementioned threshold, intersection volumes are computed under a methodology similar to [25]. Two of them are of particular interest with respect to the clinical applications: the activated volume of the target area and the activated volume outside the target area. The topology of the target area is taken from an atlas of potential regions for therapeutical stimulation and can be assumed to be convex. Whether the electric field points are inside of the convex hull of the target area or not is checked by an additional function [26].

C. Optimization scheme

In order to optimize the stimuli, the following minimization problem is defined:

$$\min_{u_i} J(u_i), \tag{4}$$

where u_i are the optimization variables (in this case, the electric potential or potentials of the stimuli) and $J(u_i)$ is a cost function to be defined that should ideally be a convex function.

The following cost function is proposed:

$$\begin{aligned} J(u_i) &= p_{\text{Spill}}(u_i)\left(\frac{100-p_{\text{Act}}(u_i)}{100-p_{\text{Th}}}\right) & p_{\text{Act}} \le p_{\text{Th}}, \\ J(u_i) &= p_{\text{Spill}}(u_i) & p_{\text{Act}} > p_{\text{Th}}, \end{aligned} \tag{5}$$

where p_{Spill} is the fraction of the activated volume that lies outside the target, p_{Act} is the fraction of the target that is activated, and p_{Th} is the minimum activation required for the target. All of them are given in percent for illustration. For this study, p_{Th} is set at 95%. It should be noted that, if several amplitudes were to be optimized, the cost function would be multi-dimensional.

The motivation behind the cost function above is that it is continuous and convex, since both p_{Act} and p_{Spill} are monotonically non-decreasing with the amplitude of the stimulus. However, there is no clear relationship between the functions p_{Act}, p_{Spill} and the stimulus amplitudes u_i. Therefore, a numerical optimization algorithm estimating the gradient of $J(u_i)$ from the computed values of p_{Act} and p_{Spill} has to be used.

An example illustrating the dependence of cost function (5) on the stimuli amplitude is given in Fig. 3.

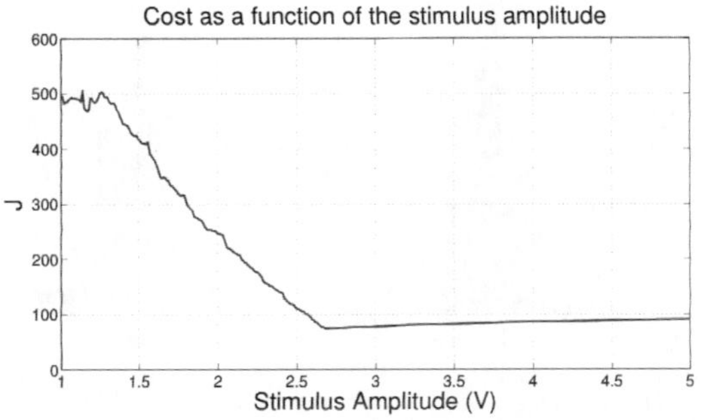

Figure 3. Example of cost function J as a function of the stimulus amplitude.

The small peaks in Fig. 3 occur because of issues with the volume computation, namely since the geometry used for both the activation volume and the target is defined in convex hulls of a discrete cloud of points. The computed volumes are thus not smooth as u changes. Although this makes the function non-convex in practice, the peaks are small enough to be skipped by increasing the step size of the optimization algorithm. A minimum step size of 0.002 V was taken.

When considering several stimulation amplitudes, extra care should be exercised then in choosing the initial guess in the optimization algorithm. It was found that the optimization gets stuck in local minima more easily than in the one-dimensional case.

III. State-of-the-art Lead

In this section, approaches to stimulation with the state-of-the-art lead are analyzed and compared.

A. Single Contact

To optimize stimulation with only one active contact, two approaches can be considered. First, the active contact position can be fixed and only the stimulus amplitude is optimized. Second, the active contact is left as an additional optimization variable, restricted to taking a single value in the set of possible positions $C_s = \{0, 1, 2, 3\}$, where contact 0 is the most distal and 3 is the most proximal. Due to the possibility of choosing the active contact at will and to illustrate the efficiency of the optimization method, the second approach is selected further.

Keeping the active contact free makes the optimization problem equivalent to four such with fixed contacts. In order to speed up the computations, best active contact could be chosen without optimization. By examining $J(u_i)$ given by (5), it can be easily seen that as long as there is an intersection between the activated volume and the target for at least one of the contacts, the cost function will be lower in general for the optimal contact no matter how large u_i is. So, it is enough to do a single evaluation of the cost function for a given value of u_i to choose the contact. Said value cannot be too low, since it might yield empty intersections, or too high, since it will take too much time to calculate due to the number of points involved. Thus, the evaluation is performed with low u_i and then, if the intersection is empty, u_i is set to a higher value.

Optimization was performed for 65 lead positions whose clinical data stated a single contact stimulation with an activation threshold of 175 and 200 V/m. Comparing the optimized results to the clinical settings is of great interest, so the fraction of configurations estimated successfully by the optimization algorithm with respect to the clinical settings was computed as well.

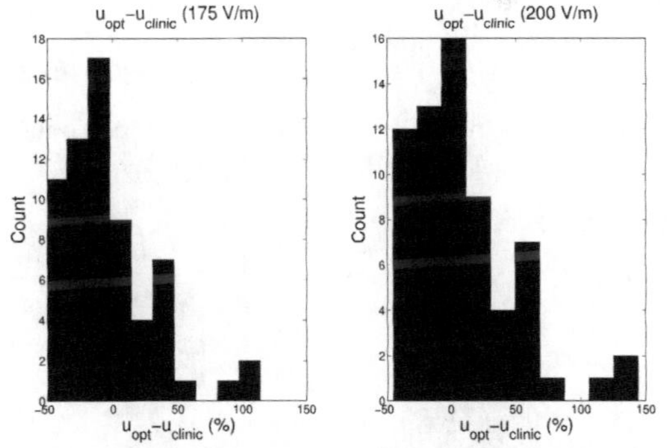

Figure 4. Discrepancy in amplitude for the optimized and clinical setting with the threshold values of 175 and 200 V/m.

As seen from Tab. I, the mathematical model predicts the clinically used contact for the defined target in roughly a half of the cases. In addition, in almost all of the cases, the predicted optimal contact is an immediate neighbor of the one specified in the clinical data. In some cases, there is no significant difference in the values of the cost function and either contact can be utilized, according to the calculated values.

TABLE I. SINGLE CONTACT OPTIMIZATION

Threshold: 200 V/m	
Correct contact (%):	53.8
1 contact error (%):	35.4
Discrepancy in amplitude (%, mean$\pm\sigma$):	9 ± 41
Threshold: 175 V/m	
Correct contact (%):	50.7
1 contact error (%):	38.4
Discrepancy in amplitude (%, mean$\pm\sigma$):	-4 ± 35

In addition, the predicted optimized stimuli amplitude is fairly close to the clinical one. The plots in Fig. 4 suggest that in most cases a threshold between 175 and 200 V/m might be sufficient. It comes though with a high standard deviation thus revealing a high variability between patients. This is to be expected, since the position of the lead has a very significant impact on the predicted optimal stimuli amplitude.

B. Multiple Contacts

Another approach to improve target coverage would be to allow for multiple active contact configurations. To facilitate the field modeling, the linearity of (1) is exploited. In particular, the field distribution for each contact stimulating with a unit stimulus while the others are grounded is computed first and denoted as $\mathbf{E}_{0,i}$ for the i-th contact. Then the interaction between active contacts and the rest in floating configuration is calculated. It follows a linear relationship and is expressed by the coefficient α_{ki}, representing the effect the i-th contact has on the k-th contact when the k-th contact is floating. This is used to transform from an active-grounded to an active-floating configuration, when the contributions are being summed.

The electric field distributions result from a sum of four contacts, with the stimuli given by the active contacts, denoted by u_i and representing the degrees of freedom and the non-active (floating) contacts contributing with the terms characterized by the corresponding α_{ki}. For instance, for a 2-contact scheme, one gets

$$\begin{aligned}\mathbf{E}_{2cont}(\mathbf{r}) &= u_1\mathbf{E}_{0,1} + u_2\mathbf{E}_{0,2} + \\ &+ (u_1\alpha_{31} + u_2\alpha_{32})\mathbf{E}_{0,3} + (u_1\alpha_{41} + u_2\alpha_{42})\mathbf{E}_{0,4}\end{aligned} \quad (6)$$

It should be noted that the numbering of the contacts above was arbitrary, and it could be any combination of them.

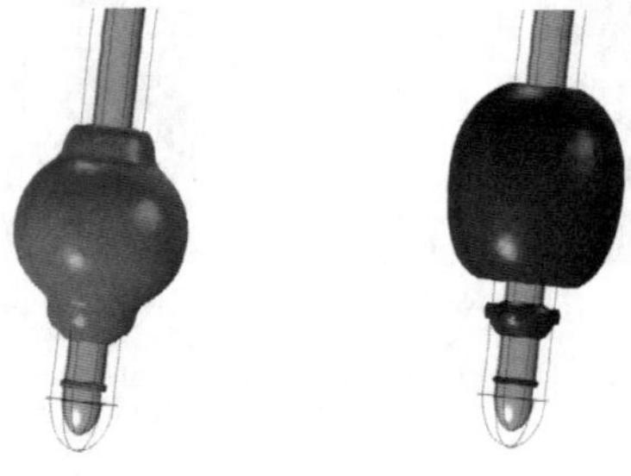

(a) (b)

Figure 5. Isolevels for $E = 200\ V/m$ for a single active contact (a) and two active contacts (b).

As Fig. 5(b) suggests, multiple active contacts might be useful to tailor the stimulation so that it achieves a similar activated volume with less overspill. The results are in principle dependent on the position of the target with respect to the active contact in the single contact approach. If the target is located next to the active contact, then it would be probably reasonable to consider just a single contact stimulation. However, if the target is located in between two contacts, shaping the stimulation with these two contacts might be beneficial.

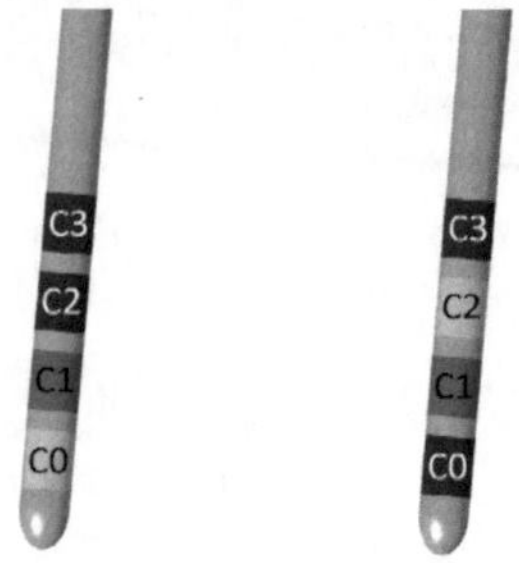

Figure 6. Example of considered multicontact configurations, with Contact 1 as the optimal (in red) and Contacts 0 (left) and 2 (right) as secondary (in green).

To speed up computations, only the configurations which involve neighboring contacts to the ones obtained in the single contact approach are considered. So, for example, if the predicted optimal contact is contact 1, only combinations involving contacts 1 and 0 and 1 and 2 are considered, see Fig. 6.

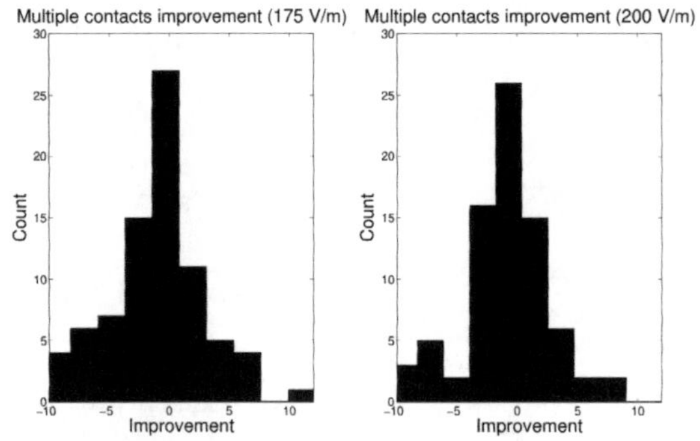

Figure 7. Improvement of overspill for 175 and 200 V/m for the multicontact approach.

TABLE II. DUAL-CONTACT OPTIMIZATION

Threshold: 200 V/m	
Improvement cases (%):	38.7
Overspill improvement (percentage points, mean$\pm\sigma$):	2.00 ± 2.28
Threshold: 175 V/m	
Improvement cases (%):	37.5
Overspill improvement (percentage points, mean$\pm\sigma$):	2.67 ± 2.83

The results are summarized in Fig. 7 and Tab. II. As expected, the improvement is situational, and appears only in a part of the cases. However, the improvement can be significant, with a decrease of up to $5-6\%$ in the absolute value of the overspill compared to the single contact approach. Note that the state-of-the-art lead considered here features a fixed distance between contacts. The effects of uneven separation between contact rows will be analyzed later in this study.

IV. Field Steering Electrode

As was investigated in [25], [27], field steering yields better results regarding overspill than the state-of-the-art radial stimulation. In this study, optimization is used in order to *in silico* confirm those findings.

Three configurations Illustrated in Fig. 1 were tested. For each configuration, the parameters to optimize are the rows where the active contacts are located and the orientation of the lead with respect to its axis. To speed up computations, the optimization followed a similar scheme to that applied with multiple contacts, taking as a baseline the results obtained with single contacts and the state-of-the-art lead. As shapes of the contacts are different, the rows at roughly the same height are considered, together with their neighbors.

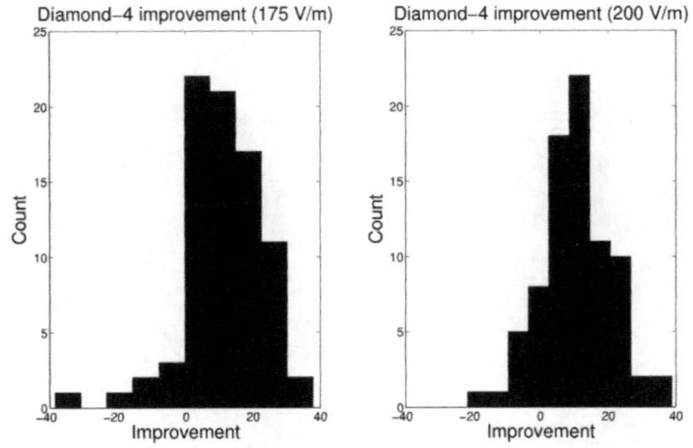

Figure 8. Improvement distribution using the Diamond 4 configuration.

TABLE III. DIAMOND 4 CONFIGURATION IMPROVEMENT.

Threshold: 200 V/m	
Improvement cases (%):	87.5
Overspill improvement (percentage points, mean$\pm\sigma$):	10.37 ± 10.52
Threshold: 175 V/m	
Improvement cases (%):	91.25
Overspill improvement (percentage points, mean$\pm\sigma$):	11.48 ± 11.63

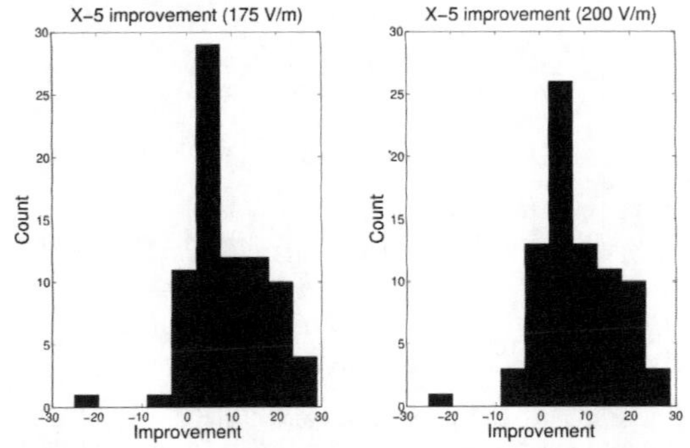

Figure 9. Improvement distribution using the X-5 configuration.

Table IV. X-5 configuration improvement

Threshold: 200 V/m	
Improvement cases (%):	88.75
Overspill improvement (percentage points, mean$\pm\sigma$):	8.65 ± 10.67
Threshold: 175 V/m	
Improvement cases (%):	92.5
Overspill improvement (percentage points, mean$\pm\sigma$):	9.76 ± 10.37

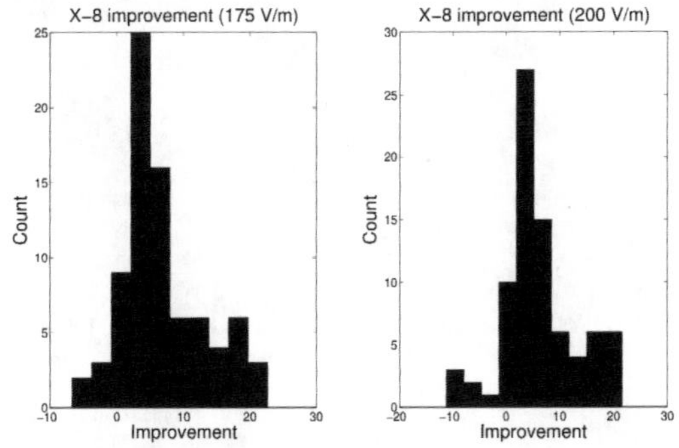

Figure 10. Improvement distribution using the X-8 configuration.

The results are summarized in Fig. 8 - Fig. 10 and Tab. III - Tab. V. In almost all cases, there is an improvement in the overspill with respect to the single contact approach. The improvement is largest in average with the Diamond-4 configuration, Fig. 8. The high standard deviation comes from the variety of geometries considered, making the improvement heavily dependent on the lead position with respect of the target. Some cases were observed where the X-5 or X-8 configurations achieved better results for a specific lead location.

V. Hypothetical Leads

Leads that have not been implemented in hardware are analyzed in this section by computing field distributions with one or several active contacts, similarly to what has been done above for the state-of-the-art and field steering lead. This part of the study is organized in three sections:

- **Row separation comparison**, where the influence of row separation on the stimulation field is evaluated, cf. Fig. 11(a) and Fig. 11(b).
- **Row segmentation comparison**, in order to understand how many contacts per row are enough to achieve good selectivity, cf. Fig. 11(b) and Fig. 11(c).
- **Configuration comparison**, to compare active contact schemes. For each lead, three active contact configurations are considered: single contact, multiple contacts

Table V. X-8 configuration improvement

Threshold: 200 V/m	
Improvement cases (%):	90
Overspill improvement (percentage points, mean$\pm\sigma$):	11.51 ± 10.63
Threshold: 175 V/m	
Improvement cases (%):	91.25
Overspill improvement (percentage points, mean$\pm\sigma$):	11.56 ± 10.41

(usually two) with the same amplitude and multiple contacts with different amplitudes.

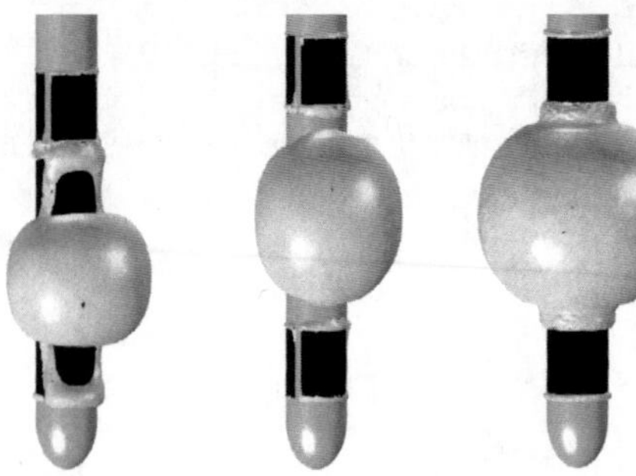

(a) (b) (c)

Figure 11. 200 V/m isolevels for a hypothetical lead with 3 contacts per row and separation between rows of 0.25 mm (a) and 1.5 mm (b). A lead of 1 contact per row and separation of 1.5 mm is depicted for comparison (c).

Due to the amount of combinations needed, this part is limited to a population of 6 leads, with only the illustrative cases being described.

A. *Row separation comparison*

The first question to look into is how the separation between rows affects the predicted stimuli amplitude, together with the overspill percentage. For the state-of-the-art lead considered in the present manuscript, the separation between the rows is 0.5 mm. Separations between rows equal to 0.25 mm, 0.75 mm, 1 mm and 1.5 mm are analyzed further.

Considering only one active contact, two trends were observed, as illustrated in Fig. 12 and Fig. 13. In some cases, increasing the distance between the contacts decreases the overspill, while in some other cases the effect is the opposite one.

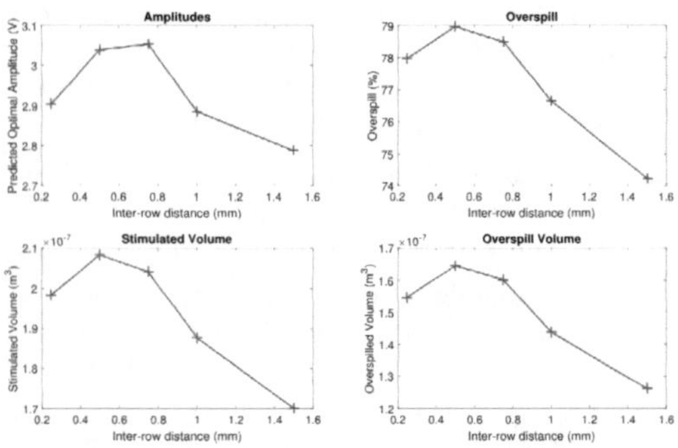

Figure 12. Predicted amplitudes, overspill percentage and activated and overspilled volumes for a single contact with 1 contact per row lead configuration.

When analyzing dual contact configurations, however, the trend is usually that the overspill is larger the more separated the contacts are, as seen in Fig. 14 and Fig. 15.

A plausible explanation for this is that the geometry of the electric field isolevels is quite different in case when the

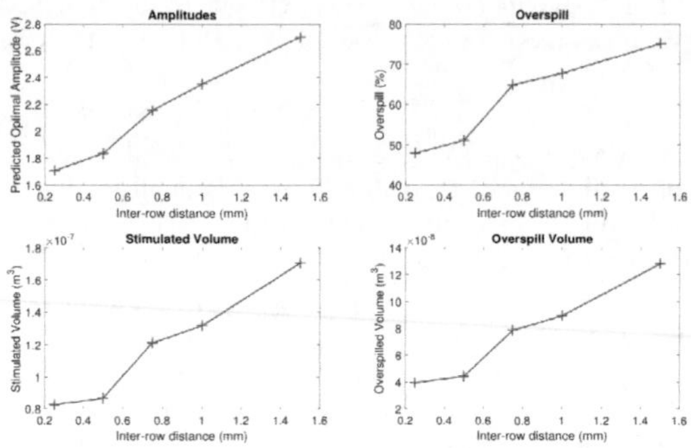

Figure 13. Predicted amplitudes, overspill percentage and activated and overspilled volumes for a single contact with 1 contact per row lead configuration.

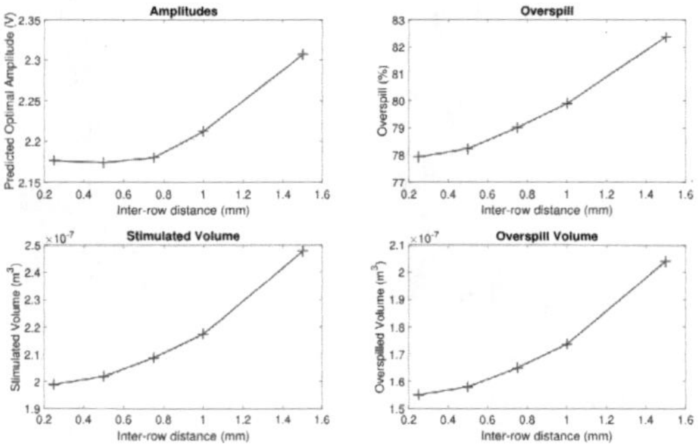

Figure 14. Predicted amplitudes, overspill percentage and activated and overspilled volumes for two active contacts in different rows with 1 contact per row lead configuration.

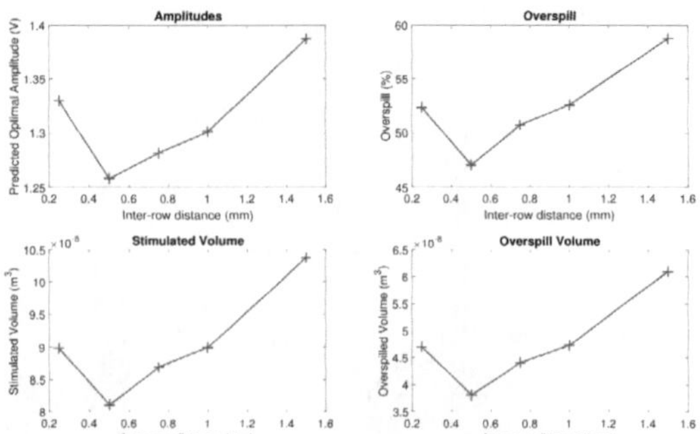

Figure 15. Predicted amplitudes, overspill percentage and activated and overspilled volumes for two active contacts in different rows with 1 contact per row lead configuration.

contacts are close to each other compared to when they are more separated. Indeed, it can be seen in Fig. 16 that the considered isolevel when contacts are close to each other is not so different from that when only one contact is active. However, when contact rows become more and more separated, the field will stretch along the lead axis. Together with the target morphology, this might explain why the overspill is larger the more separated the contacts are.

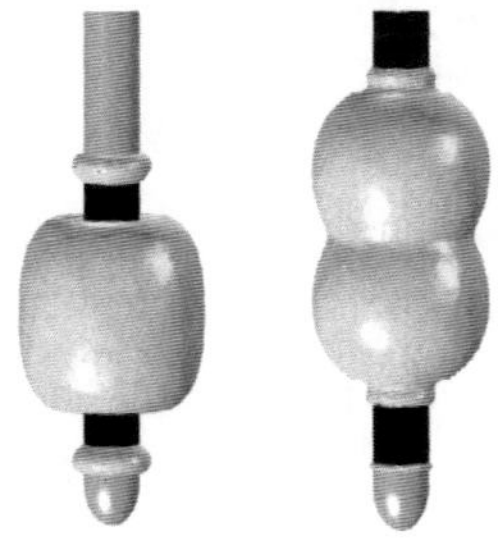

(a) (b)

Figure 16. 200 V/m isolevels for a hypothetical lead with 1 contact per row and separation between rows of 0.25 mm (a) and 1.5 mm (b).

It can be concluded that having a large separation between contacts does not facilitate achieving a good stimulation performance although it might be benefitial in some cases when considering only one active row.

B. Row Segmentation

Another possibility to look into is segmented contact rows that also offer field steering, but for a lead shape different from that analyzed in Section IV. For this part of the study, field distributions were optimized for 1, 2, 3, 4 and 6 contacts per row, see Fig. 2. From the results obtained above for the field-steering electrode, it is expected that the overspill will decrease to some extent for the segmented state-of-the-art lead, as the segmentation increases.

In the population considered, an overspill reduction is achieved in most cases when the contacts are segmented, as illustrated in Fig. 17. However, in some other cases, probably due to the location of the lead, segmentation can actually lower the predicted performance, as shown in Fig. 18. However, these cases are less frequent.

It can be seen as well that the optimal stimulation amplitude increases as the rows become more segmented. In the population considered, the lower values of overspill with a moderate amplitude are usually obtained at three or four contacts per row at most. Going beyond that implies a large stimulation amplitude that could be potentially harmful to the patient and will consume more energy in a battery-driven device, while the benefits could be minimal or non-existent. In addition, more complex hardware could be needed in order to achieve it, thus making the product more costly and difficult to handle in the clinic.

It is worth mentioning that even in cases when segmentation does not improve the overspill, multiple contacts could be active in the same row yielding something similar to a symmetric stimulation in the state-of-the-art lead. Indeed, as can be seen in Fig. 19, there is no significant difference

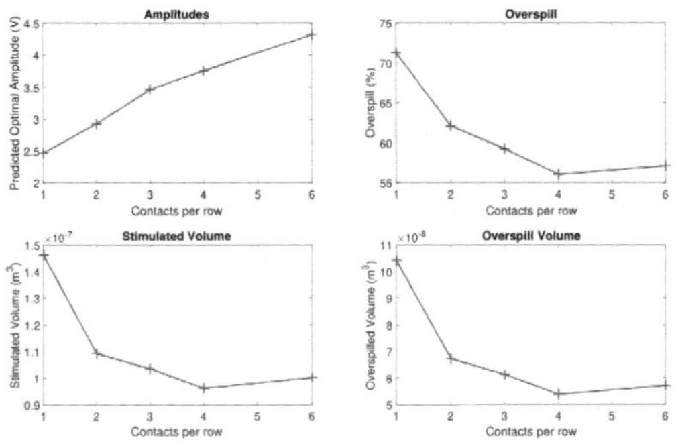

Figure 17. Predicted amplitudes, overspill percentage and activated and overspilled volumes for one active contact and different number of contacts per row. Distance between rows of 0.5 mm.

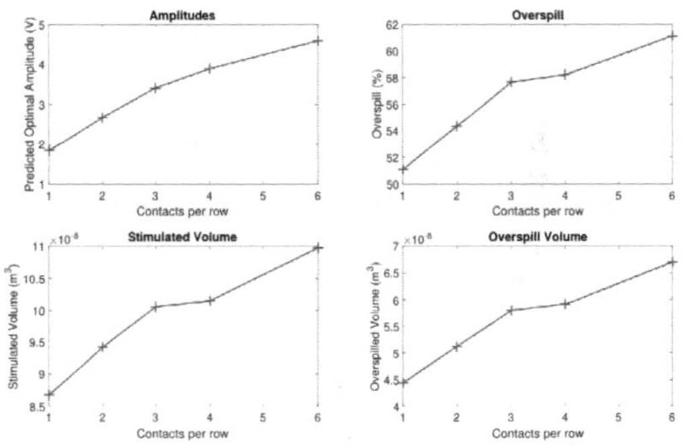

Figure 18. Predicted amplitudes, overspill percentage and activated and overspilled volumes for one active contact and different number of contacts per row. Distance between rows of 0.5 mm.

between them. Thus, even in cases when field steering does not produce an improvement, a good stimulation result can still be achieved with these hypothetical leads by making all the contacts in a given row active.

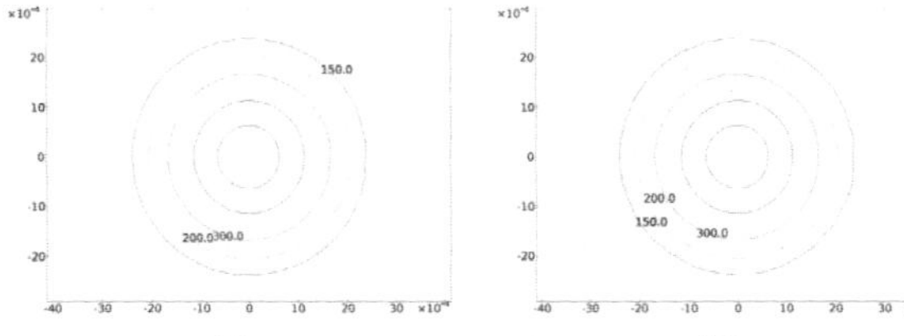

(a) (b)

Figure 19. Contour plots taken at the middle point of a contact for a 1 contact per row lead (a) and a 6 contacts per row lead (b) for levels of electric field of 150 (blue), 200 (green) and 300 (magenta) V/m.

C. Contact configurations

Contact configurations can be analyzed, in pursuit for an improvement over the single contact ones. A more thorough analysis than in Section III is conducted here, taking into consideration row segmentation and separation as well, since it was previously done only with the state-of-the-art lead specifications.

Dealing with active contact settings, the first issue to consider is the number of independent electrical sources present in the pulse generator. If only one is present, the same amplitude must be used for all active contacts. With several sources, different amplitudes can be assigned to the active contacts, making the setting more versatile. However, having several sources makes the hardware more complex and expensive. Further, the number of active contacts should be considered, along with their location that can be in the same row or in different rows.

For this study, configurations with both the same amplitude and different amplitudes are considered. In all cases, the active contacts neighbor each other.

1) Same amplitude: When considering active contacts with the same amplitude, two options can be analyzed: when the active contacts are in the same row or when they are in different rows. The former however is not interesting since it should yield the same stimulation as an electrode with contacts per row and one active contact. For example, if a six contacts per row lead is considered, having two active contacts in the same row means that the stimulation will be the same as with three contacts per row. Thus, only different rows are considered in this part of the study.

Some results are depicted in Fig. 20 and Fig. 21. The observed trends included the following:

- Significantly reduced amplitude in all cases, with the effect being more pronounced the more segmented the contacts are.
- When considering overspill, there is an improvement in some cases. Said improvement varies however significantly between leads: in some cases the improvement is larger with a very segmented lead and vice versa in others. As mentioned before, this is not an issue since segmented electrodes can behave as non-segmented ones, see Fig. 19.

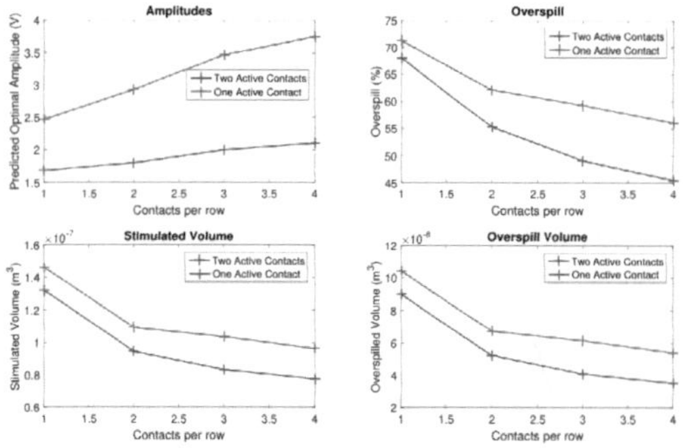

Figure 20. Predicted amplitudes, overspill percentage and activated and overspilled volumes for two active contacts (same amplitude) and different number of contacts per row. Distance between rows of 0.5 mm.

Although in some situations the overspill difference is not significant and it could be even worse when considering

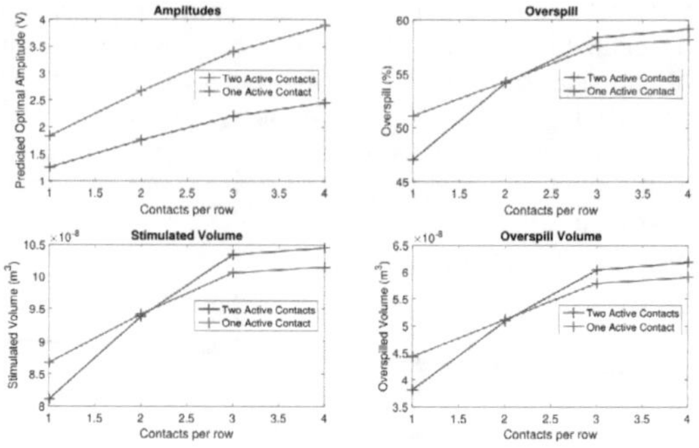

Figure 21. Predicted amplitudes, overspill percentage and activated and overspilled volumes for two active contacts (same amplitude) and different number of contacts per row. Distance between rows of 0.5 mm.

different rows, the predicted amplitude is much lower, which property translates into less power consumption and higher patient safety.

2) Different amplitude, different rows: As with the state-of-the-art lead, different stimulation amplitudes can be considered as well under the same methodology as in Section III. Different segmentation schemes will be considered as well.

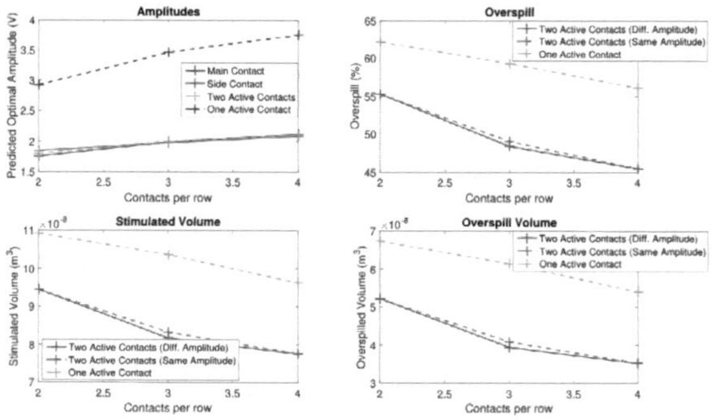

Figure 22. Predicted amplitudes, overspill percentage and activated and overspilled volumes for one and two active contacts and different number of contacts per row. Distance between rows of 0.5 mm.

From the plots in Fig. 22 and Fig. 23, the following can be concluded:

- The predicted amplitudes when different amplitudes are allowed are similar to the ones obtained with the same amplitude at all active contacts.
- Compared to the case of two contacts with the same amplitude, taking different amplitudes improves in some cases the overspill.
- In all considered cases, the amplitudes are lower than the ones obtained with only one active contact.

Thus, as expected from the results in Section III, it could be relevant in some cases to consider different amplitudes assigned to two different contacts. However, in principle, the gains might not justify the extra costs in hardware and in programming time.

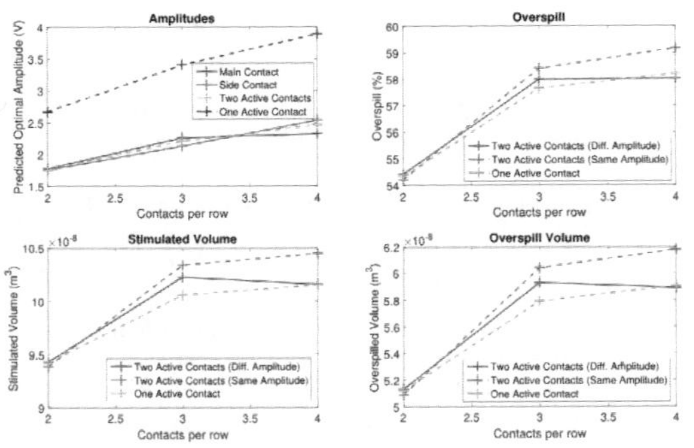

Figure 23. Predicted amplitudes, overspill percentage and activated and overspilled volumes for one and two active contacts and different number of contacts per row. Distance between rows of 0.5 mm.

3) Different amplitude, same row: It could be interesting to see how using different stimulation amplitudes can help when the active contacts are taken in the same row. For this part, only the lead with six contacts per row is considered since it makes the most illustrative case. Configurations of three and five active contacts positioned are considered, as in Fig. 24.

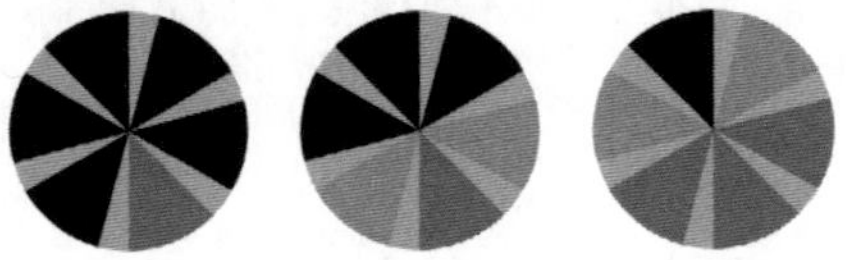

(a) (b) (c)

Figure 24. Configurations considered in the different amplitude, same row case. Active contacts are depicted in red and blue, with the red contacts having different amplitude than the blue contacts. Non-active contacts are depicted in black.

The results obtained show that there is indeed a possibility of using these configurations, as can be seen in Fig. 25 and Fig. 26. Surprisingly, with five active contacts, the predicted amplitudes are mostly the same for all contacts, which does not happen when three active contacts are considered. However, the results exhibit large variability and can make this alternative not worth the effort. Nevertheless, it could clearly improve the performance in some cases.

VI. Discussion

Using optimization schemes in order to scale the stimulus amplitude of the active contact or contacts could yield an activation volume that better covers a given target while limiting, as much as possible, stimulation beyond the target. This study has compared two available leads: a state-of-the-art lead and a field steering lead. In addition, several hypothetical leads have been tested in order to gain an insight into lead design. Different contact configurations have been tested as well.

Selecting the active contact freely for a given target with the state-of-the-art lead while using the single-contact approach, a simple model predicts the clinically used contact in roughly a half of the times in the considered lead population.

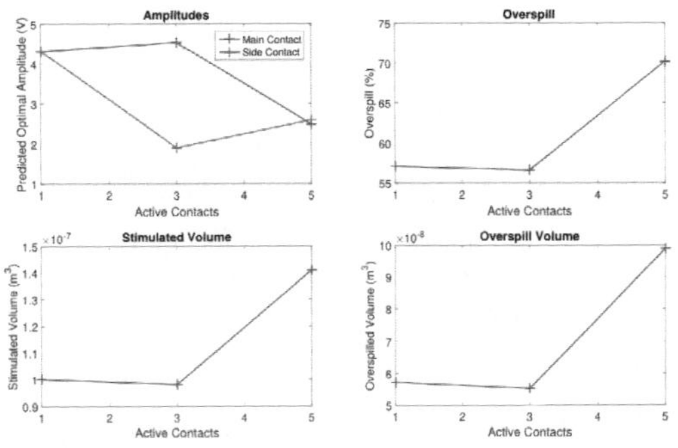

Figure 25. Predicted amplitudes, overspill percentage and activated and overspilled volumes for different active contacts. Distance between rows of 0.5 mm.

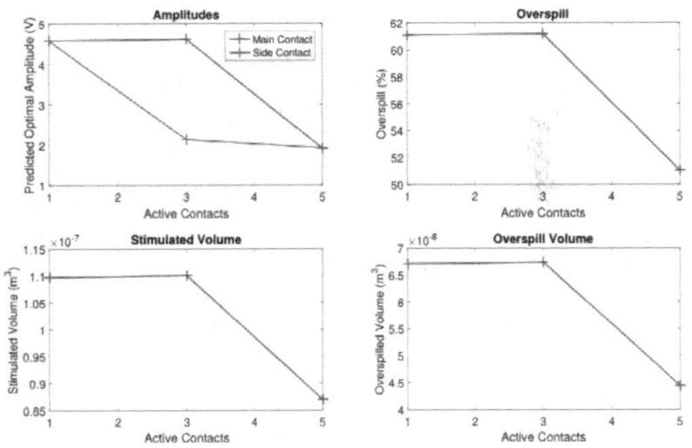

Figure 26. Predicted amplitudes, overspill percentage and activated and overspilled volumes for different active contacts. Distance between rows of 0.5 mm.

Furthermore, in some cases, there is no significant difference between the amplitudes or performance of the clinical and the optimal configurations. Extending the stimulation to multiple contacts allowed for an improvement of the overspill in around 38% of the cases.

The obtained results were compared to field steering configurations. A significant improvement of the overspill with a decrease of 10 percentage points on average was found in all cases, with an average decrease of 18 percentage points for the Diamond 4 configuration.

The analysis was then extended to hypothetical leads with a significant variability in results observed. Even so, some trends have been discerned: higher vertical separation between the contacts usually leads to a higher overspill, in particular when multiple active contacts are considered. Segmentation yields good results in most of the analyzed leads, with the best performance achieved with three or four contacts per row. Even if no improvement is present, it is still possible to stimulate in a way similar to the state-of-the-art lead by making all contacts in the same row active. Care should be exercised however, since the amplitudes needed to achieve satisfactory coverage are higher than in the non-segmented leads, an effect that becomes more profound the more segmented the leads are. When stimulating with several contacts, the amplitude is significantly lower but a higher overspill could occur. Nevertheless, variability among patients makes it difficult to achieve general conclusions in many cases.

The results obtained in this study apply under some limitations. First, the brain tissue was assumed to be homogeneous, when this is not the case and significant (patient specific) differences may arise [14]. Furthermore, the encapsulation layer surrounding the lead has uncertain physical properties, such as the conductivity and the thickness, both of which might be time varying [28]. In addition, considering the electric field as a predictor of whether a neuron is stimulated or not is an approximation. A more thorough analysis would need a complete neuron population model. Finally, the results obtained assume a certain target structure, which may be patient specific as well. Results should be verified against therapeutic outcomes, but the latter were not available for this study.

Despite the mentioned limitations, this study highlights the use of optimization schemes and geometric arguments to choose optimal DBS stimuli and facilitate the comparison between different lead designs and contract configurations. These optimization schemes could be used as a benchmark for other optimization algorithms such as in [29] and the intersection algorithms can be used in order to assess a set of stimulation settings as in [25].

ACKNOWLEDGMENT

RC and AM were partially supported by funding from the European Research Council, Advanced Grant 247035 (ERC SysTEAM). The authors of this article would like to thank the Pitié-Salpêtrière University Hospital, Paris and Medtronic Neuromodulation, Medtronic Eindhoven Design Center for providing the clinical data used in this study.

REFERENCES

[1] R. Cubo, M. Åström, and A. Medvedev, "Model-based Optimization of Lead Configurations in Deep Brain Stimulation," 1st IARIA International Conference on Smart Portable, Wearable, Implantable and Disability-oriented Devices and Systems (SPWID), Brussels, Belgium, 2015, pp. 14-19.

[2] J. A. Obeso, C. W. Olanow, M. C. Rodriguez-Oroz, P. Crack, and R. Kumar, "Deep-Brain Stimulation of the Subthalamic Nucleus or the Pars Interna of the Globus Pallidus in Parkinson's Disease," N. Engl. J. Med., vol. 345, no. 13, 2001, pp. 956-963.

[3] N. Suthana et al., "Memory enhancement and deep-brain stimulation of the entorhinal area," N. Engl. J. Med., vol. 366, 2012, pp. 502-510.

[4] C. Hamani, S. S. Stone, A. Garten, A. M. Lozano, and G. Winocur, "Memory rescue and enhanced neurogenesis following electrical stimulation of the anterior thalamus in rats treated with corticosterone," Exp. Neurol., vol. 232, no. 1, 2011, pp. 100-104.

[5] R. Gross and A. M. Lozano, "Advances in neurostimulation for movement disorders," Neurol Res., vol. 22, 2000, pp. 247-258.

[6] J. Kuhn et al., "Deep Brain Stimulation in Schizophrenia," Fortschritte der Neurol. Psychiatr., vol. 79, 2011, pp. 632-641.

[7] V. Vandewallea, C. van der Lindena, H. J. Groenewegena, and J. Caemaert, "Stereotactic treatment of Gilles de la Tourette syndrome by high frequency stimulation of thalamus," Lancet, vol. 353, no. 9154, 1999, p. 724.

[8] A. Williams et al., "Deep brain stimulation plus best medical therapy versus best medical therapy alone for advanced Parkinson's disease (PD SURG trial): a randomised, open-label trial," Lancet Neurol., vol. 9, no. 6, pp. 581-591, 2010.

[9] J. Dams et al., "Cost-effectiveness of deep brain stimulation in patients with Parkinson's disease," Mov. Disord., vol. 28, no. 6, pp. 763-771, 2013.

[10] C. D. Marsden, "Problems with long-term levodopa therapy for Parkinson's disease," Clin. Neuropharmacol., vol. 17, Suppl 2, 1994, pp. S32-44.

[11] G. Deuschl et al., "Stimulation of the subthalamic nucleus at an earlier disease stage of Parkinson's disease: concept and standards of the EARLYSTIM-study," Parkinsonism Relat. Disord., vol. 19, no. 1, Jan. 2013, pp. 56-61.

[12] J. L. Alberts et al., "Bilateral subthalamic stimulation impairs cognitive-motor performance in Parkinson's disease patients," Brain, vol. 131, no. Pt 12, Dec. 2008, pp. 3348-3360.

[13] J. Gardner, "A history of deep brain stimulation: Technological innovation and the role of clinical assessment tools," Soc. Stud. Sci., vol. 43, no. 5, pp. 707-728, 2013.

[14] A. Chaturvedi, C. R. Butson, S. Lempka, S. E. Cooper, and C. C. McIntyre, "Patient-specific models of deep brain stimulation: influence of field model complexity on neural activation predictions," Brain Stimul., vol. 3, no. 2, 2010, pp. 65-67.

[15] H. C. Martens et al., "Spatial steering of deep brain stimulation volumes using a novel lead design," Clin. Neurophysiol., vol. 122, no. 3, Mar. 2011, pp. 558-566.

[16] J. Buhlmann, L. Hofmann, P. a Tass, and C. Hauptmann, Modeling of a segmented electrode for desynchronizing deep brain stimulation., Front. Neuroeng., vol. 4, no. December, p. 15, Jan. 2011.

[17] C. R. Butson, C. C. McIntyre, "Role of electrode design on the volume of tissue activated during deep brain stimulation," J. Neural Eng., vol. 3, 2006, pp. 1-8.

[18] B. Howell, W. M. Grill, "Model-Based Optimization of Electrode Designs for Deep Brain Stimulation," 6th Annual International IEEE EMBS Conference on Neural Engineering San Diego, California, 2013, pp. 154-157.

[19] M. Åström, J. Lemaire, and K. Wårdell, "Influence of heterogeneous and anisotropic tissue conductivity on electric field distribution in deep brain stimulation," Med. Biol. Eng. Comput., vol. 50, no. 1, 2012, pp. 23-32.

[20] D. Andreuccetti, R. Fossi, and C. Petrucci, "Dielectric properties of body tissue," 2005. Available: http://niremf.ifac.cnr.it/tissprop/ [accessed: 2016.06.05].

[21] W. M. Grill, J. T. Mortimer, "Electrical properties of implant encapsulation tissue," Ann. Biomed. Eng., vol. 22, no. 1, 1994, pp. 23-33.

[22] D. R. McNeal, "Analysis of a model for excitation of myelinated nerve," IEEE Trans. Biomed. Eng., vol. 23, no. 4, Jul. 1976, pp. 329-337.

[23] F. Rattay, "Analysis of models for external stimulation of axons," IEEE Trans. Biomed. Eng., vol. 33, 1986, pp. 974-977.

[24] M. Åström, E. Diczfalusy, H. C. Martens, and K. Wårdell, "Relationship between Neural Activation and Electric Field Distribution during Deep Brain Stimulation," IEEE Trans. Biomed. Eng., vol. 62, no. 2, 2015, pp. 664-672.

[25] R. Cubo, M. Åström, and A. Medvedev, "Target coverage and selectivity in field steering brain stimulation," 36th Annual International Conference of the IEEE EMBS, Chicago, IL, 2014, pp. 522-525.

[26] J. D'Errico. Inhull function. MATLAB File Exchange. URL: http://mathworks.com/matlabcentral/fileexchange/10226-inhull [accessed: 2016.06.05].

[27] R. Cubo, M. Åström, and A. Medvedev, Stimulation field coverage and target structure selectivity in field steering brain stimulation, Mov. Disord., vol. 29, no. S1, pp. S198-S199, 2014.

[28] S. F. Lempka, M. Johnson, S. Miocinovic, J. L. Vitek, and C. C. McIntyre, "Current-controlled deep brain stimulation reduces in vivo voltage fluctuations observed during voltage-controlled stimulation," Clin. Neurophysiol., vol. 121, no. 12, 2010, pp. 2128-2133.

[29] R. Cubo, M. Åström, and A. Medvedev, "Electric Field Modeling and Spatial Control in Deep Brain Stimulation," 54th IEEE Conference on Decision and Control, Osaka, Japan, 2015.

Reverse Paternalism in Medical and Clinical Engineering Practice

Sarah Hammoud and Ali Hage-Diab
Department of Biomedical Engineering
Lebanese International University
Beirut, Lebanon
11031556@students.liu.edu.lb, ali.hagediab@liu.edu.lb

Hassan Khachfe
Center for Quality Assurance, Institutional Assessment & Scientific Research (QAIASR)
Lebanese International University
Beirut, Lebanon
hassan.khachfe@liu.edu.lb

Bassam Hussein
Department of Industrial Engineering
Lebanese International University
Beirut, Lebanon
bassam.hussein@liu.edu.lb

***Abstract*— Professions are governed by ethical frameworks of moral standards each within its own professional boundaries. So often boundaries intersect and present ethical challenges that should be dealt with from the point of view of more than one professional ethical standard; such as in the case of engineering and medicine. As the medical profession depends more and more on medical equipment for treatment and diagnosis, medical decisions present ethical dilemmas to the clinical engineer responsible for the purchase and maintenance of medical equipment. One of these challenges is related to paternalism, which is the act of interfering with a person's autonomy by making decisions for them claiming that it is "for their own good". This problem is universal, but more prevalent in countries where education and awareness are limited. In Lebanon for example, ethical challenges are widely spread in the field of medicine, though they are not being discussed in the literature. Throughout time, paternalism has found its way to turn into a new form, termed and defined for as reverse paternalism. Reverse paternalism refers to the act of sacrificing one's autonomy and self-determination and giving another person or group the right for making decisions on their behalf. Some people, as moral agents, are giving up their autonomy and rights for making decisions to medical practitioners because "they know better". The main focus of this paper is therefore on reverse paternalism that will be investigated as an ethical dilemma. What exactly is Reverse Paternalism? What might be possible causations that led to its emergence? Are there regulations that restrict such kind of paternalism? Are medical practitioners to blame in such cases? And to what extent is there auditing over what happens in hospitals, private clinics, medical centers and institutions? We focus on Lebanon and try to compare it to other countries. Our aim is to shed light on this ethical dilemma and highlight how serious and wide spread it has become, by providing statistical data we have collected.**

Keywords – professionalism; medical ethics; engineering ethics; medicine; paternalism; reverse paternalism.

I. INTRODUCTION

This study builds on a previous study identifying reverse paternalism in engineering and medicine as an emerging dilemma in developing countries such as the case of Lebanon [1]. This paper introduces more information about the issue.

Different occupations and professions exist within different societies across the globe. People get educated to major in their fields of interest and decide to become professionals or not, because not all occupations are professions. A profession is a group of individuals in the same occupation voluntarily organized to earn a living by openly serving a moral idea in the most morally permissible way [2]. Lawyers, professors, scientists, doctors, engineers, etc., are all considered to constantly find themselves in positions of taking decisions that most probably will have impact on many people's lives. They are said to be the ones that 'know better', because people tend to turn to them when it comes to life-dependant decisions. Consideration of professional or non-professional occupations differ from country to country. In Lebanon, they are considered professionals, and to govern their behavior, moralities or moral standards have been arisen as guidelines for those professionals throughout their career. In general, we find three types of ethics, common morality, personal morality, and professional ethics. Common morality is the set of standards that is shared by almost everyone, personal ethics are those that contribute to moral beliefs that each person hold, whereas professional ethics are the set of standards adopted by professionals with certain characteristics [2]. These characteristics include formal codes, focus on important profession's issues, can have negative/positive dimensions, and most importantly, they take precendence over personal morality [2]. Each profession should include a basic methodology for deciding what is morally right and what is morally wrong in one's professional conduct to qualify whether an action is right or wrong. This normative ethics and principles lead to the codes of ethics that demonstrate the accountabilities and duties of each profession and when these codes are followed, the field flourishes and brings changes to the field and the world as well. Unfortunately, there is no existence of an ideal society, despite the presence of ethical codes and rules, society will face deviations from the proper standards that are stated. What would be even more challenging is the normalization of deviance from what is allowed. Normative ethics and standards are ought to be embedded in societies, so that professions are built on an ethical base. An ethical base is important in order to practice the profession in all legal, ethical and safe ways possible. As mentioned previously, there is a difference between personal ethics and professional ethics. What might be ethical in a person's opinion must not be allowable at one's workplace. Therefore, it is very important to differentiate and control one's ability to have both. Ethical codes of standards are thus an obligation to protect workers, and society as a whole, as well as keep them as safe as possible. As a professional, one has the duty to keep their society prioritized. Deviations from those standards are proportional to risk and inversely proportional to safety. The increase of deviation from proper standards increases safety risks. Whenever one is in a position of doing harm, normalizing this act by doing it numerously, there is automatically an increase of risk on themselves, as well as the ones surrounding them. This increase compromises safety mandated by the codes of ethics and standards. In other words, ethical challenges are based on deviation from ethical norms and codes. First of all, one has to identify a profession, then categorize each occupation as a profession or not, which is normally dependent on the country. In some countries, engineering counts as a profession and therefore has specific norms of codes of ethics governing this profession, and in other countries such as in the USA, engineering is not a profession. We mainly find two models of professionalism, the business model, and the professionalism model. Priorities of the business model are mainly monetary, making profit within the boundaries set by law, gain monopoly over certain services to increase profit, and persuade governmental regulators that a great deal of autonomy is granted in the workplace [2]. The professional model includes an implicit trust relationship with the public known as the Social Contract, which will be focused on thoroughly throughout our study regarding the field of engineering and medicine. The model focuses on holding paramount the public's safety, health and welfare, and might only seek monopoly if it is to protect the public from incompetent providers for example [2].

In countries where engineering and medicine are professions governed by codes of ethics, it is important to shed light on what is known as a "Social Contract". This contract is conducted between a professional such as an engineer, and between the public. It is a contract that ensures trust between them, maintaining the public's health, safety and welfare prioritized, and guaranteeing that those professionals will do what is possible to keep that trust built up. "What does an engineer do when no one is looking?" Everyone in the society that has the ability of rational decision-making is a moral agent. Moral agency leads to acts of responsibility, and a moral agent is therefore responsible for their actions. An engineer, furthermore, is ethically responsible for the decisions that are taken concerning the society. The social contract is of high importance because it includes the engineer and the people they work with. This does not only apply to engineers, but also to the individuals in society. They are moral agents as soon as they take rational decisions and take responsibility for them. Engineers are concerned with accountability for what they have done in the past, present and will do in the future. They are not only obliged to adhere to regulatory norms and standard practices of engineering, but also to satisfy the standard of reasonable care. Since there are two types of responsibility, professional responsibility and liability [2]. Any intentionally, recklessly or negligently caused harm will have consequences, and they will be held accountable or legally responsible for them. The standard of care is a demanding norm that goes beyond what is asked to be done by professional engineers. Safety, competency, efficiency, quality and responsibility are all examples included in the standards that need to be applied by professionals. What might hold engineers back from acting responsible can be attributed to self-interest, self-deception or ignorance. Engineers should not have egocentric tendencies, a microscopic vision or an uncritical acceptance of authority, that decreases the engineer's sense of personal accountability for consequences for the public. Engineers are expected to respect professional confidentiality and honesty at work, and any form of dishonesty is an impediment to being professionally trustworthy as they should be according to the social contract. Engineers have responsibilities and duties of providing what is best and beneficial for the public. In any case of harm, they are obliged to announce and make sure everyone is aware of that harm. They are in charge of alerting, informing and advising the public and avoid any conflict of interest, because health and safety of the public always comes first.

This brings us to a point where we realize how much of unethical behavior occurs in our societies. Every field faces ethical challenges, and in fields like engineering or more precisely, clinical engineering, where technology and medicine merge, ethical challenges are more likely to be encountered because of the implications that are present. Implications might be personal, such as the impediments we have mentioned previously, meaning that engineers or doctors are causing these challenges. On the other side, there can be cultural and religious factors, depending on the type of society and the traditions or religious rules they are used to or abide by, meaning that ethical challenges arise from the public itself. Despite the fact that both parties might be causing ethical dilemmas, there is no excuse of not trying to solve these dilemmas, by finding at least middle-way solutions. This means that we must first identify the problem we are addressing. The interaction of clinical engineers heads towards patients as well as doctors, and their duties regard their responsibility to the safety of both by gaining more insight of the nature of the doctor-patient relationship. A doctor-patient relationship is said to have certain characteristics governed by ethical guidelines. These guidelines are important in order to prevent ethical dilemmas that might turn into legal problems. When mentioning this kind of relationship, paternalism is often an obvious challenge noticed in many societies, areas and fields. Paternalism, the act of interfering with a person's autonomy by making decisions for them, claiming for them to be "for their own good" [2]. Paternalism can be defined but is very difficult to be outlined, as it not only depends on medical facts, but also needs to take patient's views and judgements into consideration. This is why one should pay attention to where the patients come from, what culture or religion they follow, what beliefs and traditions they have. Doctors are ought to have more knowledge about the medical conditions of their patients, but this does not imply for them to act as if "they know better" when it comes to decision-making, especially life decisions. Paternalism is therefore an act that makes doctors take decisions for their patients regardless of the reasons. Ideally speaking, doctors would make decisions for their patients in order to save their lives, but in a materialistic world reality, there is a high chance of finding people that are more self-centered and aim for their own benefits. Instead of looking out for what is best for their patients, they search for a way of benefiting from each case. Patients who put trust in their doctors, believe in the fact that they know better and are afraid of not doing as they have been told to do. This gives rise to a phenomenon that has barely being addressed and has never been identified as an ethical challenge. Reverse paternalism, is when a patient sacrifices their autonomy and self-determination by giving another person (most probably their doctor) the right for making decisions for them. There are many reasons that might make patients act that way, and it is very important to shed light on it and categorize it as being something ethically wrong. Even though physicians might have wider knowledge about the medical status, it is not for sure that their priority is always the patient. Medicine, in general, is a field that is subject to ethical complications and this makes it more vulnerable to ethical dilemmas. Reverse paternalism is another way of saying that patients are encouraging paternalism and making it easier for physicians to act that way, without even considering the possible consequences. Sometimes it is done intentionally, and very often it is done unintentionally, without noticing that one is actually compromising his or her moral agency.

A study performed in Lebanon, a rather developing country, has shown substantial proof of the presence of reverse paternalism [1]. It is of high interest, to study the phenomena in other areas as well, including developed, developing, and underdeveloped countries. It is also important to focus on the fact that engineers have responsibility towards the awareness of the public of such dilemmas, especially when related to medical fields, such as the doctor-patient relationship.

In Section II of this paper, an overview of the diversities in ethics from around the world is presented. There are differences in the background and history of the way ethical regulations have been made in Africa, Asia or the Western countries. In Section III, the main types of professional ethics are identified, starting from Bioethics, to Medical ethics, and Engineering ethics, and of course the definition and duties of the Bioethical Engineer. It provides an explanation of the responsibilities of engineers in general, and clinical engineers in specific. Section IV defines paternalism and explains the extent of it in our society, by giving examples of a specific topic (the increase of C-sections in Lebanon). It shows how paternalism is encouraged by the patients themselves and triggered clinical engineers to identify a new ethical dilemma named reverse paternalism. It also explains the decision-making process and how it should be applied. Section V comprises the importance of informed consents in the fields of medicine and engineering. The survey we have done is represented in part B of that section, which shows the quantification of reverse paternalism in Lebanon. Section VI finally closes the article with a conclusion we have made about this issue concerning Lebanon. We also gave recommendations and mentioned what our aim as future work will be considering this subject in other societies.

II. DIVERSITY OF ETHICS WORLDWIDE

Ethics can be classified into a variety of categories, either as professional and non-professional, or with respect to cultural, religious, or even moral interpretation. It is a group of principles, values, rules and regulations, beliefs, morals and rules of conduct [2]; a group that organizes either the goals, or the actions that need to be implemented for certain achievements. Ethics can also be described as a system of moral principles that differentiate between what is right and wrong, a norm of conducts that recommends concepts of acceptable and unacceptable behavior [1].

When analyzing ethics one must take into consideration that it has diverse perspectives depending on each area of interest, because ethics are local. There are countries that have their ethical norms of conduct affected by their culture rather than the main religion governing their area. Then there is western and non-western ethics, each having different ethical expectations. Studies that compare ethics in countries with different cultural dimensions, show that these dimensions could serve as predictors of the ethical standards desired in a specific society [3]. National culture plays a fundamental role in forming cultural values [3]. Ethics is the discipline that examines one's moral standards as well as the moral standards of a society. Whenever a subject is to be analyzed based on ethical standards or rules in any society, one has to take into account on what these rules are based on. As mentioned before, each society has their ethics embedded in different ideas or beliefs.

In most of Africa, a group of societies has evolved ethical systems to guide social and moral behavior. African philosophers have been evolving their values for the last three decades in order to make some contribution to the understanding of African ethical thinking. So, through their critical analyses and arguments, philosophers try to explain, sharpen, clarify or even enlarge the understanding of the concepts and issues of morality. In order to do so, one has to approach this subject by an inquiry into African moral language and search for the word 'ethics' in the different African languages. It is interesting to mention that most of the languages in Africa do not have a direct equivalent of the word 'ethics' or 'morality' [4]. But what is even more interesting and relevant to our study, is that, the African religion is traditionally characterized to be a mystic religion [4]. This indicates that African ethics is independent of religion and makes it an autonomous moral system. It is correct to regard African religion as ethical, instead of regarding African ethics as religious.

Moral personhood is attained in the later years by carrying out the obligations that transform one from the it-status of early childhood into the ethical mature personhood. Thus, one can say that the concept of a person in African thought embodies ethical presuppositions. Different societies in Africa have various definitions for the word "person". A central notion would be that an individual can be a human being without being a person. Even though, it must be noted that not being a person does not withdraw any right as a human being. Not being a person simply implies not having a good character, if used normatively. But how is a good character defined? How do African ethics define and differentiate between right and wrong? According to traditional thinkers, a good character is built up by deeds, habits, and behavior patterns considered by the society as worthwhile because of their consequences for human welfare [4]. Generosity, truthfulness, faithfulness, respect, justice, honesty, hospitality, etc., are examples of the goods that give a person a good character in the African society. This again implies that African ethics is humanist, thus a moral system that is preoccupied with human welfare. According to Monica Wilson, "The basis of morality was fulfillment of obligation to kinsmen and neighbors, and living in amity with them" [5]. An important statement was made by Edwin Smith claiming that the norm of right and wrong is custom; that is, the good is that which receives the community's approval; the bad is that which is disapproved. The right builds up society; the wrong tears it down. One is social; the other is anti-social [6].

Criteria for ethics vary from culture to culture and field to field. The south Asian culture is a conglomeration of many religions such as Hinduism, Buddhism, Islam and Christianity [7]. Thus, ethical principles are mainly influenced by the culture of the south Asian countries. In terms of medical ethics, these civilizations have been influenced lately by the Western medicine during the colonial period [7]. When mentioning the field of medical ethics, the Hippocratic Oath is what comes first to the mind as well as the tenets of the early religious healing traditions of the West. There are also several Asian traditions of ethical tenets governing the physician-patient relationship [8]. In the field of contemporary medical ethics, the first codes of ethics were developed by the doctors in the USA in the first meeting of the American Medical Association (AMA) in 1846. The impact of science and technology increased and traditional ethics have changed into an interdisciplinary field involving lawyers, historians, theologians, social scientists, physicians and other health professionals [9]. Every culture is ought to have an ordered moral system or set of norms to guide the behavior of their citizens. These are most of the times a reflection of both the nature of morality and the culture's own moral repertoire, mostly religion and theology. They have also played an active role in the enterprise of the early Greek approaches and trajectories for the ethical life and vice versa [10]. Certain movements in recent times have sought to return to the ancient Greek insights (from Aristotle to the Stoics) to avert the crisis that some writers, such as Foucault following Nietzsche, argument has been precipitated by the codes of Christian moralism and rationalism of the (European) Enlightenment [10].

When intellectuals in India come together to talk about ethics in Indian tradition they ask one very important question: "Has there ever been ethics in India?" [10]. Indian thinkers recognized morality's pervasiveness throughout human life and culture, and did not shy away from inquiry into the nature of morality of 'right' and 'wrong' or 'good' and 'bad'. In Indian philosophy, one begins with the practices that are embedded in all human cognitive and aesthetic forms [10]. Thus, we can assign India as a civilization whose roots recede into antiquity, and expected to have a variety of ethical systems within the Indian tradition. The notion of ethics in India has undergone significant shifts in meaning and emphasis over the long history of Indian philosophical speculation [10]. This is also true with respect to Western ethics where classical and modern moral philosophy is sharply distinguished by the work of Henry Sidwick in his *Method of Ethics* [10].

As a conclusion, ethics can be identified as well as defined in many ways, but what Western, Asian, African and most of the countries agree on, is that ethics serves as guidance for their societies to distinguish between 'good' and 'bad'. Differences lie in the extent of how much of culture and religion influences these ethical regulations.

III. THE MAIN TYPES OF PROFESSIONAL ETHICS

This study's main objective concerns ethics in the field of biomedical/clinical engineering, the intersection of bioethics, medical ethics and engineering ethics (Figure 1). In order to understand bioethics from an engineering perspective, the three will be defined, as well as a description of the relationship and contribution to one another, will be provided. An explanation of the ethical frame of each field will be given by demonstrating their standards and codes of ethics, in addition to their ethical relation in a clinical engineering context. Most importantly, there will be a detailed description of the applications, principles, and ethical requirements of clinical engineering, leading to the profession of bioethical engineering. A principle is a basic truth that is used as a basis for ethical reasoning by guiding a specific action or behavior, also helping in assisting moral agents in making moral decisions where its guidance is more general than that of laws [2].

A. Bioethics, Medical Ethics, and Engineering Ethics

Bioethics is an activity; it is a shared, reflective examination of ethical issues in health care, health science, and health policy [1]. It is a discussion and a relatively new field that emerged due to new medical technologies and legal cases that have thrown up ethical issues. Bioethics is a multidisciplinary, where bioethicists are clinically, legally and philosophically 'informed', by learning from doctors and other scientists that work in clinical and research areas of biomedicine. It is a discussion that is often sparked with new developments due to the enhancements of clinical scientific technologies, but also raises new questions about old issues.

Medical ethics is a system of morals and principles being applied to situations that are specific to the medical world and the practice of medicine. It formally considers the morality of medical decision-making and addresses thereby the wide ethical principles that impact not only physicians and healthcare providers, but also the patients [11]. It also concerns the code of ethics of healthcare providers. Ethics in general can be seen as systemic rules or principles that point out the right and wrong of actions, in addition to the good and bad of the motives and ends of these actions. Another definition would be that it is a moral construct focused on the medical issues of medical practitioners, stating the principles of proper professional conduct concerning the rights as well as the duties of physicians, patients, and fellow practitioners, in addition to the care of patients and in relation to their families [12]. What is important to notice, is that the history of medical ethics goes far beyond that of bioethics, since it began with the Hippocratic Oath. "To treat the ill to the best of one's ability, to preserve a patient's privacy, to teach the secrets of medicine to the next generation, etc…" [13]. Medical ethics has four commonly accepted principles excerpted from Beauchamp and Childress [14]:

1. Principle of respect for autonomy
2. Principle of non-maleficence
3. Principle of beneficence
4. Principle of justice

Engineering ethics stands for the set of ethical standards and principles ruling the behavior of engineers in their title role as professionals. The term profession has been presented in the introduction and will be explained later on in relation to bioethical engineers. A profession is motivated by either economic self-interest (business model) that makes the social practice or occupation concerned with making profits, or by ethical commitment (professional model), which makes professionals agree to regulate their practice in accordance to promoting the public good [2]. In the engineering profession, the ethical commitment of an engineer must overshadow the business and profit model because they agree to regulate themselves by high standards of technical competence and ethical practice so that their main goal remains in the area of the decent and fair of the public. The professional codes of ethics have been documented by several professional engineering societies, such as the National Society of Professional Engineers (NSPE), the American Society of Mechanical Engineers (ASME), the Institute of Electrical and Electronics Engineers (IEEE) and the American Society of Biomedical Engineers (ASBME). The idea of ethical codes is rather uniform. A code can be general "Using their knowledge and skill for the enhancement of human welfare", or specific "Engineers shall hold paramount the safety, health, and welfare of the public" [2]. According to the preamble stated in the National Society for Professional Engineers (NSPE) [2]:

Engineering is an important and learned profession. As members of this profession, engineers are expected to exhibit the highest standards of honesty and integrity. Engineering has a direct and vital impact on the quality of life for all people. Accordingly, the services provided by engineers require honesty, impartiality, fairness, and equity, and must be dedicated to the protection of the public health, safety, and welfare. Engineers must perform under a standard of professional behavior that requires adherence to the highest principles of ethical conduct.

All of which brings us back to the point that an engineer should always be devoted to the protection of public health, safety and well-being; even if it takes going beyond what an engineer is expected to do.

B. Biomedical Engineering Ethics

Biomedical Engineering is the application of engineering principles and techniques on medicine and biology [1]. It comprises design and problem solving skills of engineering with medical and biological sciences to enhance the quality of people's life, by evolving medical health care and technology. Clinical Engineering is a sub-specialty of Biomedical Engineering concerned with healthcare delivery. Each profession is ought to include a basic methodology for deciding what is morally right and what is morally wrong in one's professional conduct [1]. Ethics is therefore a central concept and not a peripheral one, because its principles guide biomedical engineers to recognize ethical problems and attempt to solve them. This is why there is a code of ethics that emphasizes the major canons for biomedical engineers, helping them to recognize, think critically and engulf the ethical problems they might face. One important thing is the responsibility they have towards their profession as well as them being fully attentive of the potential for their professional knowledge and skills to affect health and human life. Public health, safety, and well-being are paramount considerations, and ethical responsibilities incorporate those of engineers and medical practitioners. Thus, the moral obligations of medicine and biomedical engineering diverge due to the specialized nature of both practices. It should be noted that biomedical engineers are considered as indirect practitioners; the technologies and techniques they advance co-determine medical practice and affect the medical field as well [15]. Three types of ethics are manifested in biomedical engineering: *Professional ethics*, *Patient ethics*, and *Natural & Human ethics* [14]. Honesty, fairness, and not publishing false data comprise professional ethics. Honesty and confidentiality are necessary to allow engineers to conduct research with patients comprising patient ethics. Preserving the standards of nature and not crossing the line between enhancing one's quality of life and changing their traits comprise natural and human ethics. Again, Respect for autonomy, beneficence, non-maleficence, and justice, are the most widely used frameworks and provide an extensive consideration for biomedical engineering ethics while analyzing bioethical issues.

C. The Bioethical Engineer

Since the professional does profess, he asks that he trusted. The client is not a true judge of the value of the service he receives; furthermore, the problems and affairs of men are such that the best of professional advice and action will not always solve them ... The client is to trust the professional; he must tell him all secrets which bear upon the affairs in hand. He must trust his judgement and skill. [16]

Everett C. Hughes

All kinds of professions are somehow related to bioethics, especially engineers and physicians who work to enhance the quality of life, care for the public's health and safety. There is a need for bioethical sensibilities in the engineering codes of ethics such as those just mentioned. What is always expected from any professional in any field, is trust and reliability. Honesty, confidentiality and many other forms of honesty are of high importance in engineers. Society demands and puts trust in the professionals they deal with, from medical to engineering to legal and other professionals. Not only are they expected to be current and capable, but also honest, especially when undergoing medical treatment and dealing with healthcare providers, such as physicians, nurses, emergency personnel and others. Society cedes a substantial amount of trust to a relatively small group of experts; the professionals in increasingly complex and complicated disciplines that have grown out of the technological advances that began in the middle of the twentieth century and grew exponentially in its waning decades [16]. There is a contiuum among science, engineering and technology, becase many health problems require interrelated harmony among doctors, clinical engineers, and technicians [16]. Ethics related to these fields, are bioethics that govern what is clearly wrong and clearly right.

All engineering projects are commnal; there would be no computers, there would be no airplanes, there would not even be civilization, if engineering were a solitary activity. What follows? It follows that we must be able to rely on other engineers; we must be able to trust their work. That is, it follows that there is a principle which binds engineering together, because without it the individal engineer would be helpless. This principle is truthflulness. [16]

Joseph Bronowski

In order to gain the trust of a whole society, an engineer should first be able to meet the trustworthiness of the engineering community, which is articulated by the engineering profession through the codes of ethics. We have mentioned before that a good engineer is an engineer that has the traits of a professional character. A character that might go beyond what is asked from, in order to posses pride in technical excellence, social awareness and environmental consciousness [2]. Since there must be professional harmony between engineers, such as clinical engineers, medical practitioners, and physicians, the relationship should be based on honesty, trust and reliability, towards themselves and the society. In order to be trustworthy, engineers also have to be professionally responsible as well as legally. Engineers should be held accountable for not only what has been done in the past, but what will happen in the future. "What does an engineer do when no one is looking?" a social contract of professional intrinsic ethics controls this kind of contract between an engineer and society as a whole. A good clinical engineer knows that he is accountable for what is happening with patients when undergoing any kind of medical procedure. They satisfy a norm called the standard of care, which goes beyond basic job resposnibilites as defined by employment terms. According to the preamble of the code of ethics of the National Society for Professional Engineers (NSPE):

Engineering is an important and learned profession. As members of this profession, engineers are expected to exhibit the highest standards of honesty and integrity. Engineering has a direct and vital impact on the quality of life for all people. Accordingly, the services provided by engineers require honesty, impartiality, fairness, and equity, and must be dedicated to the protection of the public health, safety, and welfare. Engineers must perform under a standard of professional behavior that requires adherence to the highest principles of ethical conduct.

IV. From Paternalism To Reverse Paternalism

Professionals possess the knowledge that qualifies them to be superior in the field they work in. For example, medical doctors have more scientific and medical information, which might be a reason for them to act paternalistically. Paternalism is by definition, the interference of a state or a person with another person's autonomy. This is motivated by the claim that this interference will provide benefit or protect from harm. Paternalism can take place in different areas of our personal and public life. It can be reasonable, when being protective and it can be unacceptable, when being beneficial for the wrong party. What we will focus on is medical paternalism, because as clinical engineers it is of our concern to provide safety and health for patients that undergo medical procedures, since there is always contact with any kind of medical equipment. In addition to that, biomedical/clinical engineers are in interaction not only with doctors, nurses, clinical research departments, etc., but also patients (Figure 1). Moreover, as professional engineers, it is of our responsibility to protect and provide safety for these patients. This is why we have to make sure that paternalism does not completely incapacitate patients' ability to make decisions. There lies an even bigger problem in the fact that paternalism is being asked for. For reasons, we are trying to find out, patients are giving physicians the opportunity to make decisions for them and thereby give up their moral agency. They do not realize that in some cases physicians might take advantage of those patients who leave their choices to the doctors they trust, blindly. Even though paternalism can sometimes be the correct way of handling patients, it is ethically unacceptable to not involve them in the decision-making process.

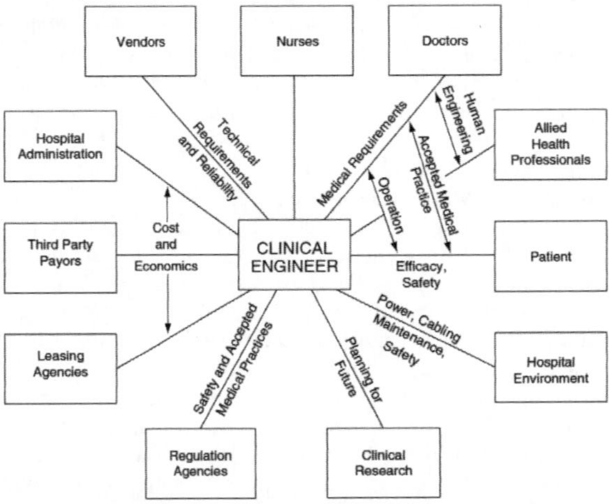

Figure 1. Diagram Illustration of the interaction of a Clinical Engineer [17].

A. *Paternalism in our society*

Every moral agent can reason, make rational decisions and form self-interested judgments based on concepts of right and wrong conduct, being therefore responsible for their actions [2]. Moral responsibility for one's actions is based on the concept of autonomy, that refers to the aptitude of a balanced individual to make an educated an un-coerced decision. In ethics, autonomy refers to a person's capacity for self-determination and decision-making in the context of moral choices and making decisions based on a course of action out of respect for moral duty [14]. Paternalistic interventions are categorized into legitimate and illegitimate. They are legitimate when the patient is incapable of making an autonomous and voluntary decision. This implies that being or acting paternalistically needs certain conditions and is not always acceptable or even allowable.

Medicine is facing a transformation for relocating the authority of decision making from physicians to patients. This is noted by a comparison of the ethical codes of the American Medical Association (AMA) in the last two centuries. In article II of the 1847 AMA ethical code entitled "Obligations of patients to their physicians", the following statement was found in Section 6 [13]:

"The obedience of a patient to the prescriptions of his physician should be prompt and implicit. He should never permit his own crude opinions as to their fitness, to influence his attention to them. A failure in one particular case may render an otherwise judicious treatment dangerous, and even fatal."

In contrast, AMA's opinion in 1990 on "Fundamental Elements of the Patient-Physician Relationship" now states a radically different position [18]:

"The patient has the right to make decisions regarding the health care that is recommended by his or her physician. Accordingly, patients may accept or refuse any recommended medical treatment."

Today, the principle of patient autonomy and self-determination has emerged as the dominant ethos in health care, threatening in many instances to totally eclipse the principle of medical beneficence [18]. Sometimes, doctors act paternalistically by not explaining in a detailed manner why a patient needs to undergo certain treatments, because in their opinion "they know better" and patients most often will not understand. A pregnant woman sees her doctor regularly and begins to trust her, so when that doctor tells her that she needs to undergo caesarean operation instead of giving birth naturally, she would believe her. What matters to her is the safety and health of her child. But what is so dangerous about this issue, is that many women do not ask why. They do not pay attention to the fact that many doctors have so much self-interest that they only care for how much money they can make out of this operation. Women do not ask for all the information they deserve to know in order to understand the advantages and disadvantages or side effects of any treatment or surgery. This kind of "marketing" is very common in Lebanon and the rate of C-section (CS) patients is rising unbelievably, most of which are performed for nonmedical reasons. C-sections were introduced in clinical practice as a life saving procedure for the mother and the baby. In 1985, the World Health Organization (WHO) stated: "There is no justification for any region to have CS rates higher than 10-15%" [19]. According to a study done in 2010, about the global numbers and costs of additionally needed and unnecessary Caesarean Sections performed per year, Lebanon is among the countries that have a rate of above 15% of C-sections with a percentage of 23.3% [20]. The rate of CSs in Lebanon is the highest in the Arab region, and the rates are still rising. In 2013, the Ministry of Public Health (MoPH) covered a rate of 44-45% CSs of total deliveries. Along with the Syrian crisis, the women delivering their babies in Lebanon by CSs is 35%, according to the UNHCR (collected data from the hospitals Syrian women were admitted to) which is higher than the rate in Syria (23%) [21]. Again showing an increase in comparison to the acceptable rate given by WHO (15%). WHO stated that they do not recommend achieving a specific rate at population level anymore, but urges focusing on the medical needs of women and the indications that necessitate the procedure [22]. In order to reduce the rate of unnecessary or repeated CSs, Vaginal Birth after CSs (VBAC) is rising in developing countries, whereas in Lebanon this rate is only of 7%. Which is relatively low and indicates that women who undergo CSs are more likely to undergo repeated CSs due to the refusal of most physicians to perform VBAC [23].

What we are addressing, is the right of patients to decide and gain full knowledge about their medical conditions. In this case, women must gain full knowledge about the risks they will have to handle when the CSs are unnecessary. They can result in major health risks and various complications for mothers and newborns, which might lead to significant burdens on health care systems [24]. Multiple systematic reviews indicated increase in adverse health consequences such as the need for antibiotic treatment [25]; neonatal intensive care unit admission [26]; blood transfusion [27]; hysterectomy [28]; and sometimes death [29], in addition to many others. Thus, acting paternalistically as a physician in such cases is not always beneficial for the patient. Factors that might lead to doctors advising women to go through CSs are the absence of national guidelines, diversity in medical schools (diversified practices and absence of unified medical standards in the field of maternal healthcare [30]), opposition of powerful stakeholders (opposing standardized regulations that aim to reduce the rate of CSs in the country [31]), absence of law, need for strengthened primary healthcare, unregulated medical practice, presence of medical insurance (women with medical insurance are more likely to have CSs than others [32]), higher hospitalization costs and benefits [33] as well as higher procedural costs and revenues [34]. Which implies that CSs ensure excessive profits for hospitals as they require higher bed occupancy and longer hospital stays that subsequently result from programmed births, which in turn increases the benefits of the hospital and leads to an increase in CSs rate. In Lebanon, the coverage of CSs by the MoPH and the National Social Security Fund (NSSF) insurance schemes and physician reimbursement are higher than those of vaginal or normal deliveries, which is of course associated with the preference of health providers to do more CSs [29]. This is evidence of practiced paternalism in Lebanon by medical physicians towards their pregnant patients of performing unnecessary CSs for the benefits of their own. Paternalism is a pure realm of applied ethics, and raises many ethical and theoretical questions. How should we think about individual autonomy and its limits? What is the trade-off, if any, between regard for the welfare of another and respect for their right to make their own decisions? When does a physician have the right to be paternalistic? We have started our study in Lebanon, which seems to be one of the countries that suffer from medical paternalism. Traditional cultures across the world empower family members and doctors alike to "protect" patients from knowing the truth of their medical condition or by ruling their medical decisions as well. In countries like Japan, Iran, Turkey, Saudi Arabia, Kuwait and Lebanon, these views are strongly tied to social norms and traditions that topple the western conception. Ethics does shed light on the concept of paternalism, in particularly medical paternalism, but what we have noticed is that there has been a reverse of paternalism coming from the patients themselves.

B. *Reverse paternalism*

As we have made clear, about the long practiced medical paternalism that doctors tend to make decisions for their patients. This is how they take advantage of their position as specialists having the power to reach personal achievements. We have identified an emerging ethical dilemma concerning the doctor-patient relationship that shows how patients themselves are encouraging the act of medical paternalism. This ethical issue is commonly present in the field of medicine in a developing country like Lebanon and termed it for the first time as reverse paternalism. Reverse paternalism refers to the act of sacrificing one's autonomy and self-determination and giving another person or group the right for making decisions on their behalf [1]. Whenever a patient refuses to choose a treatment procedure among other procedures and trusts the doctor to choose for them, we have a case of reverse paternalism. People have not acknowledged the severity of the presence of such an ethical issue yet. One of the reasons is the absence of regulations that specific to reverse paternalism that restrict physicians' unethical behavior towards their patients. There is a lack of auditing and supervision over what happens in hospitals or clinics. What was also been noticed is that patients are not usually advised to seek second opinions about certain medical treatments or procedures. Many patients show too much dependence on their doctors, others are ignorant and do not seek information about their medical situation other than what their doctors tell them, some do not have the courage to question their doctors. Most of the time patients trust their physicians too much. Patients' educational level, medical experiences, financial status and psychological states can play a huge role in enhancing this kind of paternalism. Paternalism induces power imbalance between health professionals and patients. Doctors have the medical knowledge that makes them superior to patients in making decisions as mentioned before. A decision-making process is a process of selecting a belief or a course of action among various alternative choices. There are seven main steps that highlight the importance of patients' moral agency. These steps also include the principle of an informed consent of which we will provide a detailed explanation in the following section. Patients are advised to surrender to an epistemic authority. It is the process of selecting a belief or a course of action among various alternative choices. It is very important for patients to be aware of how this can be done.

The main steps are:

1. Identify decision to be made
2. Gather relevant information
3. Identify alternatives
4. Weigh evidence
5. Choose among alternatives
6. Take action
7. Review decision and consequences

Table I, as shown in the Appendix, illustrates a comparison between how decision-making should occur and how it is done in Lebanon [1]. The description of how it is happening in Lebanon is only generalized and points out how important it is to solve this issue and search for recommendations. Reasons are numerous and most of them might be of cultural sources. There is a strong relationship of dominance and affection between the way decisions are made and the cultural perceptions. Each culture brings its own views and values to the health care system, which alters health care beliefs, health practices and the nature of doctor-patient relationships [1]. Mutual respect and appreciation of roles is the basic guideline on which a healthy relationship should be based on. Professionals should not abuse their position by manipulating or coercing patients against their will, so patients must not coerce professionals to go against their fundamental ethical convictions and professional values [35].

C. Clinical Engineers and Reverse Paternalism

Biomedical/Clinical engineers have a range of interactions in which they might be required to engage in a hospital setting (Figure 1). In cooperation with doctors, they share duties towards patients. They are involved in medical operations and accepted medical practices between the doctor and the patient to ensure efficacy and safety. Thus, clinical engineers must act in a patient-centered manner and apply engineering principles in managing medical systems and devices in the patient setting. Since engineers are responsible for decisions taken about particular designs that will affect the lives of patients and financial well-being of many people, give professional advice, they are obliged to regard responsibility towards the health and safety of patients. As biomedical engineers, the current doctor-patient relationship presented in our society, has triggered our sense of responsibility. Reverse paternalism is an ethical dilemma that interferes with decisions taken by engineers working in the medical field. It is of our duty to alert and inform the public so their moral agency keeps protected. We have therefore, for the first time, identified, quantified and discussed this issue by assigning it as a problem in various fields such as medical diagnosis and treatment in Lebanon. So if someone asks if engineers really do have patients, in order to be concerned for an issue such as reverse paternalism, the answer is clearly yes. In the fourth canon of the National Society of Professional Engineers (NSPE) it is stated that [16]:

Engineers shall act for each employer or client as faithful agents or trustees.

Furthermore, the preamble to the NSPE code affirms:

Engineering has a direct and vital impact on the quality of life for all people. Biomedical engineers design and test devices to be used to treat diseases and to ameliorate the quality of life of individual patients [16]. Thus, the real clients of clinical engineers are physicians, but being the trustee of the public, the devices and systems must hold paramount health, safety, and welfare. This makes patients indirect clients of engineers. At a minimum, engineers are part of the team that supports the physician, who in turn treats the patient. Which implies that the clinical engineer is held responsible to both, the client (physician) and patient (recipient of the engineered system) [16].

V. Informed Consents in Medicine and Engineering

We will take a closer look at the concept of informed consents. The principle of autonomy implies that a patient has the capacity to act intentionally, with understanding and without controlling influences that would influence a free and voluntary act. Which is the basis for the practice of informed consents. It is important to shed light on this process after noticing that this is still unknown to a large portion of the public.

A. The importance of informed consents

Informed consents can be defined as the process that gets a patient's permission before being subjected to healthcare interventions. The patient is requested to consent before receiving therapy, or a clinical researcher asks a research applicant before signing them up into clinical trial. It comprises a clear appreciation and understanding of the facts, implications and consequences of the specific therapy, surgery or trial as well as providing all relevant facts. A physician is obliged to give a detailed explanation of every step of a treatment, the reason and possible side effects to the patient, and thereby get their permission. This is done by providing them with a document that contains all the information the patient has to be aware of in order to accurately go through the decision-making process we have explained previously. Certainly, no information should be kept from the patient so that they are able to form a rational decision and avoid severe ethical issues arising from the lack of sufficient data. Informed consents are another way of respecting moral agents' autonomy and right of taking decisions related to their health. Mental disability, sleep deprivation, Alzheimer's disease or being in a coma, or immaturity are cases of limited moral agency, implying that other individuals are certified to give consent on their behalf, such as parents, siblings, or legal guardians of a child. Informed consents can be divided into two parts, one containing the information specific to what type of medical intervention, and the second one comprising the consent. The information component refers to disclosure of information and comprehension of what is disclosed giving the patient the chance to consider its contents in their decision-making process. The consent component refers to that the decision about to be made is voluntary and permission is given to proceed. Note that informed consents are collected according to guidelines from the fields of medical and research ethics [1]. We have found out by a survey we will provide in the next section, that many patients have never received neither heard of informed consents. Which is ethically unacceptable especially in cases of surgeries and medical interventions with probable side effects. Permission is often taken verbally if not paternalistically, and what we are concerned about is the fact that many of those cases are results of reverse paternalistic cases.

Informed consents should be seen as protection not only for the patients, but physicians as well. It is evidence that patients are aware of all the possible outcomes and have fully understood what and why is going to happen, and what might happen if this intervention is not taking place. However, many practitioners believe that patients may thus be better served if efforts are directed instead of finding ways of minimizing hard paternalism without too great of compromise on patient's freedom [18]. This argument is yet to be validated from an ethical perspective.

B. Survey

The purpose of this survey is to examine the presence of the suggested phenomenon in Lebanon and to assess to what extent it is present in the field of medicine. Based on the results we may indicate if it is spread out in the Lebanese society. If yes, the next step would be to alert the public and suggest some regulations to restrict this kind of paternalism.

The following hypothesis is formulated to achieve the objectives of the present study: A new kind of paternalism is emerging in the field of medicine in Lebanon, termed as reverse paternalism.

The study was conducted on a representative sample of 85 patients in the region of Beirut. The patients are a selection of males and females with diversity in age and education (Table II, as shown in the Appendix, represents a sample of the questionnaire). The questionnaire consists of 20 items each with five alternative responses: strongly disagree, disagree, neutral, agree, and strongly agree. The items are related to the following concepts:

1. Patient's autonomy
2. Decision making process

The questionnaire comprises a variety of questions that refer to a paradigm of reverse paternalism or the absence of reverse paternalism, as well as a neutral point of view. Figures 2-a, b, c and d give an illustration of the age, gender, marital status, employment status, and educational level. A total of 85 patients have answered 20 questions. Each question was analyzed in order to categorize it. Questions 1, 4, 6, 7, 8, 10, 12 and 15 are direct questions referring to reverse paternalism. The questions can be separated into two types, 10 positive questions and 10 inversed questions (meaning the opposite of the positive ones). As Figure 3 shows, 85% ask their doctors about suggested treatments or procedures. 70% disagree with their doctors not involving them in decisions about their treatments (Figure 4). Only 53% do not allow their doctors to choose on their behalf, 19% have a neutral opinion, which means that 28% allow their doctors to take decisions for them (Figure 5). Figure 6 illustrates how 38% agree that it is ethically permissible for doctors to act paternalistically with their patients. When asked if they refuse to let their doctors take decisions for them, 44% disagreed (Figure 7).

As most of the results indicate the existence of weak reverse paternalism, we took a closer look at the age and educational level of those who showed tendency towards reverse paternalism. Some of the patients that are in the age of 40-60 years have a lower educational level, due to the complications of war Lebanon has faced, also showing tendency towards reversing paternalism. We chose a patient to ask about his last visit to a doctor. This patient is a married employed male of age between 40-60 years with an elementary educational level. He was asked about how much he trusts his doctor and how much he believes in what he prescribes as treatments. He agreed on telling us what his problem was and what was prescribed, and when we asked him if he knows what each drug is for he said no: "He is a very good doctor and I am sure he knows what is best for me to get better." Again, we took a closer look at the questionnaire this patient has filled and noticed that they do not quite match the way he really acts.

Figure 2. An overview of the personal information and diversity of surveyed patients, in terms of (a) age, (b) gender, (c) marital status, and (d) employment status.

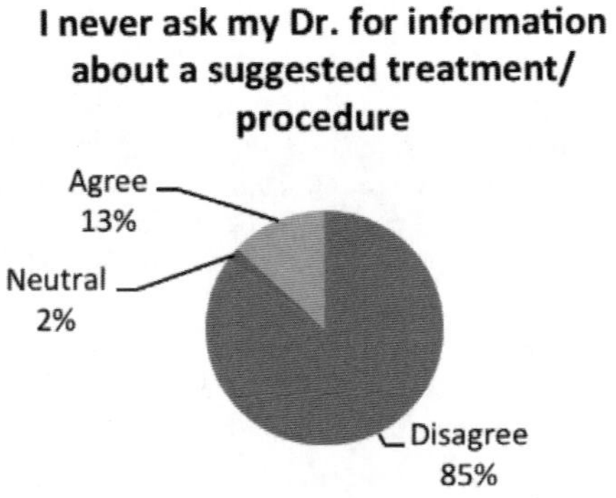

Figure 3. Illustration of how many patients ask for information.

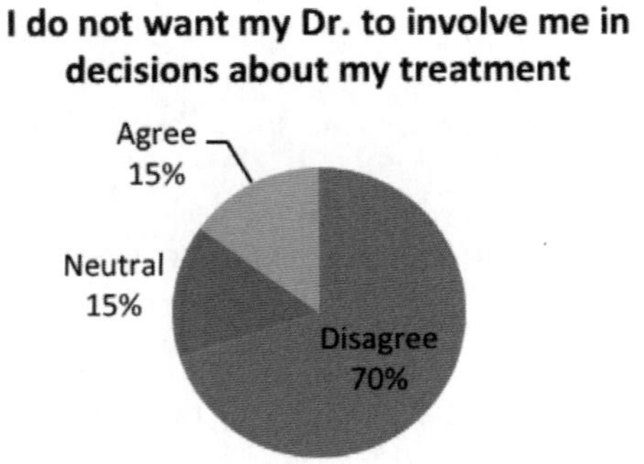

Figure 4. Patients' disagreement on not being involved in decision-making processes.

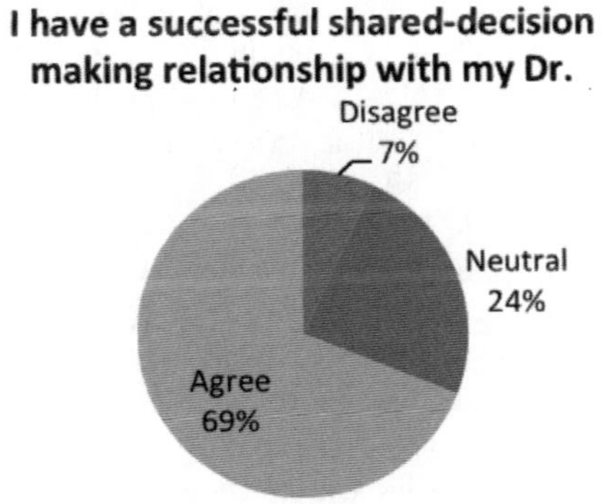

Figure 5. Patients that say that they have a successful shared-decision making process with their doctors.

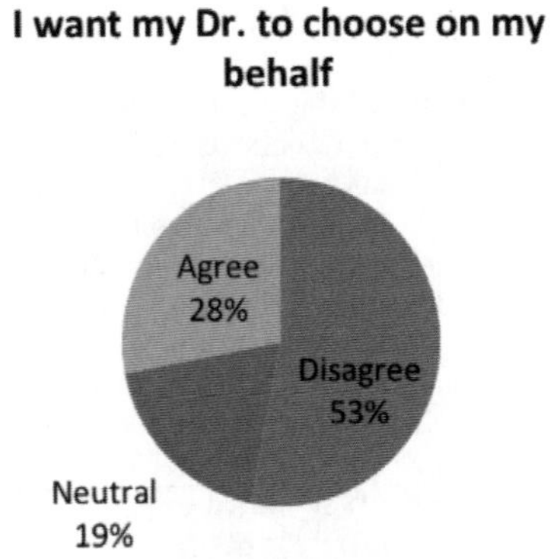

Figure 6. Patients that want their doctor to choose and decide on their behalf.

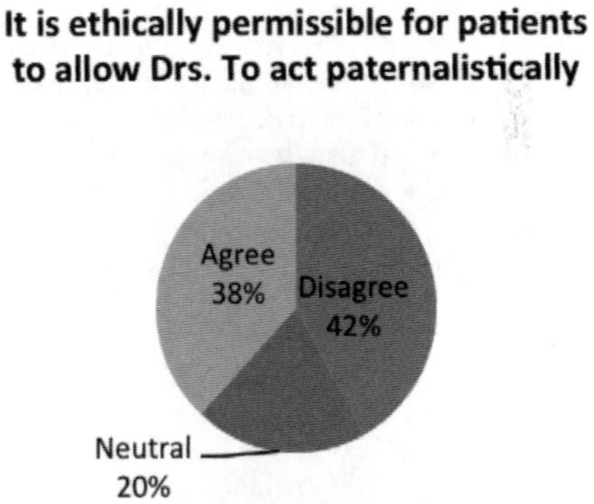

Figure 7. The percentage of patients who find it ethically permissible for doctors to act paternalistically.

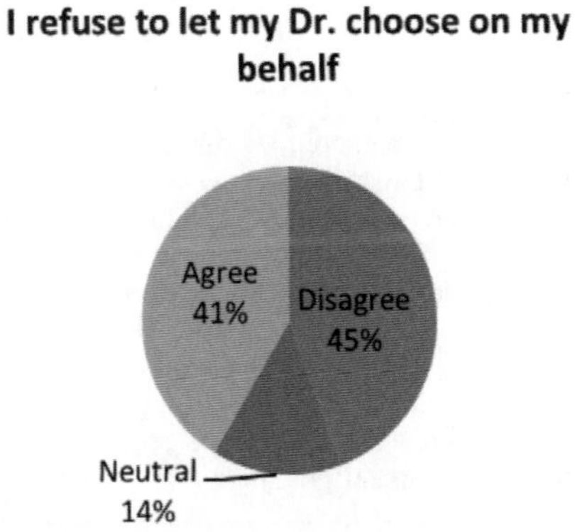

Figure 8. The patients that refuse to let their doctors choose on their behalf.

VI. Conclusion

Living in a society where professionals are considered trustees most of the public believe in, it is important to pay attention to the ethical issues that may be encountered. Patients trust their doctors sometimes too much, which can lead to the emergence of new phenomena in ethics. Paternalism is a long-known dilemma that is widely present in our society, but what we have noticed is that paternalism is not as hard to be achieved as it once was. By encouraging it by patients, doctors find it much easier nowadays. This refers to a new ethical dilemma termed as reverse paternalism. When we first started to gather information we had to look at it from different perspectives. Is paternalism a bad thing to do? Can reverse paternalism have positive aspects? So, we started interviewing people and with time had a clearer view at what is really happening in our society. We already knew that culture and religion play major roles in the decision-making processes patients undergo, and knew that there are many other factors that make patients tend to trust their doctors sometimes blindly. Though the outcome of the survey showed weak reverse paternalism according to the collected and compiled data, but the fact that it is present is a problem itself. The absence of regulations specific to reverse paternalism that can restrict the physicians' unethical behavior towards patients is one of the most difficult problems to solve. It is difficult in a society where doctors take advantage of their patients, thus a society that is losing faith in humanity. In Lebanon, medical practitioners lack the sense of responsibility due to the lack of auditing and supervision over what happens in hospitals/clinics. Who is to blame? The doctors or the patients themselves? A good test for their responsibility is the question "Do physicians commit to ethical or legal standards when there is no supervision?" and it seems that most doctors fail this test! It is the same question that must trigger the consciousness of engineers when asking "What does an engineer do when no one is looking?" we must always remember that there is a social contract between the public and us, promising health, safety and welfare.

As we have mentioned, some of the questions were answered in ways patients only "wish" to act in real life. But what is important is that it started to raise awareness and open the eyes of those who were involved in our study. As engineers it is our responsibility to alert people, inform them about the challenges they might face and advise in order to help. We have started in Lebanon and wish to reach other countries where this might be happening too.

Thus, recommendations must be provided to control this ethical dilemma. Ethical guidance that governs the behavior of doctors and patients in cases of reverse paternalism should be developed. Highlighting the importance of consent before any medical intervention is another recommendation. This can be done by the organization of seminars for patients to raise their awareness of having the right to get all the information they need as moral agents. This is an ethical issue that should not only be acknowledged in Lebanon, but in all societies that suffer from reverse paternalism.

References

[1] S. Hammoud, Z. Shahrour, A. Hage-Diab, H. Khachfeh, and B. Hussein, "Emerging Issues in Medical and Engineering Ethics: The Case of Reverse Paternalism in Lebanon," *The Fourth International Conference on Global Health Challenges (GH 2015) IARIA*, Nice, France, 2015.

[2] C. Harris, M. Pritchard, and M. Rabins, Engineering Ethics Concepts & Cases, California: Wadsworth-Cengage Learning Belmont, 2009.

[3] R. Alas, "Ethics in countries with different cultural dimensions," *Journal of Business Ethics,* no. 69, pp. 237-247, 2006.

[4] K. Gyekye, "African Ethics," The Stanford Encyclopedia of African Philosophy, 2011.

[5] M. Wilson, Religion and Transformation of Society, Cambridge: Cambridge: Cambridge University Press., 1971.

[6] M. J. McVeigh, God in Africa, Cape Cod, MA: CLade Stark, 1974.

[7] R. A. Kallivayalil and R. K. Chadda, "Culture, ethics and medicine in South Asia," *International Journal of Person Centered Medicine,* vol. 1, no. 1, pp. 56-61, 2011.

[8] P. N. Desai, "Medical Ethics in India," *Journal of Medicine and Philosophy,* vol. 13, no. 3, pp. 231-255, 1988.

[9] R. M. Veatch, "How philosophy of medicine has changed medical ethics," *Journal of Medicine and Philosophy: A Forum for Bioethics and Philosophy of Medicine, 31 (6),* pp. 585-600, 2006.

[10] P. Bilimoria, J. Prabhu, and R. M. Sharma, Indian ethics: Classical traditions and contemporary challenges (Vol. 1), Ashgate Publishing, Ltd, 2007.

[11] "American Medical Association and New York Academy of Medicine, Code of medical ethics," in *Ludwig & Company*, 1848.

[12] K. D. Clouser, "What is Medical Ethics?," *Annals of Internal Medicine,* vol. 80, no. 5, pp. 657-660, 1974.

[13] L. Edelstein, The Hippocratic Oath, text, translation, and interpretation., 1943.

[14] T. L. Beauchamp and J. F. Childress, Principles of biomedical ethics, Oxford University Press, 2001.

[15] P. Brey and J. P. Berg-Olsen, Biomedical Engineering Ethics, Blackwell, 2009.

[16] D. A. Vallero, Biomedical Ethics for Engineers: Ethics and Decision Making in Biomedical and Biosystem Engineering, Elsevier, 2007.

[17] J. Dyro, Clinical Engineering Handbook, Academic Press, 2004.

[18] J. J. Chin, "Doctor-patient relationship: from medical paternalism to enhanced autonomy," *Singapore medical journal*, vol. 43, no. 3, pp. 152-155, 1996.

[19] World Health Organization, "Appropriate technology for birth," in *Lancet 2(8452): 436-7*, 1985.

[20] R. Jurdi and M. Khawaja, "Cesarean Section Rates in the Arab Region: a cross-national study," *Health Policy and Planning,* vol. 19, pp. 101-110, 2004.

[21] K. Huster, "Cesarean Sections among Syrian refugees in Lebanon since December2012/January 2013 to June 2013: Probable causes and recommendations," *The Yale journal of biology and medicine,* vol. 3, no. 87, pp. 269-88, 2013.

[22] F. Richard and V. De Brouwere. Non-clinical intervention for reducing unnecessary caesarean section: RHL commentary. The Who Reproductive Health Library; Geneva: World Health Organization. 2012.

[23] World Health Organization - WHO, 2010. The Global Numbers and Costs of Additionally Needed and Unnecessary Caesarean Sections Performed per Year: Overuse as a Barrier to Universal Coverage. [Online]. Available from: http://www.who.int/healthsystems/topics/financing/ 2016.5.30.

[24] R. Porreco and J. Thorp, "The cesarean birth epidemic: Trends, causes, and solutions," *American Journal of Obstetrics and Gynecology,* pp. 369-374, 1996.

[25] F. M. Smaill and G. M. Gyte, Antibiotic prophylaxis versus no prophylaxis for preventing, The Cochrane Library., 2010.

[26] M. E. Hannah, W. I. Hannah, S. A. Hewson, E. D. Hodnett, S. Saigal, and A. R. Willan A. R., "Planned caesarean section versus planned vaginal birth for breech presentation at term: a randomised multicentre trial," *The Lancet, 356 (9239),* pp. 1375 - 1383, 2000.

[27] D. J. Rouse, C. MacPherson, M. Landon, M. W. Varner, K. J. Leveno, A. H. Moawad, and B. M. Mercer, "Blood transfusion and cesarean delivery," in *Obstetrics & Gynecology, 108(4)*, 2006, pp. 891-897.

[28] C. M. Rooney, A. T. Crawford, B. J. Vassallo, S. D. Kleeman, and M. M. Karram, "Is previous cesarean section a risk for incidental cystotomy at the time of hysterectomy?: A case-controlled study," *American journal of obstetrics and gynecology, 193(6),* pp. 2041 - 2044, 2005.

[29] J. Kacmar, L. Bhimani, M. Boyd, R. Shah-Hosseini, and J. Peipert, "Route of delivery as a risk factor for emergent peripartum hysterectomy: A case-control study," *Obstet Gynecol, 102(1),* pp. 141-145, 2003.

[30] M. Carayol, A. Zein, N. Ghosn, C. Mazaubrun, and G. Breart, "Determinants of caesarean section in Lebanon: Geographical differences," in *Paediatric and Perinatal Epidemiology*, 2008, pp. 136 - 144.

[31] T. Kabakian-Khasholian, R. Shayboub, and F. El-Kak, "Seeking Maternal Care at Times of Conflict: The Case of Lebanon," *Health care for women international, 34(5),* pp. 352-362, 2013.

[32] H. Tamim, S. El-Chemali, A. Nassar, A. Aaraji, O. Campbell, A. Kaddour, and K. Yunis, "Cesarean Delivery Among Nulliparous Women in Beirut: Assessing Predictors in Nine Hospitals," in *Birth* , 2007, pp. 14-20.

[33] K. H. Hsu, P. Liao, C. Hwang, "Factors Affecting Taiwanese Women's Choice of Cesarean Section," *Social Science and Medicine; 66(1),* pp. 201-9, 2008.

[34] A. Mroueh, "The maternal healthcare in Lebanon: Problems and challenges," (Arabic version of a published report), 2011.

[35] J. Wyatt, "Medical Paternalism and the Fetus," *Journal of Medical Ethics,* vol. 27, no. 2, pp. ii15 - ii20, 2001.

APPENDIX

Table I. The Decision Making Process.

Process steps	Description of each step	Application of each step in Lebanon
Identify decision to be made	Go through an internal process of trying to define clearly the nature of the decision you must make.	Patients realize that there is a decision to be made but instead of going through an internal process, they immediately ask their physicians for advices and what to do.
Gather relevant information	Most decisions require collecting pertinent information. The real trick in this step is to know what information is needed the best sources of this information, and how to go about getting it. Some information must be sought from within you through a process of self-assessment; other information must be sought from outside yourself-from books, people, and a variety of other sources. This step, therefore, involves both internal and external "work".	Many people do now know where to look for information or whom to ask. Others try to get information from people with similar experiences instead of researching properly. The process of self-assessment is sometimes not clear to certain patients.
Identify alternatives	Through the process of collecting information you will probably identify several possible paths of action, or alternatives. You may also use your imagination and information to construct new alternatives. In this step of the decision-making process, you will list all possible and desirable alternatives.	Many patients ask their physicians for alternatives, but do not know where to look for information other than their healthcare practitioners, which is the same problem found in step 2.
Weigh evidence	You draw on your information and emotions to imagine what it would be like if you carried out each of the alternatives to the end. You must evaluate whether the need identified in Step 1 would be helped or solved through the use of each alternative. In going through this difficult internal process, you begin to favor certain alternatives, which appear to have higher potential for reaching your goal. Eventually you are able to place the alternatives in priority order, based upon your own value system.	The challenge in this step is that many patients do not even reach this step. But helping them reach this point would make it easier for them to be able to imagine themselves in certain situations.
Choose among alternatives	Once you have weighed all the evidence, you are ready to select the alternative, which seems to be best suited to you. You may even choose a combination of alternatives.	What is done here, is that most patients only take into account the alternatives their physicians have told them, so when left with a number of alternatives they are lost when confronting decisions on their own. (Only if physicians haven't been paternalistic when implying what alternative to choose).
Take action	You now take some positive action, which begins to implement the alternative you chose in Step 5.	This is where patients return to reverse paternalism and let their health care practitioners choose what alternative to choose and implement.
Review decisions and consequences	In the last step you experience the results of your decision and evaluate whether or not it has "solved" the need you identified in Step 1. If it has, you may stay with this decision for some period of time. If the decision has not resolved the identified need, you may repeat certain steps of the process in order to make a new decision. You may, for example, gather more detailed or somewhat different information or discover additional alternatives on which to base your decision.	This depends on what type of decision was made. If the decision has not resolved the identified need, if a surgery has not been successful, patients often blame their physicians. These physicians however, have been told to decide for them, which is why shared-decision making is of highest importance.

Table II Patient Questionnaire

Answer questions as they relate to you.
Check the box(es) that are most applicable to you.

8) About You

a) 1. Your Age

- ☐ Below 15
- ☐ 15-20
- ☐ 20-40
- ☐ 40-60
- ☐ Above 60

b) 2. Your Gender

- ☐ Female
- ☐ Male

c) 3. Your Marital Status

- ☐ Single
- ☐ Married
- ☐ Previously Married

d) 4. Your Employment status

- ☐ Student
- ☐ Employed
- ☐ Unemployed
- ☐ Retired

e) 5. Your Educational level

- ☐ Elementary
- ☐ Intermediary
- ☐ High School
- ☐ College

9) Doctor-patient Relationship

Please complete the following questionnaire by circling the appropriate answer.

	Strongly Disagree	Disagree	Neutral	Agree	Strongly Agree
I never ask my doctors for information about a suggested treatment/procedure	1	2	3	4	5
I seek multiple opinions before selecting a surgery/treatment	1	2	3	4	5
I am confident that my doctors provides me the best treatment	1	2	3	4	5
I don't want my doctor to involve me in decisions about my treatment	1	2	3	4	5
I have a successful shared-decision making relationship with my doctor	1	2	3	4	5
I want my doctor to choose on my behalf	1	2	3	4	5
Doctors know best for patients and they have to decide for them	1	2	3	4	5
It is ethically permissible for patients to allow doctors to act paternalistically	1	2	3	4	5
In critical cases I prefer my doctor to choose on my behalf	1	2	3	4	5
I trust my doctor in everything he/she says because he/she is well-known to be the best in his/her field	1	2	3	4	5
I always ask my doctor for information about a suggested treatment/procedure	1	2	3	4	5
It is not necessary to seek multiple opinions before selecting a surgery/treatment	1	2	3	4	5
I don't trust my doctor's ability to provide the best treatment for me	1	2	3	4	5
I want my doctor to involve me in decisions about my treatment	1	2	3	4	5
My doctor-patient relationship lacks a successful shared-decision making process	1	2	3	4	5
I refuse to let my doctor choose on my behalf	1	2	3	4	5
Even though doctors know better, they don't have the right to choose for patients	1	2	3	4	5
It is not ethically permissible for patients to allow doctors to act paternalistically	1	2	3	4	5
I prefer to take all my medical decisions by myself	1	2	3	4	5
I don't trust my doctor completely just because he/she is known to be the best in his/her field	1	2	3	4	5

Monitoring the use of impaired hand by a new low cost device during daily life activities with a real-time visual feedback

Nour El Hoda Saleh[1], Ali Hage-Diab[2], Ghadir Salhab[1], Bilal Debs[2], Mohamad Hajj-Hassan[2], Hassan Khachfe[2], Ahmad Rifai Sarraj[1], Soha Saleh[1,2,3]

1. Department of Physical Therapy, School of Public Health, Lebanese University, Beirut, Lebanon
2. Department of Biomedical Engineering, Lebanese International University, Beirut, Lebanon
3. Human Performance and Engineering Lab, Kessler Foundation, West Orange, NJ

nour_40@hotmail.co.uk, ali.hagediab@liu.edu.lb, ghadirsalhab@hotmail.com, bilal.debs@hotmail.com, mohamad.hajjhassan@liu.edu.lb, hassan.khachfe@liu.edu.lb, ahmadrifai@optl.org, saleh.soha@gmail.com

Abstract— **Hand movement tracking devices are important for monitoring impaired hand function during daily life activities. The study presented the design of a finger movement evaluation device to record hand movement in daily life activities. It also investigates the feasibility of using such a device for rehabilitative purposes. Finger Movement Evaluation Device (FMED) was developed for these purposes and was tested on six stroke subjects who used it at home for two days. Based on the results of the feasibility study and subjects' request to use the device for a longer period of time, a home-based therapy pilot study on one stroke survivor was done. Results show a high patient acceptance of using FMED, with an ability to use acquired data to extract quantitative information about finger movement. However, the clinical training trial shows that the first version of the device is not practical for intensive use and its performance is not stable. These preliminary findings lead to designing a new version of the device using different hardware and setup.**

Keywords- data glove; finger movement impairment; stroke; home-based therapy

I. INTRODUCTION

This paper extends the paper [1] that was presented at the 4th International Conference on Global Health Challenges. In the field of stroke rehabilitation, evaluating the use of impaired hand during daily life requires the use of motion-tracking devices like the Finger Movement Evaluation Device [1]. Stroke is the major cause of neurological disability all over the world [2]. However, upper extremity (UE) motor impairment, specifically hand paresis, is the most disabling and persisting residual impairment after this event [3] and it is evident that it limits basic activities of daily living [4]. For this reason, the role of stroke rehabilitation is to promote independence in daily life activities [5]. Moreover, the use of outcome measures (OMs) in neurological physical therapy is essential to evaluate the improvement of function during rehabilitation [6]. Therefore, an essential issue in the assessment after stroke is to determine how much of the impairment of upper extremity is the source of loss of function, and if the selected rehabilitation intervention improves the daily activities of stroke survivors.

Numerous standardized clinical measures are available for clinicians to evaluate UE function after stroke. However, these measures are rarely used in clinical practice because of time constraints, high level of difficulty, lack of equipment, and lack of knowledge regarding OMs [7]. Besides, most of these measures do not collect information about the use of UE in the Activities of Daily Living (ADLs) and do not provide clinicians with quantitative and objective information about patients' use of impaired limb during the day [8]; thus, they do not reflect how patients function in their daily life and real world [9]. The use of assessment tools at home and community is essential for evaluating UE function during daily activities in order to improve therapeutic intervention and avoid having patients stop using the impaired limb due to pain or absence of confidence and eventually lose the ability to use it due to the learned non-use phenomena [10].

Wearable measurement devices and home monitoring devices provide clinicians with additional assessment opportunities such as collecting hand posture and movement data when individuals perform daily activities outside the clinic [11]. Despite their importance in measuring the fingers' range of motion during dynamic tasks [12], different limitations exist; they are expensive, heavy and uncomfortable to be worn in daily life outside the clinic [13, 14] and do not provide long-term monitoring [13].

Preliminary research in the area of hand glove devices has focused primarily on testing protocols that evaluated the characteristics of glove devices [11-13, 15-18]. None of them has explored the use of these devices to monitor the impaired UE function during daily life.

This article describes the design, development, and testing of a low-cost device for the assessment of finger movement during daily activities. Section II provides a review of evaluation measures of hand function after stroke. Section III describes the device design and implementation. Section IV describes the methodology of feasibility studies. Sections V elaborates on the results and discussion, followed by conclusions provided in Section VI.

II. Evaluation Measures of Hand Function After stroke

A. Clinical Measures

Numerous clinical measures are disposable to clinicians to use in clinic for the evaluation of upper extremity functions after stroke, that either measures self-report or performance. Performance measures include different clinical tests, frequently the Action Research Arm Test (ARAT) [20], Box and Blocks Test (BB) [21], Chedokee Arm and Hand Activity Inventory (CAHAI), Jebsen-Taylor Hand Function Test (JTT), [22] Nine- Hole Peg Test [23], and the Wolf Motor Function Test (WMFT) [24]. The most cited self-report measures include the Stroke Impact Scale (SIS) and the Motor Activity Log (MAL) [25]. However, all of these clinical measures of motor function do not provide data about how the person functions in their daily life [19].

B. Quantitative Tools

Probably the most commonly used evaluation procedure is the measurement of joint range of motion (ROM) [26] using mechanical or electronic goniometers [12]. The range of motion (ROM) is defined by the ability to move the joint(s), and can be evaluated as active and passive ROM [27].

The complex structure of the hand negatively affects the accuracy of any hand ROM measurement [12]. Moreover, goniometry is related to static ROM measurement, but the hands are used principally in complex and dynamic tasks. Hence, ROM measurements using goniometry cannot predict the effective ability of the hand to perform functional tasks [28].

C. Hand Movement Data Gloves

In order to overcome the limitations of clinical measures and traditional goniometry and understand how individuals interact in the real world, there is a need for quantitative measures of finger joint motion over longer periods of time at home [15]. In principle, the use of glove devices establishes an objective procedure to measure hand function independent of examiner subjective interpretation [29]. Examples of data gloves in the market include the Cyber glove (Immersion Corporation, San Jose, CA), the Data-Glove Family (Fifth Dimension Technologies (5DT), Irvine, CA), the SIGMA Glove [16], the Human Glove [12], the shadow monitor [28, 30, and 15], the Wü glove [31], the Smart Glove [17], and the Neuro-Assess Glove [18].

Some of these systems are commercially available but are not feasible for use on individuals with severe hand and finger impairments and/or neurological disorders. This is due to the fact that these devices are too complex (some require specific software and extra accessories) and unaffordable to be owned by individuals for personal use [15].

D. Home-Based Therapy and Visual Feedback

Home-based upper limb therapy can be more beneficial than conventional therapy used in rehabilitation centers. Theoretically, home-based rehabilitation permits a repeated practice of occupationally embedded tasks in the individual's own environment [35]. This is perhaps more advantageous than hospital-based or outpatient treatment in accordance with the "specificity of learning" principle [36], which predicts that the learning of a new skill is improved when conditions of practice match those of the task in real life [35]. The FMED can be used in home-based upper extremity rehabilitation therapy as it is low-cost, lightweight, and easy to use. Furthermore, it is equipped with LEDs, which provide visual feedback, and can track the use of the subjects' upper extremities during daily life and supervise the exercises of home-based therapy to ensure that patients are following the instructions of their clinicians at home. The FMED's entity with visual feedback makes it suitable for home rehabilitation in that it engages and motivates patients during their exercises at home. Thus, an evaluation of the efficacy of the use of this device in the home-based stroke rehabilitation is essential. In addition, there are indications that integration of augmented feedback and exercises can stimulate the learning process in rehabilitation therapy by making patients more conscious of their performance [37]; hence, the use of FMED in therapy programs can add an important value to the rehabilitation efficacy.

III. The FMED : An overview

A. Finger Movement Evaluation Device (FMED)

Figure 1 represents the first prototype of the Finger Movement Evaluation Device worn by a volunteer. FMED was designed to act as an offline electronic goniometer that measures the angle of finger flexion of two joints simultaneously. The device includes two bending sensors (SpectraSymbol®, UT, USA) that can be placed on two fingers' MCP joints (for example, index and middle fingers) at a time using Velcro™. Only two joints can be tracked with this prototype in order to reduce the cost of the device and allow the user to focus on two fingers at a time with the freedom of choosing which joints to track. Figure 2 illustrates the main parts of the device. A voltage regulator circuit was implemented to downregulate the power from a 12V (6500 mAh) rechargeable battery to 5V. A dc-dc converter was implemented before the voltage regulator to convert 12V to 7V in order to avoid too much heat dissipation in case the voltage regulator downregulates from 12V to 5V. A charged battery (12V, 6500 mAh) was used to power the device for more than 48 hours.

The microcontroller (ATmega2560) reads input from the bending sensors through a voltage divider signal conditioning circuit. This is a low-power complementary metal–oxide–semiconductor (CMOS) 8-bit microcontroller that supports a real Read-While-Write Programming mechanism. The microcontroller processes the data and saves the values of each joint on an SD card in real time. The raw data is saved on the SD (Secured Digital) card in addition to the fractionation angle (difference in angle between the two joints). This device also gives patients a real-time feedback of their movement using a set of light-emitting diodes (LEDs) indicating the level of finger flexion in increments of 10° (10°, 20°, 30°, 40°, 50°, 60°, 70°, 80°, 90°) and fractionation in increments of 5° (5°, 10°, 15°, 20°,

25°). The main electronics of the device were chosen to be surface-mounted in order to reduce the size and weight of the device.

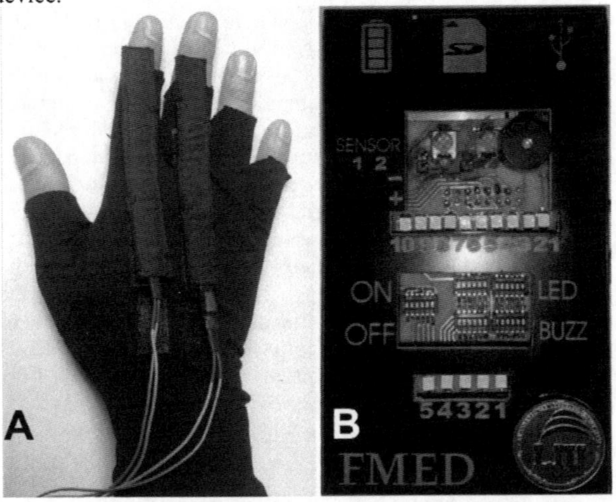

Figure 1. Prototype 1 of FMED. A. Glove worn by a volunteer. B. Device Hardware showing the set of LEDs for real-time visual feedback

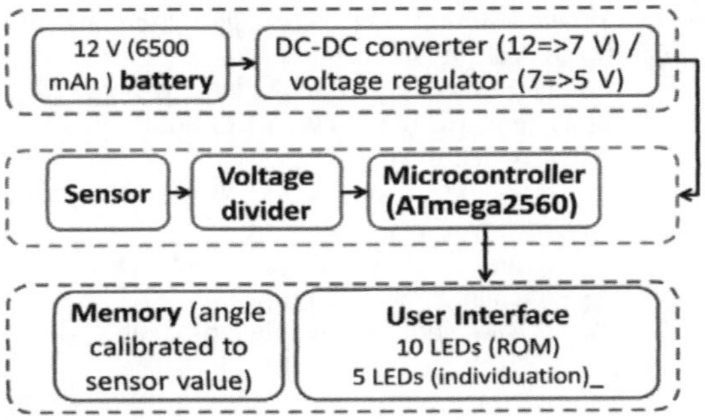

Figure 2. Main components of FMED

B. Characterization of the Sensor

Two procedures were used to characterize the linear relationship between sensor resistance and bending angle.

Test 1: A healthy volunteer who bends his index finger 0 degrees, 45 degrees, and 90 degrees wore the glove. Data was collected 4 times at each angle to check the repeatability of resistance at these fixed positions. The results show instability in the performance of the sensor on the mid-angle, which is around 45^0 (See Figure 3).

Test 2: A customized hardware was developed to read the accuracy and repeatability of the bending sensors. The device specifically mimics the actions of a finger joint. A stepper motor, a protractor, and flex sensor holder were assembled to work as a hinge. Materials needed for this testing device (Flex Sensor Testing Device, FSTD) are stepper motor, stepper motor driver, sensor and motor holder, and a protractor. The stepper motor uses a 12V input and allows rotation with increments of 2 degrees. The stepper motor driver is a driver board that includes three main electronic chips: the NE555 precision timer, L297 stepper motor controller, and L298 dual full-bridge driver.

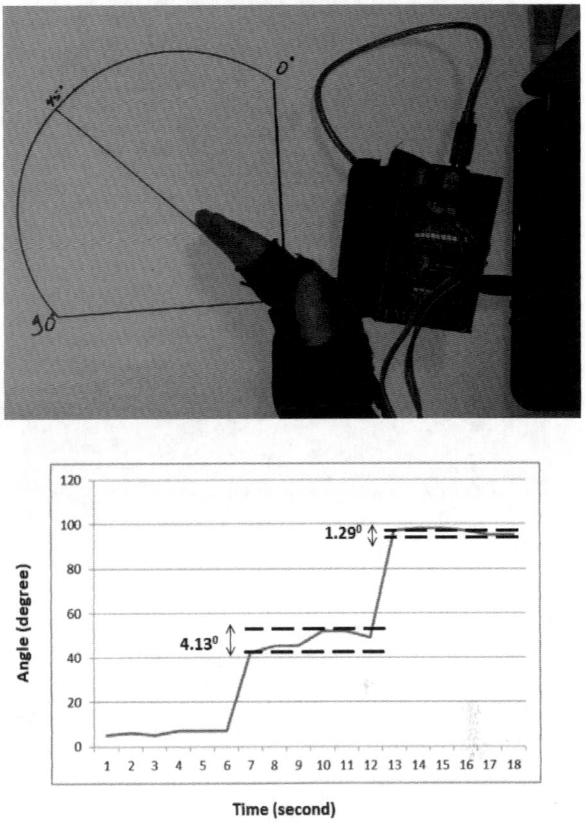

Figure 3. Top. Testing the sensor values at three angles, 0, 45, and 90 degrees. Bottom. Results of one trial, the sensor's values are not stable at 45 degrees.

The device was used to test the relationship between collected data (based on sensor impedance) and bending angles from 0 to 90 degrees (by increments of 2 degrees). Five trials were collected for the two used sensors. The results with the best and worst linear relationship between recorded value and protractor angle are shown in Figure 4.

C. Collected Data

As mentioned in the last section, the device saves the values of the variation of angles for sensors 1 and 2 (for example, index and middle finger flexion angles) and the individuation (difference between angles recorded by sensor 1 and sensor 2) versus time. A customized Matlab® algorithm was written to process this data. The first processing step was to calculate differences between adjacent elements of the dataset in order to detect movement

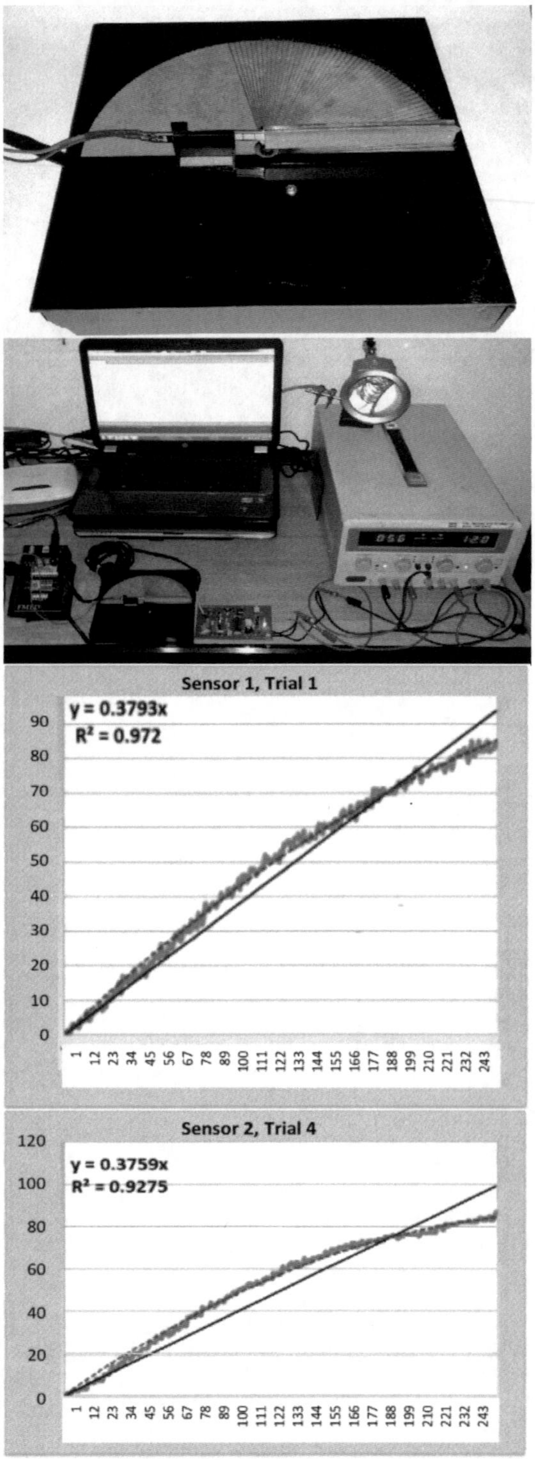

Figure 4. Upper two panels show the customized device and the setup to record sensor value at a range of 0 to 90 degrees. Lower two panels show the results of two trials.

episodes (change in flexion and individual angles). A threshold of 2° was used to count the episodes of movement (flexion angle exceeding a 2° predefined threshold). The ratio of counted samples over the whole dataset provided the Ratio of Movement (RaOM) values of each finger.

The other parameter that was calculated is related to the mean of difference in the angle between the two fingers (individual finger movement). Episodes or consecutive samples where there was a difference in flexion between index and middle fingers were reported. These episodes were counted to derive the Integral of Individuated Movement (IIM) episodes value. This value indicates how much the subject was moving the index finger independent of the middle finger and vice versa. IIM reflects how much the patient is capable of controlling one finger independent of the other during executing a functional task. The parameters (RaOM and IIM) calculated based on the recorded data were used as the main outcomes of FMED to effectively quantify the amount of movement during the day.

IV. Using FMED in Home-Based Therapy

A. *Feasibility Study*

1) *Subjects*

Subjects with stroke were recruited from multiple rehabilitation centers in Beirut, Lebanon. Six individuals with a clinical diagnosis of stroke in the chronic phase (three males and three females, mean age 49.33±8.1 > 6 months post-stroke) participated in the study.

Subjects were included because they had residual upper extremity impairments (Upper extremity Fugl-Meyer [FM] scores, with a range of (45-56)/66; and with mean flexion fingers ROM ± 1 standard deviation: 73.3 ±7.4 degrees). Table I presents the patient's demographic and clinical characteristics.

Inclusion criteria were chosen to give a nearly homogenous group of subjects between 40-60 years, with a similar representation of both sexes, and approximately same degree of hand function deficit. The participants signed informed consents approved by the Lebanese University, school of health ethical review board.

2) *Protocol of the Feasibility Study*

Subjects were trained for a few minutes on how to wear the device at home and turn it on or off. Subjects wore the glove at home for two days. In the first day, they were instructed to wear the glove in the impaired hand and use it like they usually do during daily activities and remove it before sleeping. In the second day, they received a call in the morning from the study personnel and were instructed to do specific activities using their impaired hand during the day in addition to their daily routine.

TABLE I. SUBJECTS' DEMOGRAPHICS AND CLINICAL INFORMATION

Subject	Age	Gender	Months since CVA	Fingers' Average ROM	CVA side	FM UL section (0-66)
S1	40	M	8	70	L	49
S2	60	F	8	60	L	50
S3	45	F	12	80	L	56
S4	53	M	48	70	L	50
S5	58	F	6	80	L	45
S6	40	M	24	80	R	52
Average	49.33			73.3		50.33
SD	8.12			7.45		3.61

SD: standard deviation. CVA: cerebrovascular accident. ROM: Range of Motion. FM: Fugl Meyer.

The list of activities is as follows:

- Stacking cups and dishes on shelves, organizing the laundry, and other different household tasks with the impaired hand (especially if subject usually does such tasks normally)
- Trying to write using the impaired hand
- Using a remote control with the impaired hand when watching TV
- Getting dressed with the use of the impaired hand like zipping and buttoning
- Combing his/her hair using the impaired hand
- Working on the PC using the impaired hand
- Tying shoes using the impaired hand
- Using the impaired hand while using the phone
- At night, removing the glove before sleeping.

The research team did not supervise the patient at home; however, the device recorded the data from the subject's movement on the SD card. After collecting the FMED device from the subjects on the following day, they completed a user feedback questionnaire, and the data was saved on the SD card were collected for offline analysis.

User acceptance of the device and patient feedback were evaluated based on a user feedback questionnaire [13] presented in Table III and an open-ended discussion, performed after using the device for two days. The participants were supposed to answer a list of 11 questions on a scale of 1 to 7; 1 meaning strongly disagree, 7 meaning strongly agree, and 4 meaning neutral. The study personnel was mainly interested in knowing whether the device was comfortable or not and if it was effective in engaging the subjects and motivating them to do more home daily activities. The recorded data was inspected for quality and movement quality parameters. The RaOM and IIM parameters were calculated in order to inspect if these parameters changed in the second day after the study personnel had asked the participants to do extra exercises.

B. *Two Weeks Home-Based Therapy*

The feasibility study shows the subjects' interest to use FMED for an extended period of time. Hence, the device was used in a two-week therapeutic intervention for a stroke survivor suffering from hand movement impairment and limited range of motion.

The training protocol is summarized in Table II. The participant was instructed to practice a set of exercises at home and use the device during exercise to receive real-time visual feedback of index and middle finger range of motion and individuation angle between the two joints. Wolf Motor Function Test (WMFT) [24] was used to evaluate functional movement before and after the training.

TABLE II. LIST OF EXERCISES TO DO AT HOME

Day	Training Type
Monday	Finger movement flexion/ extension exercising
Tuesday	Drawing, connecting dots
Thursday	Mirror training
Friday	Weight lifting
Saturday	Functional exercises (door unlocking brushing teeth, tying shoes, etc)

V. RESULTS

A. *Feasibility Study*

This section presents the results of the user feedback questionnaire and the recorded data. Table III lists the questionnaire questions and mean responses to each question, the results of the t-test performed between the mean responses to each question, and a hypothesized mean of 4 [neutral score]. Results show a significant difference from the neutral score ($p<0.001$). In the open-ended discussion, subjects expressed high satisfaction and reported that the visual feedback by the LEDs was engaging and motivating in moving the impaired hand more than usual. They also expressed their willingness to use FMED at home

TABLE III. USER FEEDBACK QUESTIONNAIRE RESULTS

Question	Average	SD	t-value	p-value
I felt comfortable as the glove was put on	6.33	0.82	7.0	
I did not feel like my fingers were put into any uncomfortable position as the glove was put on	6.33	1.21	4.7	
I felt any restriction to movement with this glove is similar to other gloves I have worn	6.67	0.52	12.6	
I would feel comfortable wearing this glove in public	6.67	0.52	12.6	
I felt comfortable performing the activities in this study	6.50	0.84	7.3	
I feel I can do most of my daily activities (except those involving water) while wearing this glove	6.67	0.52	12.6	
The glove did not feel too tight (it did not make my hands or fingers tingle)	6.83	0.41	17.0	< 0.001
I feel like I can bend my fingers just like I can without wearing the glove	6.83	0.41	17.0	
The glove did not feel too hot or too cold	6.50	1.22	5.0	
I did not feel like my fingers were put into any uncomfortable position as the glove was removed	6.33	1.03	5.5	
I felt comfortable as the glove was removed	6.33	0.82	7.0	
Average	6.55	0.20		

for a therapeutic intervention because of their desire to intensify their hand use during daily activities.

Figure 3 shows the results of data collected from the device during the two days. The subjects were doing a minimal impaired hand finger flexion activity around 60% of the time while using the device (RaOM range 0.55 – 0.61 over the two days). It should be noted that the activity initially reported on day 1 might not accurately reflect the regular activity of the participants without using the device; while wearing FMED, subjects might be moving more than they usually do, knowing that they are being watched. This is known as the Hawthorne effect [32]. However, the participants are stroke survivors and have movement impairments; thus, the Hawthorne effect will not increase the movement score beyond a subject's true functional capability, although it might increase the amount of his or her movement in comparison to regular days. This brings us back to another argument: it is helpful if these individuals with movement impairments feel they are being watched so that they move more according to their functional capability. In addition, by being watched and getting positive visual feedback of their movement (like the feedback by the LEDs in FMED), the subjects become more engaged in daily life functional activities, more than they averagely do. This is believed to be helpful in avoiding the learned non-use phenomena in stroke survivors in which the less the individuals use their impaired limbs, the harder it gets for them to recover their motor skills due to brain remodeling over time [9]. These results are promising due to their effect in validating the use of the FMED and other similar devices in patients with an impaired hand. This study demonstrates that the FMED can be a useful tool to track and monitor the use of paretic hand during daily activities in home environment. It was demonstrated that it is feasible as well as accepted by patients with stroke. Additionally, it was shown that this device could be used to motivate patients to improve their hand movements in daily life and more importantly, do exercises at home (home-based therapy).

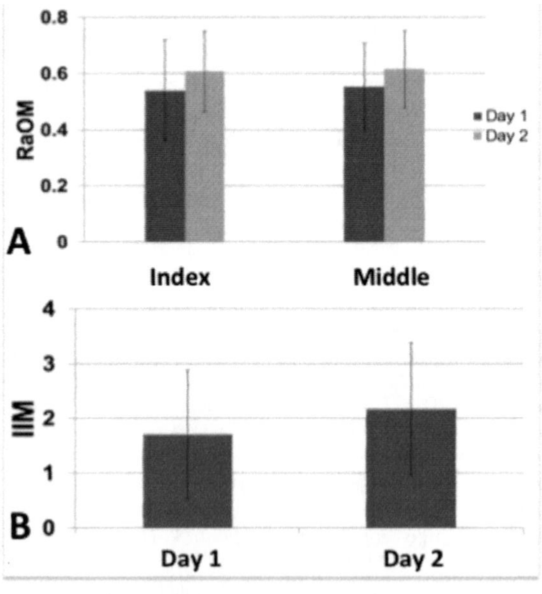

Figure 5. A. Average value of range of motion angle for each joint during the two days, B. Integral of Individuated Movement for the two days of testing the device

B. Two Weeks Home Based Therapy Study

This section discusses the results of a two-week clinical intervention on one stroke subject. The subject was given a set of exercises to do at home based therapy on pre-assigned schedule and she was asked to use FMED while performing the exercises.

1) Clinical Assessment

The WMFT clinical score for the set of 17 timed exercises was 174 seconds before the two-week training and 106 seconds after training, showing a 61% improvement in clinical score.

2) Movement Kinematics Assessment

The data shown in Figure 4 was collected in 1 week of using the device during the home-based therapy. The collected data corresponds to 12326 seconds of device use, which is equivalent to around 205 minutes (3.4 hours). This concludes that the subject complied with therapist's instructions to do 30 minutes of exercising, 5 days a week. RaOM values were 0.89 and 0.87 for index and middle fingers respectively, implying that the subject was exercising while using the device. Values 0.89 and 0.87 imply that during exercising, the subject was performing a minimal impaired hand index finger and middle finger flexion during 89% and 87% of the exercising time, respectively.

Visual observation of the data also shows that the subject was active during the recording time. The average flexion angle of the recorded data was 61.5 degrees for the index finger and 60.5 degrees for the middle finger. However, in the first half of the dataset, the sensors' values ranged from 50 to 100 degrees while in the second half, the sensor's values were between 0 and 50. This indicates that the linear relationship between bending angle and recorded voltage changed during the experiment, either due to a change in the setup or location of the sensor on the finger joint, or damage to the sensor's material due to excessive use for hundreds of repetitions. In this design of the device, a routine calibration was not required, which was a major limitation that leads to corruption in the data collection during the experiment.

At the end of 2nd week of training, the subject reported that the LEDs were not always flashing. Inspection of the data shows improper functioning of the device during training. Similar to week 1 data, the sensors' values were abnormally high due to an error in calibration. However, the device did not record all episodes of movement. The total recorded time was less than 30 minutes so the conclusion was that this set of data was corrupted and was discarded from further analysis. Another conclusion was that the flex sensor is not useful for this application; monitoring the use of an impaired hand in a home-based therapy protocol, due to the uncertainty in the performance of the sensor after an excessive number of repetitions. This conclusion led to recommendations for the creation of a new design of FMED and the termination of the clinical study to avoid wasting time and resources.

Based on the results of the feasibility and clinical study, a new vision for the device was deduced. Accordingly, a new prototype of the device was designed using customized stretch sensors, 5 of which were used to track the 5 MCP joints of the hand in addition to a 6th sensor used to track thumb abduction/adduction movement. The sensors were designed by Dr. Ali Hage-Diab at LIU using graphite powder as conductive material soaked in alcohol and laminated with oil.

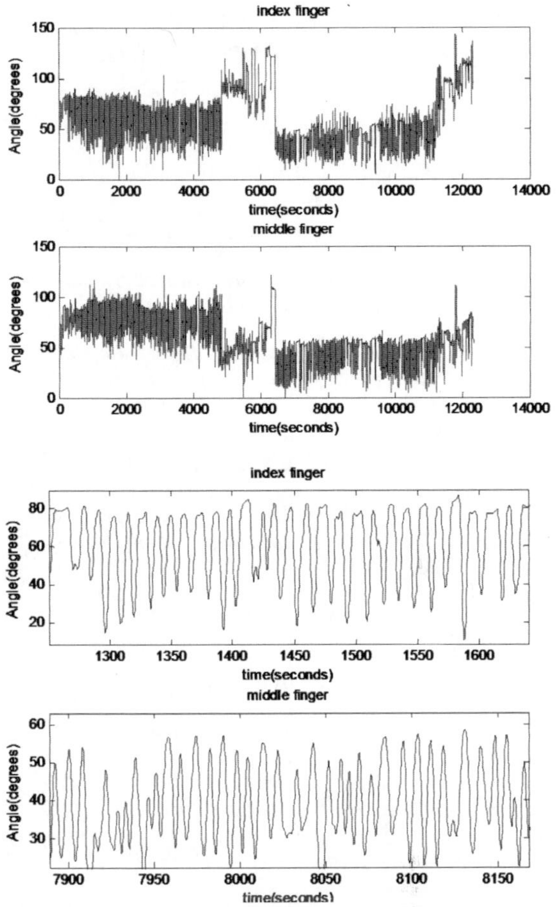

Figure 6. . A. Data recorded by FMED during the first week of training. B. Segment of data shown in A. scaled out to show details of finger flexion.

The sensor was named GO (Graphite Oil) sensor. Detailed description of its design is published in [33]. Sensor performance was tested and results show a linear relationship with bending angle (see Figure 7A). Testing also shows sharp step response of the sensor with bending (see Figure 5B) and repeatability that is better than the flex sensor used in the first prototype (see Figure 8). In this version of FMED, Atmega2560 microcontroller was replaced with Atmega328. In addition, the real-time visual feedback circuitry was replaced with an LED bar that displays the average of whole-hand finger flexion angle during movement. A calibration procedure was designed so that every time a user would wear the device, he or she had to keep the hand in a flat position for 5 seconds and then switch it to a fist position for another 5 seconds. FMED saves the numbers (for each finger) that are acquired during flat position as minimum input values corresponding to 0 degrees bending angle. Similarly, the values recorded during the fist position are assigned to 90 degrees bending angle. Input data between the minimum (during flat) and maximum (during fist) values are mapped to a range of angles between 0 and 90 degrees.

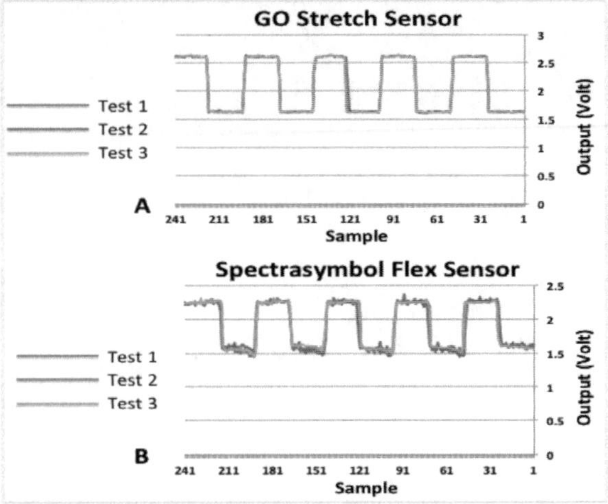

Figure 7. Repeatability test of the sensors. A. Results of the GO sensor. B. Results of the Flex sensor used in the 1st prototype of FMED

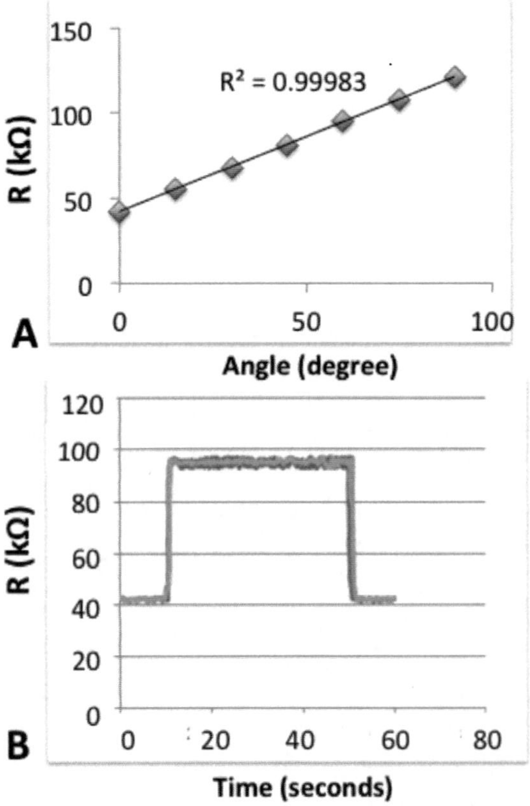

Figure 8. A. Results of Linearity test of the GO sensor. B. Results of the step response test of the GO sensor where the sensor is flexed from 0 to 60 degrees.

VI. CONCLUSION AND FUTURE WORK

FMED allows clinicians to evaluate the improvement of hand function in the context of home environments. It can be a useful tool to complement the role of standardized outcome measures by assessing hand use in real life so that clinicians are not limited to the clinical setting. The high rate of acceptance of FMED by the participants in this study and high enthusiasm of patients to continue using it for therapy due to the presence of visual feedback, suggested the need to test the usability of FMED in a home-based rehabilitation therapy intervention so that it can be produced with a very low cost (~$100).

Home-based therapy study on one stroke survivor showed high competency; however, it revealed issues in the initial design of FMED, specifically weakness in the performance of the flex sensors. An important modification on FMED, as stated previously in this study, involves replacing the flex sensors with the stretch (GO) sensor and increasing the number of sensors to 6 to measure all five finger flexion angles and thumb abduction/adduction movement. The new prototype of the FMED is under development and testing.

Low cost, user-friendly, and low weight are the main advantages of FMED in comparison to other hand motion tracking gloves that are available in the market. The presence of visual feedback setup also allows FMED to be useful as a therapeutic tool and not just as a movement recording device. In the future, the second version of the device will be used in a clinical study to evaluate FMED as both an assistive tool during home-based therapy of impaired hand function and as a research tool that captures significant data of stroke survivors' use of their impaired hands during daily life activities.

ACKNOWLEDGMENT

Authors would like to thank Ali Ibrahim and Ashraf Fouani for their help in testing the GO sensors. Authors would also like to thank Houssam Yassine for his help in recruiting the subjects for the first feasibility study.

REFERENCES

[1] N. H. Saleh, B. Debs, E. Mousallem, A. Hage-Diab, M. Hajj-Hassan, H. Khachfe, H. Yassine, A. Rifaii Sarraj, S. Saleh, "A Finger Movement Evaluation Device to Monitor the Use of Paretic Hand During Daily Life Activities," The Fourth International Conference on Global Health Challenges, Nice, France, pp. 45-50, 2015.4th

[2] G. R. Williams, "Incidence and characteristics of total stroke in the United States," BMC Neurol, vol. 1, p. 2, Dec 18 2001.

[3] A. Turolla, M. Dam, L. Ventura, P. Tonin, M. Agostini, C. Zucconi, et al., "Virtual reality for the rehabilitation of the upper limb motor function after stroke: a prospective controlled trial," J Neuroeng Rehabil, vol. 10, p. 85, 2013.

[4] R. Goulding, D. Thompson, and C. Beech, "Caring for patients with hemiplegia in an arm following a stroke," Br J Nurs, vol. 13, pp. 534-539, May 13-26 2004.

[5] J. C. Chen and F. Z. Shaw, "Progress in sensorimotor rehabilitative physical therapy programs for stroke patients," World J Clin Cases, vol. 2, pp. 316-326, Aug 16 2014.

[6] K. Potter, G. D. Fulk, Y. Salem, and J. Sullivan, "Outcome measures in neurological physical therapy practice: part I. Making sound decisions," J Neurol Phys Ther, vol. 35, pp. 57-64, Jun 2011.

[7] D. U. Jette, J. Halbert, C. Iverson, E. Miceli, and P. Shah, "Use of standardized outcome measures in physical therapist practice: perceptions and applications," Phys Ther, vol. 89, pp. 125-135, Feb 2009.

[8] D. Rand, J. J. Eng, P. F. Tang, J. S. Jeng, and C. Hung, "How active are people with stroke?: use of accelerometers to assess physical activity," Stroke, vol. 40, pp. 163-168, Jan 2009.

[9] R. G. Navarathnam, I. Y. Wang, D. Thomas, and P. L. Klineberg, "Evaluation of the transcutaneous electrical nerve stimulator for postoperative analgesia following cardiac surgery," Anaesth Intensive Care, vol. 12, pp. 345-350, Nov 1984.

[10] R. P. Allred, S. Y. Kim, and T. A. Jones, "Use it and/or lose it-experience effects on brain remodeling across time after stroke," Front Hum Neurosci, vol. 8, p. 379, 2014.

[11] L. K. Simone, N. Sundarrajan, E. P. Elovic, X. Luo, Y. Jia, and D. G. Kamper, "Measuring finger flexion and activity trends over a 25 hour period using a low cost wireless device," Conf Proc IEEE Eng Med Biol Soc, vol. 1, pp. 6281-6284, 2006.

[12] L. Dipietro, A. M. Sabatini, and P. Dario, "Evaluation of an instrumented glove for hand-movement acquisition," J Rehabil Res Dev, vol. 40, pp. 179-189, Mar-Apr 2003.

[13] L. K. Simone, N. Sundarrajan, X. Luo, Y. Jia, and D. G. Kamper, "A low cost instrumented glove for extended monitoring and functional hand assessment," J Neurosci Methods, vol. 160, pp. 335-348, Mar 15 2007.

[14] S. Wise, W. Gardner, E. Sabelman, E. Valainis, Y. Wong, K. Glass, et al., "Evaluation of a fiber optic glove for semi-automated goniometric measurements," J Rehabil Res Dev, vol. 27, pp. 411-424, Fall 1990.

[15] L. K. Simone and D. G. Kamper, "Design considerations for a wearable monitor to measure finger posture," J Neuroeng Rehabil, vol. 2, p. 5, Mar 1 2005.

[16] N. W. Williams, J. M. Penrose, C. M. Caddy, E. Barnes, D. R. Hose, and P. Harley, "A goniometric glove for clinical hand assessment. Construction, calibration and validation," J Hand Surg Br, vol. 25, pp. 200-207, Apr 2000.

[17] K. Li, I. M. Chen, S. H. Yeo, and C. K. Lim, "Development of finger-motion capturing device based on optical linear encoder," J Rehabil Res Dev, vol. 48, pp. 69-82, 2011.

[18] N. P. Oess, J. Wanek, and A. Curt, "Design and evaluation of a low-cost instrumented glove for hand function assessment," J Neuroeng Rehabil, vol. 9 : 2, 2012.

[19] C.E Lang, M.D Bland, R.R Bailey, S.Y Schaefer, R. L Birkenmeier, "Assessment of upper extremity impairment, function, and activity following stroke: Foundations for clinical decision making," J Hand Therapy, vol. 26, no 2, pp. 104-114, 2013.

[20] M. McDonnell, "Action research arm test," Aust J Physiother, vol. 54, no 3 : 220, 2008.

[21] M.J Mulcahey MJ, S. Kozin, L. Merenda, J. Gaughan, F. Tian, G. Gogola, "Evaluation of the box and blocks test, stereognosis and item banks of activity and upper extremity function in youths with brachial plexus birth palsy," J Pediatr Orthop. vol. 32, suppl 2, pp. S114-122, 2012.

[22] S.R Barreca, P.W. Stratford, C.L Lambert, L.M. Masters, D. L. Streiner, "Test-retest reliability, validity, and sensitivity of the Chedoke arm and hand activity inventory: a new measure of upper-limb function for survivors of stroke," Arch Phys Med Rehabil, vol. 86, pp. 1616-1622, 2005.

[23] G. K. Oxford, K. A. Vogel, V. Le, A. Mitchell, S. Muniz, M. A. Vollmer, "Adult norms for a commercially available Nine Hole Peg Test for finger dexterity," Am J Occup Ther, vol. 57, no. 5, pp. 570-573, 2003.

[24] D. M. Morris, G. Uswatte, J. E. Crago, E. W. Cook, E. Taub, "The reliability of the wolf motor function test for assessing upper extremity function after stroke," Arch Phys Med Rehabil, vol. 86, no. 6, pp. 750-755, 2001.

[25] G. Uswatte, E. Taub, D. Morris, K. Light, P. A. Thompson, "The Motor Activity Log-28: assessing daily use of the hemiparetic arm after stroke," Neurology, vol. 67, no. 7, pp. 1189-1194, 2006.

[26] R. L. Gajdosik, R. W. Bohannon, "Clinical measurement of range of motion. Review of goniometry emphasizing reliability and validity," Phys Ther, vol.67, no. 12, pp. 1867-1872, 1987.

[27] R. M. de Carvalho, N. Mazzer, C. H. Barbieri, "Analysis of the reliability and reproducibility of goniometry compared to hand photogrammetry," Acta Ortop Bras, vol. 20, no. 3, pp. 139-149, 2012.

[28] L. K. Simone, N. Sundarrajan , X. Luo, Y. Jia ,G. Kamper, "A low cost instrumented glove for extended monitoring and functional hand assessment," J Neurosci Methods, vol.160. pp. 335-348, 2007.

[29] S. Wise, W. Gardner, E. Sabelman, E. Valainis, Y. Wong, K. Glass, "Evaluation of a fiber optic glove for semi-automated goniometric measurements," J Rehabil Res Dev, vol. 27, no. 4, pp. 411-424, 1990.

[30] L. K. Simone, N. Sundarrajan, P. Elovic, X. Luo , Y. Jia , G. Kamper, "Measuring Finger Flexion and Activity Trends over a 25 Hour Period using a Low Cost Wireless Device," Conf Proc 28th IEEE Eng Med Biol Soc. New York, NY, 2006.

[31] R. Gentner, J. Classen, "Development and evaluation of a low-cost sensor glove for assessment of human finger movements in neurophysiological settings," J Neurosci Methods, vol. 178, no. 1, pp. 138-147, 2009

[32] F. Sciurba, S. I. Rennard, "Rationale for a redundant formulary. The hawthorne effect and the art of medicine," Am J Respir Crit Care Med, vol. 191, pp. 1224-5, Jun 1 2015.

[33] O. Osman, I. H. Ahmad, A. Hage-Diab, "Thoracic Kyphosis Alert System," Conf Proc Adv in Biomedical Engineering, Beirut, Lebanon, 2015.

[34] AJ. Turton, P. Cunningham, E. Heron, F. Van Wijck, C. Sackley, C. Rogers, et al. "Home-based reach-to-grasp training for people after stroke: study protocol for a feasibility randomized controlled trial," Trials. 2013 April; 14(109).

[35] L. Proteau, R. G. Marteniuk, L. A. Lévesque, "Sensorimotor basis for motor learning: evidence indicating specificity of practice". Q J Exp Psychol A. 1992 April; 44(3)(557-75).

[36] A. Henderson, N. Korner-Bitensky, M. Levin, "Virtual reality in stroke rehabilitation: a systematic review of its effectiveness for upper limb motor recovery," Stroke Rehabil. 2007 March-April; 14(2)(52-61).

[37] M. K. Holden, "Virtual environments for motor rehabilitation: review," Cyberpsychol Behav. 2005 June; 8(3)(187-211).

Comparability, Availability and Use of Medication eHealth Services in the Nordic Countries

Heidi Gilstad, Berit Brattheim and Arild Faxvaag
Health Informatics Research Group, Dept. of Neuroscience
Norwegian University of Science and Technology, NTNU
Trondheim, Norway
heidi.gilstad@ntnu.no, berit.j.brattheim@ntnu.no and arild.faxvaag@ntnu.no

Christian Nøhr and Sidsel Villumsen
Dept. of Development and Planning
Aalborg University
Aalborg, Denmark
cn@plan.aau.dk and sidvil@plan.aau.dk

Maarit Kangas
FinnTelemedicum, Research Unit of Medical Imaging, Physics and Technology (MIPT)
University of Oulu
Oulu, Finland
maarit.kangas@oulu.fi

Jarmo Reponen
FinnTelemedicum, Research Unit of Medical Imaging, Physics and Technology (MIPT)
University of Oulu. and
Raahe Hospital
Oulu, Finland
jarmo.reponen@oulu.fi

Guðrún Audur Harðardóttir
Dept. of Health Information Management
Directorate of Health
Reykjavik, Iceland
audur@landlaeknir.is

Hege Andreassen
Norwegian Centre for Integrated Care & Telemedicine
Tromsø, Norway
hege.andreassen@telemed.no

Sabine Koch
Health Informatics Centre, Dept. of Learning, Informatics, Management and Ethics
Karolinska Institutet
Stockholm, Sweden
sabine.koch@ki.se

Hannele Hypponen
Dept. of Information
National Institute for Health and Welfare
Helsinki, Finland
hannele.hypponen@thl.fi

Abstract— **A prescription and medication service that is optimised to protect against unnecessary harm is an essential component of a safer healthcare system. To this means, the Nordic countries have put considerable efforts in digitizing their prescription and dispensing processes and making medication related eHealth services available for clinicians, pharmacists and patients. As these e-services are being established and applied, there is a need to monitor and learn from their use. This paper reports from a sub-study of a larger activity on developing indicators for monitoring eHealth services in the Nordic countries. We describe different medication eHealth services and compare their availability to professionals and patients in the Nordic countries and the usage rates. We found that an ePrescription service is available for clinicians and patients in all Nordic countries, but services that enable renewal or viewing of prescriptions by patients are not commonly available yet. The usage rate of the e-services was not systematically registered in all the Nordic countries at the time of the study, so a comparison between the countries was impossible. A major challenge when comparing medication eHealth services is the fact that definitions of the service itself as well as the indicators used to monitor it vary between countries. The main output is a knowledge-based discussion from the Nordic context on indicators for monitoring eHealth services, evaluated by the potential outcome in terms of comparability and benchmarking.**

Keywords- medication eHealth services; medication data; ePrescription; medication list; indicators; monitoring; comparability, availability and use.

I. INTRODUCTION

Elaborating on the paper "Challenges of Comparing Medication eHealth Services in the Nordic Countries" [1] from the IARIA conference "GLOBAL HEALTH 2015, The Fourth International Conference on Global Health Challenges", this paper offers a discussion on the availability and use of medication related eHealth services in the Nordic countries.

Most nations now devote large resources in digitizing their healthcare systems and in building eHealth services for healthcare professionals and patients. These technological solutions allow better quality and exchange of information between health professionals as well as between health professionals and patients. The assumption is that a better flow of information will consequently lead to better health outcome. As such services have been built and taken into use, there is a need for monitoring and assessing the use of these services for mutual learning and improvement [2].

Although it is acknowledged that eHealth solutions are key measures to handle the challenges in modern healthcare [3][4][5][6][7] acquisition, implementation and development of eHealth has not been systematically based on evaluations and monitoring of the everyday ehealth communication practices in the healthcare services or the health communication of the citizens. Assessments of existing eHealth services' availability, use and usability can contribute to improve healthcare services. In a meta-review of eHealth implementation studies, Mair et al. [8] showed that while some eHealth evaluation studies are used to influence utilization and future eHealth implementations, other studies deal with patient safety and efforts made to avoid clinical errors. Sound eHealth evaluation studies can inform strategic planning and improve eHealth activities and communication for different stakeholders [3][9][10].

What characterizes eHealth services that are available at a national level in the Nordic countries and how are these being used? These are some of the questions that the Nordic eHealth Research Network (NeRN) has posed in an inter-Nordic collaboration on developing indicators for monitoring eHealth. NeRN is a research group [11] reporting to the eHealth group of the Nordic Council of Ministers, and is working with development, testing and assessment of a common set of indicators for monitoring eHealth in the Nordic countries (Finland, Sweden, Norway, Iceland and Denmark), plus Greenland, the Faroe Islands and Åland. The focus has been on developing and subsequently testing indicators for monitoring eHealth, and the test results have been evaluated by potential outcome in terms of comparability and benchmarking. This kind of benchmarking work can support political decision-making in healthcare as well as the development of existing and new eHealth services.

The data referred to in this paper are based on the study Nordic eHealth Benchmarking - Status 2014 [12] conducted by NeRN and reported to the eHealth Group of the Nordic Council of Ministers. The focus is on the comparability of indicators across different healthcare systems.

This paper addresses the issue of availability and use of ePrescription related *eHealth services* and offers a comparison of the availability and use for patients and healthcare professionals in the Nordic countries. The following research question guides the work presented in this paper:

What are the availability and usage rates of ePrescription and eMedication list services in the Nordic countries?

The research question encompasses information from indicators identified by NeRN. The indicators relevant for this paper are: 1) Availability of a national ePrescription service, 2) Availability of a national electronic medication list of prescribed and dispensed medication, 3) Availability of electronic medication renewal, 4) Availability of electronic viewing of patient's own medication data.

The research question is seperated into the following sub-questions:

- Is an ePrescription service available?
- Is a national electronic medication list comprising prescribed and dispensed medication available?
- Is it possible for patients to renew their prescriptions electronically?
- Is it possible for patients to view their ePrescriptions?
- Given that the eHealth service is available, what is the usage rate?

The subsequent part of Section I discusses notions concerning Medication eHealth Services. Section II describes the methods used in the project. Section III offers a presentation of the results, and Section IV includes the discussion. Section V comprises concluding remarks.

Medication eHealth Services

Medication eHealth services include a variety of different systems and e-services related to medication management for patients, pharmacists and healthcare professionals. In this paper we cover the national ePrescription service. We have not included the closed-loop for medication management processes in hospitals. Access to information about medication is crucial for high quality healthcare and patient safety [13]. Viewing an up-to-date list of current medications is a prerequisite when prescribing a new drug, administering medications or assessing potential side effects, decreasing errors when dispensing medications, for preventing medication errors and adverse drug events in the healthcare system [14], as well as for control of financial aspects for prescription products. From the patient's perspective, having an updated list of their medications is an effective means of ensuring that the healthcare professionals they encounter on their path through the health system are kept aware of some of the most important aspects of their health.

A central element of medication eHealth services is the electronic recording of prescriptions. The representations of prescriptions can be described as involving four different characteristics: Prescription as the *decision to medicate*, Prescription as assigning a *right to collect a medication* (Prescribe), Prescription as a *collection action* (Dispensing of a medication) and Prescription as an *administered action*. Administering medication includes several actions: The doctor decides whether the medication should be injections (i.e., intramuscular, depot etc), tablets, etc., and the nurse or the patient (if self-care) administers in line with the prescribed instruction. For health professionals the administering also comprises the task of documentation. As such, the *decision to medicate, prescribing*, *dispensing* and *administering medicine,* are different aspects of a Medication eHealth service in form of an ePrescription service, as suggested in Figure 1.

The decision to medicate is the first step, where the healthcare professional decides when and how the patient should be medicated. ePrescribing is the electronic prescribing of medicine by a healthcare professional to a patient and making it electronically available to a pharmacy, where the medicine can be dispensed and picked up by the patient. The prescription is a signed artifact (document) that describes the medication and how it shall be taken. It gives the patient the right to pick up the medication at the pharmacy and use it according to the description. In a hospital, the healthcare professional does not need to send the prescription to an external server and can proceed directly from deciding to medicate to dispensing of the medicine, thus passing by one step shown in the general process in Figure 1.

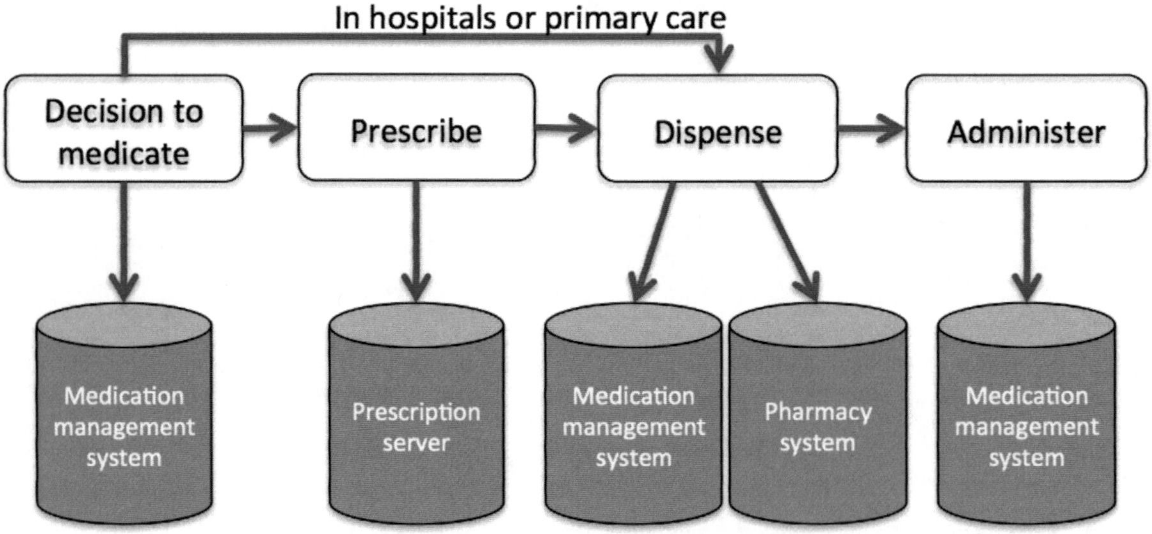

Figure 1. Overview of the process of a prescription from decision to medicate to administration of the medicine and local and national repositories where the data is stored during different phases of the project.

Dispensing is the retrieval of a prescription and the dispensing of the medicine to the patient. The patient consequently administers the medication, when consuming it, or in the case of intravenous medication being administered by healthcare professionals, the administration lies upon the healthcare professional. A prescription list is an overview of the prescription artefacts (the right to collect the medication) of the patient, whereas the medication list is the overview of the medications that are prescribed *and* dispensed to the patient. An eMedication list service allows for both patients and professionals to access it. However, the medication list is not a complete list of all medications of the patient since over the counter medication will not be included in this list. The medication list completeness is one indicator, and only when over-the-counter and herbal medications are included, can it be said to be complete.

Figure 1 illustrates worktasks related to medication, and the storage of data related to each task. The decision of medication is noted by the healthcare professional in a Medication Management System (MMS), which is the Medication section in the EPR-system. In most Nordic countries, the vast majority are sent as an electronic order. Outside hospitals a prescription can be issued on a sheet of paper, telephoned to a pharmacy or sent as an electronic order to a prescription server, where it can be accessed by pharmacies. When a drug is dispensed at a hospital it will be documented in the MMS, if it is dispensed at a pharmacy it will be documented in a pharmacy system – in some countries at a national level. Health care professionals at hospitals, in clinics and long term care facilities store information about the administration of drugs in a MMS, where the system has been implemented. Information on administration performed by the patient himself outside the clinical setting is not recorded in any official health information system in the Nordic countries. Only health care administered (or observed administered) medication will be noted, whereas self-administered medication outside hospitals, clinics etc. are usually not noted anywhere. Although there are some recent emerging mHealth tools, which can follow the act of self-administration [15], it is not possible to register systematically whether dispensed medications are actually administered at home.

II. Material and Methods

The indicators used in "The Nordic eHealth Benchmarking. Status 2014."- study [12] were derived from a rating survey performed in 2013, constructed on the basis of national survey questionnaires in the Nordic countries, an OECD model survey developed in 2012 [16], eHealth policy analysis performed in 2013 and variables presented in the eHealth evaluation literature [17].

Data about the indicators for ePrescription and eMedication list services arose from discussions in a series of workshops with participants from all the Nordic countries arranged by the NeRN and a summary of the national survey questionnaires in the Nordic countries performed from 2010-2014 [12]. The results are presented as proportion of public healthcare organisations having the functionality within each of the Nordic countries.

The study was conducted through four main tasks:

Task 1: Prioritizing functionalities, for which common indicators are needed, and defining measures for availability, usage rate and usability (Responsibility: National Institute for Health and Welfare (THL), Finland, main authors Hannele Hyppönen and Sabine Koch).

Task 2: Collection and reporting of results of eHealth functionality availability, usability and benefit-variables from national surveys (Responsibility: University of Oulu, Finland, main authors Maarit Kangas and Jarmo Reponen).

Task 3: Defining the availability of common usage rate variables from log files (Responsibility: University of Aalborg, Denmark, main authors Christian Nøhr and Sidsel Villumsen).

Task 4: Reporting of results of Nordic eHealth access, usage rate, usability and benefits (Responsibility: THL, Finland).

In the first task, three methodological approaches were applied: 1) content analysis of the existing national eHealth monitoring surveys for listing of existing measures (variables), 2) quantitative and qualitative analysis of rating survey results for key stakeholders to prioritize the measures and 3) analysis of the Nordic eHealth policies. The second task was based on the rated list of variables from the first task and survey data in each of the countries. The methods for harvesting log data for task 3 were different in the various countries due to different systems and practices.

The study has methodological limitations that should be identified. There were some differences in the data collection methods of the Nordic national surveys, sampling and response rates. National surveys were targeted to different professional groups (either availability or usability and experienced benefits of services). Some included private practice while other did not. For example, the proportion of public health organizations where electronic prescription renewal was available for patients was either based on expert knowledge (Denmark, Iceland, Norway, Sweden) or data colleted from the organizational national survey where the Chief Medical Officers (CMO) and Chief Intelligence Officers (CIO´s) were the target population (Finland). Furthermore, there is a limitation in the comparability of usage rate as the availability and granularity of log data varied between the countries. The variation in methods limits comparison of the results. However, the availability of an ePrescription service was a national functionality within all the Nordic countries, and therefore comparability was not an issue for that indicator.

III. Results

When presenting the results, the sub-questions are addressed separately. Each Section shows the results of the availability and the usage rate from the 2014 status of Nordic eHealth benchmarking [12]. The overall results indicate that the availability of electronic medication services varies, and there are differences between the countries in how systematic they are in registering usage rate.

A. Availability and use of a national ePrescription service

This Nordic indicator is identical to an OECD indicator (Availability of making prescriptions electronically available to pharmacies outside of own organization – answer option "Yes, any pharmacy outside of my organization"), but measured at a national level.

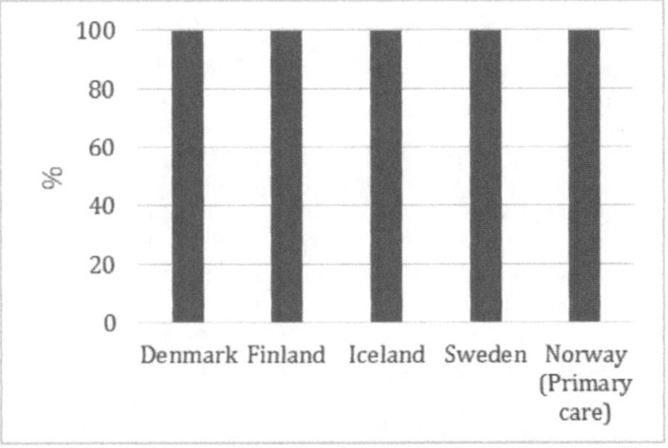

Figure 2. Availability of an ePrescription service.

Figure 2 shows that by the end of May 2014, all the Nordic countries had a national ePrescription service in place. In Finland, Denmark, Iceland and Sweden, ePrescription is available at a national level, i.e., at all public hospitals within the country, for all GP´s, and at every pharmacy in the country. In Finland, the roll-out of ePrescription to the private sector health care providers is currently almost complete (including dentists), but at the time of the data collection in 2014, only public sector physicians and dentists had access to the service. In Norway, all pharmacies, general practitioners, private specialists, dentists and emergency doctors, and all (non-hospital) doctors allowed to prescribe drugs have access to ePrescription.

The professional usage of electronic list of patient prescriptions was measured as a proportion of viewings of prescriptions by professionals (nominator) by all prescriptions (paper based, telephone based or electronic prescriptions) made per year (denominator). This included both national and available regional data on electronic and paper-based prescriptions outside own organisation. In the study, there were differences in the availability of data, as the Nordic countries have different practices for how to log data. In Denmark, there is 100% viewing, in Finland 37%, in Iceland 6%, in Sweden 0.3%. At the time of the study there was no data available from Norway on this indicator.

When looking at the proportion of viewings by professionals by electronic prescriptions made per year (denominator), we see that the results are similar. In Denmark, the number is 100%, in Finland 60%, in Iceland 9%, in Sweden 0.3%. There was no data available from Norway on the nominator.

The results above demonstrate that although the ePrescription service is available in all the Nordic countries, the knowledge and systematic collection about the usage rate vary. One needs to keep in mind that in Iceland the viewing of prescriptions by doctors across healthcare institutions was in a pilot phase at the time of the study. Hence, only a small portion of physicians within Iceland had access to this service. In Finland the national ePrescription had just become available for many organisations in the public sector and was not available for private practitioners. Lower availability as well as requirement to change work practices in order to view the prescriptions from the national database may account for lower usage rates in Finland. Denmark has had this functionality available on a national level to healthcare professionals since 2010 – excluding professionals in the municipalities who were fully implemented by 2014.

B. Availability of a national electronic list of prescribed and dispensed medication

The indicator is identical to the OECD indicator (Availability of information on dispensing status by the pharmacist, answer option "Yes, for most or all of my patients"). It measures the availability of information about medication that has been previously prescribed and dispensed (including prescriptions from other institutions).

However, the contents of this indicator vary in the Nordic countries. A national list of prescribed and dispensed medication is not necessarily the same as the patient´s current medication list, since for example, the medication dispensed while admitted to a hospital or purchased without a prescription may not be included.

Figure 3 shows the availability of national electronic lists of prescriptions and dispensings in the Nordic countries. In Denmark, the medication list has been 100% available since 2010, including all types of prescriptions made outside hospitals as well as all medications prescribed on discharge from the hospital.

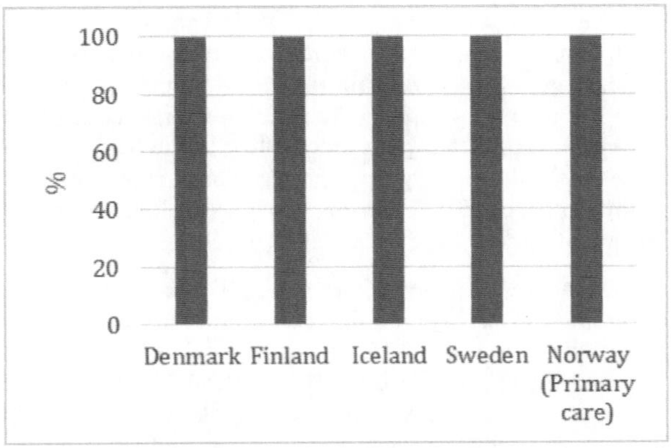

Figure 3. Availability of a national level list of prescrptions and dispensings in public sector.

In Finland, the national prescription database shows prescribed and dispensed medication, which in public sector was 100% in 2014, but not those administered during hospital stay. Patients can preclude health professionals from accessing the data. The national list of prescriptions does not include prescriptions on paper or prescriptions that have been made by phone, nor prescriptions related to social care. From 2015, the national KANTA-system generates a comprehensive list of current medication for the patient from the prescription database and data from individual electronic patient record (EPR) systems, including medication administered during hospital stay and all types of prescriptions.

In Iceland, the availability of the national list of prescribed and dispensed medicine is 100%, since 2014, and includes all ePrescriptions, both prescribed and dispensed, as well as some paper and telephone prescriptions. All paper and telephone prescriptions are available in 2015. As in Finland, the medication list does not include the medication administered during hospital stay. By law the medication is made viewable for the past three years within the national pharmaceutical database. Patients cannot opt out of this service, meaning that the doctor treating the patient does not need the patient´s permission to access his/her medical history.

In Norway, the availability of list of prescriptions and dispensings in primary care is 100 %. A national medication list is to be found in the "Kjernejournal" (Summary care record), and it may also be accessed via the national portal "helsenorge.no". "Kjernejournal" is running as (in 2014) a pilot implementation in two regions. "Kjernejournal" contains a list of the medicines the patient has been prescribed (both ePrescriptions and paper prescriptions) in Norwegian pharmacies. Medicines the patient purchased without a prescription, received at an emergency department, hospital / nursing home or purchased abroad will not appear. Prescriptions that have been dispensed are stored in the "Kjernejournal" for three years.

In Sweden, the list of medications that have been dispensed to the patient has been available since 2012. The patient decides if the doctor is allowed to see the information in the database. A consent is needed from the patient. Very few patients i.e., 3-4.000 patients out of 9 million actually choose to hide their information.

In the NeRN Status Report, the frequency of use of electronic prescriptions is not monitored. However, NeRN suggests that the indicators "proportion of dispensed prescriptions of electronic prescriptions made" and "proportion of dispensing list viewings by professionals (excluding pharmacists) of electronic prescriptions made" could be additional indicators for monitoring usage rate of electronic prescriptions.

C. *Availability of electronic medication renewal*

This indicator shows the availability of services that enable electronic medication renewal for patients at the national level. The indicator is identical to the OECD indicator. Some countries have data for local functionality while other countries have data for the national functionality.

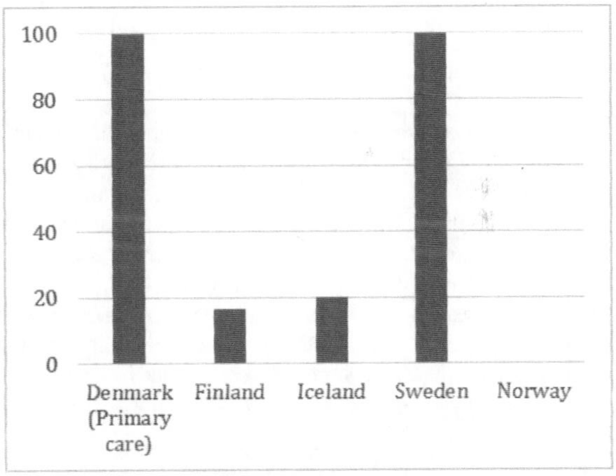

Figure 4. Availability of electronic medication renewal.

Figure 4 shows the availability of electronic medication renewal services in the Nordic countries. In Denmark, there is 100% availability of electronic medication renewal in primary care at a national level.

In Finland, this was an organizational activity in 2014, but currently this functionality is available for citizens via the national patient portal. In 2014, the patient needed to contact the pharmacy or primary health care centre to ask for a renewal, although some organizations provide an electronic web portal to mediate the request as depicted in Figure 4.

In Iceland, only a few healthcare institutions offered this service in 2014. The functionality was in the form of patients sending an e-mail from the healthcare organization´s web site that offered these services for their patients and requesting the medication renewal. However, this is currently being implemented via a national patient portal and is expected to be at a national level before end of 2015.

In Norway, this service has not been established at the national level. General practitioners can offer service functionalities for patients depending on what portal provider they have chosen.

In Sweden, electronic medication renewal has been available since 2012 in the national service "My healthcare contacts (MVK)". MVK is a citizen web portal that enables secure communication between patient/ consumer/ customer and healthcare and long-term care. The patient can book and rebook appointments, renew prescriptions, order a copy of his patient record and in some county councils also access it.

The usage rate of electronic medication renewal is based on the OECD variable "frequency of use of electronic prescription renewal requests by patients". The results show that none of the Nordic countries have a systematic overview, in log data, of electronic renewal requests made by patients.

D. *Availability of electronic viewing of patient's own prescription*

This indicator concerns electronic services that enable patients to view their own medication data. We present data for services at a national level. The indicator is not completely identical to the OECD indicator (Availability for patients to remotely access the Medication lists from their provider-maintained electronic record): The OECD indicator focuses on local level availability of a medication list, the Nordic indicator on a national level availability of a list of prescriptions and dispensed medication.

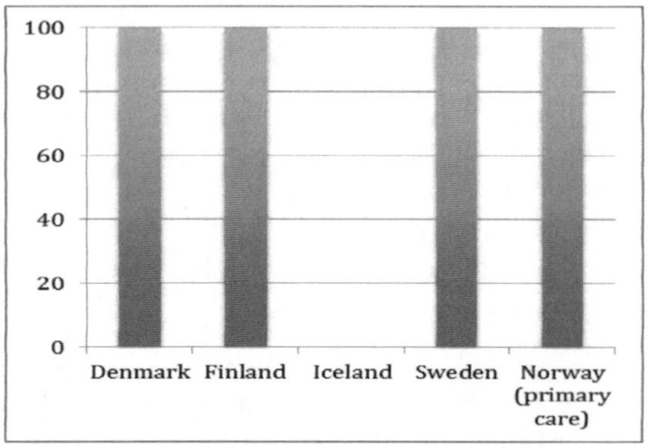

Figure 5. Availability of patients' viewing electronically of own medication data

Figure 5 shows the availability of patients` viewing electronically of own medication (prescriptions and dispensings, where available) data. In Denmark, patients have had the opportunity to view their own medication data covering the past two years since 2009. In the beginning, it only enabled viewing of prescriptions made outside hospitals. Since 2013, viewings of prescriptions made by hospital personnel have been included.

In Finland, all patients have since 2013 had access to all prescriptions that are in the prescription database. This service did not exist in Iceland until October 2014 via the national patient portal. Currently in Iceland, it only includes ePrescribed medications. However, plans are already underway to enhance these services to include also paper- and telephone prescribed medication.

Norway established this service in 2012-2013 via "My prescriptions" in helsenorge.no [18]. The service currently enables viewing of the most recent prescriptions made by general practitioners.

In Sweden, this service has been available since 2012 as a national service through "My healthcare contacts" (MVK).

Availability for patients viewing of medication data that (public/private sector) professionals have prescribed is a feature of national health information systems in the Nordic countries.

The usage rate of electronic viewing of patients own prescriptions, i.e., active list of patients current medication via the national information system, was not benchmarked within the Nordic countires. However, OECD covers this in the model survey ("Availability for patients to remotely access the Medication lists from their provider-maintained electronic record – usage rate").

IV. Discussion

The Nordic countries seem relatively homogeneous and comparable in terms of political systems, infrastructure, culture, and educational, social- and healthcare systems; however providing comparable eHealth indicators from surveys across the Nordic countries involves a number of challenges. The eHealth functionalities, albeit spoken of in same terms (e.g., ePrescription service), were not identical in different Nordic countries. The samples of the survey varied: in Denmark, a representative sample of clinical end users participated whereas in the other countries leaders in health care institutions were approached. The survey questions were formulated in the language of the respective countries, and the time and frequency of the surveys varied. Detailed discussions of these differences settled most of the variance they introduced, and the results obtained on the medication issues were quite comparable.

Comparable e-services regarding the ePrescription include availability of the prescriptions for pharmacies and patients on national level, and availability of the list of medicines prescribed to the patient on national level, i.e., the proportion of public and/or private organizations where prescribed medicine outside their own organization, are available in all the Nordic countries. Although ePrescription is available, it is still possible to issue prescriptions on paper or by telephoning the pharmacy. This proportion has not been measured, but it is assumed that it is neglectable given the high number of prescriptions made electronically.

The ePrescription services are well established and mature in all of the Nordic countries. However, the availability of viewing the prescriptions on a national level is still in a pilot phase in some of the countries.

The availability of a national list of prescribed and dispensed medication has also, by 2014, reached a level of saturation. In Denmark, this service has been available for some years. Although the service is available in all the countries, the architecture of the systems behind the services differ significantly, but a detailed analysis of these differences has not been targeted in this study. A special feature for patients to hide specific medications is available in all countries for similar ethical reasons.

An e-service to renew medication is not available in all of the Nordic countries. In Denmark and Sweden the service has been available for some years; however it is implemented in different ways. In Denmark, this service was implemented for all patients to use as part of the agreement between the general practitioners and the regions who pay them fee for service. In Sweden, the service is available to all citizens through a national portal. In other countries this service was available only through dedicated organizations.

The service that enables patients to view their own prescriptions has been implemented in all of the Nordic countries.

However, when going into detail about the content of the indicators, the NeRN group realized that characteristics of the eHealth functionalities as well as the monitoring data provided in national surveys and logs varied somewhat between the respective countries.

The availability of a national ePrescription service was saturated, but the content measured was different between the Nordic countries. In the definitions of the indicators, the fact whether the medication was prescribed, dispensed or administered was not clearly specified, or the data were not available because the question was not asked specifically in the surveys. It became apparent that the content of ePrescriptions and the measurements of them varied between the countries making detailed explanation in the presentation of the results necessary for each indicator and each country.

Another point, which makes comparison difficult, is the fact that ePrescription does not cover paper-based prescriptions per se, which are regulated in another way than electronic prescriptions. It has different consequences in the respective countries. In Denmark, for instance the paper-based prescriptions will be synchronized with the electronic overview of the patient's own prescriptions once the medicine has been dispensed in a pharmacy. A related issue is that while ePrescriptions are 1-1 prescribed and dispensed medication, where the paper based prescription can hold prescriptions of several different medicines on the same piece of paper.

The e-services in this paper more specifically referred to as *the medication eHealth service*, may have different scopes, i.e., intended coverage area. While some e-services are accessible at a national level, others are either limited geographically to a regional level, administratively to the hospitals or the organizations, or to specific roles, for example to healthcare professionals and not to patients. The focus in this study was the availability of medication eHealth services at a national level and availability at a more granular level was therefore not presented.

The study also showed that there are different practices in the Nordic countries whether they are systematically logging usage rate of the electronic medication services. It was possible to retrieve log data about use of ePrescription viewing in Denmark, Finland, and Iceland but due to different systems in the countries, the definitions of the denominators varied slightly. Furthermore, as this service was in a pilot phase in Iceland at the time of the study, comparability was an issue. There were no systematic log data available on the usage of the other e-medication services discussed here, i.e., viewing by patients and electronic renewal requests by patients.

V. CONCLUSION

The study showed that the availability of patients' prescription information and ePrescriptions made available to any pharmacy is acknowledged via the national ePrescription systems in each Nordic country. Moreover, the availability of medication renewal requests as well as the availability of electronic viewing of patients' own prescriptions is comprehensive on national level in some countries (Sweden, Denmark, Finland and Iceland). Patients' access to view their prescription data electronically is also broad. However, the NeRN findings demonstrated that the services are carried out differently in the respective countries and also definitions of indicators vary between countries hampering comparison. The implementation of eHealth services within healthcare is expected to enhance patient safety and quality of healthcare delivery. Prerequisites to access this goal are that the complete list of patient's current medication is available and systematically used to inform clinical decision-making. The results of this study provide valuable information to guide decision making at the healthcare and eHealth policy level with regard to development, acquisition, implementation and assessment of eHealth services. Furthermore, it highlights the need for implementation of standardized, accessible systems for monitoring and benchmarking eHealth services.

Benchmarking is important in order to detect possible benefits in use of the eHealth service and to identify best practise in the respective countries that consequently could inform the development in other countries. Benchmarking is also important for detecting possible problems and risks. Despite of the limitations of the work, NERN succeeded in benchmarking availability and use of several Health Information Exchange (HIE) and Patient Health Record (PHR) functionalities, including the eHealth services related to medication. The results show that the Nordic countries advances in eHealth services for healthcare professionals and for citizens. The results and the experiences from the study generate the following recommendations [12]:

1) The Nordic countries should agree on common indicators, in order to monitor the same aspects and consequently exchange knowledge and "best practise" in eHealth service provision; 2) The Nordic countries should provide access to log data for monitoring and research (and not only key numbers or forecasts); 3) Since there are great differences in the national architectures, there is a need for more detailed comparison of the data retrieval processes and outcomes in the respective countries; 4) As the utilities health record systems in Iceland shows, usability and expected utility is not just about high eHealth budgets, but about wise practices; 5) One anticipated impact of ePrescription /comprehensive medication list is the reduction of medication errors. Although the results from Denmark showed the opposite (the patient safety reporting system is more comprehensive than in the other countries, and is thus detecting more medication errors and near-misses), it is assumed that if a systematic and coordinated practice with registration of prescriptions and medication lists is introduced, the proportion of medication errors will fall.

The data in this paper are based on the results from the Nordic eHealth Benchmarking status 2014. It must be noted that the eHealth services are continuously under development in the Nordic countries. This study would assumingly have different results if conducted in 2016.

ACKNOWLEDGMENT

Our gratitude goes to the eHealth Group of the Nordic Ministry of Health who supports the work of the Nordic eHealth Research Network in our efforts to define common indicators for monitoring eHealth in the Nordic countries. Thanks also to Lars Jervall and Thomas Pehrsson for support during the indicator work and data collection.

REFERENCES

[1] H. Gilstad, A. Faxvaag, C. Nøhr, S.Villumsen, J. Reponen, H. Andreassen, L. Jervall, T. Pehrsson, M. Kangas, G.A. Hardardottir, S. Koch, and H. Hypponen, "Challenges of Comparing Medication eHealth Services in the Nordic Countries," GLOBAL HEALTH 2015, The Fourth International Conference on Global Health Challenges. [Online]. Available from http://toc.proceedings.com/27226webtoc.pdf 2016.05.30

[2] The European Federation of Medical Informatics "Working Group Proposal, adopted by the EFMI board," [Online]. Available from http://iig.umit.at/efmi/ 2016.05.30

[3] WHO National eHealth Strategy Toolkit, 2012. [Online]. Available from http://www.who.int/ehealth/publications/overview.pdf 2016.05.30

[4] OECD. "Improving Health Sector Efficiency. The Role of Information and Communication Technologies," [Online]. Available from http://www.keepeek.com/Digital-Asset-Management/oecd/social-issues-migrationhealth/improving-health-sector-efficiency_9789264084612-en#page2

[5] EU-European Union. "eHealth Action Plan 2012-2020," [Online]. Available from http://ec.europa.eu/health/ehealth/docs/com_2012_736_en.pd

[6] H. Hyppönen, A. Faxvaag, H. Gilstad, G.A. Hardardottir, L. Jerlvall, M. Kangas, S. Koch, C. Nøhr, T. Pehrsson, J. Reponen, Å. Walldius and V. Vimarlund, "Nordic eHealth Indicators: Organisation of Research, First Results and Plan for the Future," MEDINFO 2013C.U. Lehmann et al. (Eds.) © IMIA and IOS Press.

[7] The Norwegian Ministry of Health, "White Paper no 9 (2012-2013)," [Online]. Available from https://www.regjeringen.no/nb/dokumenter/meld-st-9-20122013/id708609/ 2016.05.30

[8] FS. Mair, C. May, C. O'Donnell, T. Finch, F. Sullivan and E. Murray "Factors that promote or inhibit the implementation of eHealth systems: an explanatory systematic review," Bull World Health Organ. 2012 May 1; 90(5):357-64.

[9] C. Codagnone and F. Lupiañez-Villanueva, "Benchmarking Deployment of eHealth among General Practitioners–Final report." European Commission. Directorate-General of Communications Networks. Content & Technology, [Online]. Available from https://ec.europa.eu/digital-agenda/en/news/benchmarking-deployment-ehealth-among-general-practitioners-2013-smart-20110033 2016.05.30

[10]H. Gilstad, A. Faxvaag, H. Hypponen, S. Koch, C. Nøhr and K. Skauli, "Multinational surveys for monitoring eHealth policy implementations - usefulness and pitfalls," Studies in health technology and informatics, 01/2014.

[11] H. Hyppönen, A. Faxvaag, H. Gilstad, G.A. Hardardottir, L. Jerlvall, M. Kangas, S. Koch, C. Nøhr, T. Pehrsson, J. Reponen, A. Walldius and V. Vimarlund. "Nordic eHealth Indicators. Organization of Research, First Results and Plan for the Future," Stud Health Technol Inform 192:273-7.

[12] H. Hyppönen, M. Kangas, J. Reponen, C. Nøhr, S. Villumsen, S. Koch, G.A. Hardardottir, H. Gilstad, L. Jerlvall, T. Pehrsson, A. Faxvaag, H. Andreassen, B. Brattheim V. Vimarlund and J. Kaipio, "Nordic eHealth Benchmarking: Status 2014," TemaNord 2015:539, Nordic Council of

Ministers, Copenhagen, Denmark. [Online]. Available from www.norden.org/en/publications 2016.05.30

[13] D. M. Benjamin, "Reducing Medication Errors and Increasing Patient Safety: Case Studies in Clinical Pharmacology," The Journal of Clinical Pharmacology. Blackwell Publishing Ltd. 2013.

[14] G.E. Karagiannis, L. Tzachani, B. R. M. Manning, V. G. Stamatopoulos, M. Roughton, A. A. Lazakidou, M. Petridou, D. Iliopoulou and M. L Rigby, "Patient-Generated EHR Input System Trials: An Analysis of Perceived Benefits Across a Range of Disease Groups," in eds D.D. Koutsouris and A.A. Lazakidou R, "Concepts and Trends in Healthcare Information Systems". Springer Link, 2014.

[15] "Proteus Ingestible Sensor for Tracking Medication Intake Receives FDA Clearance," [Online]. Available from http://www.medgadget.com/2012/07/proteus-ingestible-sensor-for-tracking-medication-intake-receives-fda-clearance.html

[16]OECD Directorate for Employment, Labour, and Social Affairs and Directorate for Science, Technology, and Industry, "Draft OECD Guide for Measuring ICTs in the Health Sector, Paris: OECD. COM/DELSA/DSTI(2013)3/FINAL", [Online]. Available from http://www.oecd.org/health/health-systems/Draft-oecd-guide-to-measuring-icts-in-the-health-sector.pdf 2016.05.30

[17] J. Adler-Milstein, E. Ronchi, G. R. Cohen, L.A. Pannella Winn, and A. K. Jha, "Benchmarking health IT among OECD countries: better data for better policy," J Am Med Inform Assoc 2014;21:111–116. doi:10.1136/amiajnl-2013-001710

[18]The Norwegian Directorate for eHealth, "helsenorge.no patient portal," [Online]. Available from https://helsenorge.no/ 2016.05.30

Toward Usable and Trustworthy Online Monitoring on e-Health Applications

Youna Jung
Department of Computing and Information Sciences
Virginia Military Institute
Lexington, Virginia, United States
e-mail: jungy@vmi.edu

Abstract — **To enable and validate their effectiveness, many e-health applications track how they are used by patients. While online monitoring can improve the accuracy and quality of e-health applications, there is the potential of serious privacy violations. As e-health applications use online monitoring services, sensitive health data could be exposed to not only the healthcare providers but also the monitoring service providers and third-parties such as advertisement companies against wishes of a user. To prevent privacy loss during online monitoring, as a preliminary work, we came up with the idea of a privacy-preserving online monitoring framework, in short PPoM, that helps both of e-health providers and users specify their own policies and enforce user privacy policies during monitoring in systematic manner. In this paper, we extend the idea of the PPoM framework by describing a motivating example of privacy violation during online monitoring and specifying each component in the framework, and demonstrating a prototype of the secure user browser, called PPoM browser.**

Keywords - e-health application; online monitoring; privacy protection; framework; secure monitoring service; secure browser; toolkit.

I. INTRODUCTION

Online monitoring and analytics are essential techniques to evaluate and enhance the performance of online applications. They help the online service providers improve the usability of online applications by collecting user/usage data and analyzing the performance of applications [1][2][3]. In general, there are three different approaches to online user monitoring: 1) log file analysis on the server side, 2) proxy-based monitoring, and 3) use of monitoring scripts provided by online monitoring/analytics services on the client side [4]. Among those approaches, we focus on the third approach because it is widely used and requires less time and effort to collect, analyze, and visualize user/usage data.

Online monitoring and analytics services, such as Google Analytics [5] and Adobe Analytics [6] have been extensively used in a variety of online application areas, such as e-health [7], e-commerce [8], information retrieval [9], and so on. These monitoring/analytics services enable the tracking and recording of user actions and characteristics, such as mouse clicks, frequency of use of an application, time spent in a particular page, media viewed, page navigation sequences, content entered into a textbox, location information, whether a mobile device is being used, and so on.

As e-health applications are becoming ingrained into the everyday life of many people, we focus on the healthcare and wellness domains among various application areas. According to Eysenbach, e-health is an umbrella term that includes a variety of online healthcare applications and systems that use information technologies, such as e-Learning for healthcare, e-Diagnosis, e-Prescribing and online health interventions [10]. It is an emerging field at the intersection of medical informatics, public health and business, referring to health services and information delivered or enhanced through the Internet and related technologies. By using advanced information technologies, including electronic data management and rich interaction skills, e-health applications are capable of 1) providing personalized services, 2) reducing healthcare cost, 3) ensuring easy access regardless of time and place, 4) ensuring consistent quality of services over time, 5) enabling automated data collection/analysis, and 6) demonstrating the potential for having more honest self-reporting by patients. Many e-health applications have been used for online healthcare education [11], healthcare research [12] and recruitment of its participants [13], collecting healthcare data for research or national healthcare purposes, and conducting healthcare interventions to facilitate disease prevention, disease self-management, and health promotion [14].

To accomplish the purposes of e-health applications, most e-health applications provide several of the following functionalities: 1) self-assessment or self-profiling to recognize individuals' health-related status and in turn provide personalized messages and/or healthcare services, 2) continuous communication with patients/users using interactive tools such as online trackers, and 3) wide dissemination of information related to health and safety, presented in text and/or multimedia format. To provide the required functionalities, on one hand, detailed monitoring is critical to confirm that e-health applications are correctly used and to validate their efficacy. In order to do so, e-health applications must collect detailed, and often identifiable, user data including health information. On the other hand, the protection of user privacy is however critical since e-health applications often deal with very sensitive private data, including health status, medical records, and family health histories. Control over the sharing of such information is of the utmost importance and urgency because indiscriminate monitoring, if inconsiderate of user privacy, may result in private health data being used for unwanted purposes and/or shared with unknown people [2][15][16]. In case of e-health applications, even generic usage data can violate privacy if disclosed. For example, disclosure of the login frequency into an online treatment application for substance abuse can unintentionally reveal a user's medical status. Consequently, it is urgent and critical for research to examine how we can simultaneously achieve these two important yet opposing goals -- monitoring identifiable user data while protecting user privacy.

To enable e-health applications to conduct trustworthy user monitoring without concern for loss of privacy, in this paper, we enhance the Privacy-Preserving online Monitoring (PPoM) framework [1]. In the PPoM framework, online monitoring services collect user/usage data based on users' policies. In addition, users can verify user/usage data being monitored in real-time and strictly enforce user policies on the client side by controlling outgoing messages set from users' browsers. To support non-IT medical staff who do not have enough knowledge and skills on Information Technologies (IT), the PPOM framework provides intuitive and semi-automatic tools that enable them to generate privacy policies and insert monitoring code into their e-health applications. The rest of this paper is organized as follows. In Section II, the limitations of existing online monitoring approaches and the necessity for a secure monitoring in e-health applications are identified with an example scenario. In Section III, the overall architecture of the PPoM framework is described and the detailed functionalities and methods for each component are specified. In Section IV, how the PPoM framework mitigates the privacy vulnerabilities described in Section II from the perspective of e-health service providers. In Section V, we present a prototype of the privacy-preserving browser as a first step in the development of the PPoM framework. The evaluation plans is described in Section VI and related work are introduced in Section VII. The conclusions and future work are presented in Section VIII.

II. MOTIVATION

A. Limitations

Existing online monitoring approaches on e-health applications have two major problems, as follows:

1) Lack of systemactic methods to verify and enforce privacy policies mutually agreed by users and providers: To protect user privacy during online monitoring, a user needs to specify his/her preference in data disclosure while the administrators of an e-health application specify their privacy policy describing what kinds of user data might be monitored, what those data are used for, who those data will be shared with, and how user data are maintained. Users and administrators can specify their policies using policy languages such as P3P [20] and WS-XACML [21]. Once a user agrees to an application's policies, the enforcement of agreed policies has been primarily relied on the honor system [17] within the application without any external verification process. To ensure user privacy, the federal Health Insurance Portability and Accountability Act (HIPAA) [18] stipulates that a healthcare component must not disclose protected health information to another component (HIPAA 164.105.(a)(ii)) with only a few exceptions (HIPAA 164.512). However, it is difficult to expose violations of HIPAA regulations within e-health applications in existing approach. If a provider embeds monitoring code and/or third-party data-collecting ads in its webpages, private data can easily be released regardless of users' wishes. Although this is an obvious violation of HIPAA rules, there are no solutions to systematically detect the application's fraud and prevent user data from undesirable use and disclosure.

To protect user privacy from undesirable use, some online applications anonymize/de-identify user data by deleting identifiers in original data but such anonymized data can often be re-identified/de-anonymized [19]. It is hence not enough to hide user identifiers and we need a new method not to share critical information based on user preferences. In addition, anonymization might not be applicable to some e-health applications that require identifiable user data for personalized services. Without a strong enforcement method, many users are unlikely to consent to online monitoring.

2) Need for professional IT knowledge and skills: At present, professional IT knowledge is needed for developing monitoring-enabled e-health applications with a privacy policy. For example, to specify privacy policies of an e-health application, an application administrator must understand privacy policy languages, such as P3P [20] or WS-XACML [21] and be able to precisely specify the application's policy in that language. In addition, to use an online monitoring service, the administrator must understand the client-side monitoring code (e.g., in JavaScript), and be able to manually insert privacy-preserving code into the original source code (possibly using different languages) for each web object or webpage being monitored. Hence, administrators of e-health applications need to understand at least one language to integrate privacy-preserving monitoring into applications. This is however an impractical expectation for many non-IT administrators, such as doctors, nurses, health educators and communicators. Not only for non-IT clinical staff who manage e-health applications but also for average users who use e-health applications, it is difficult to exactly specify privacy policies and enable their applications/browsers to protect user privacy. The lack of IT knowledge of administrators and users of e-health applications significantly increases the need for easy-to-use tools for a privacy-preserving framework.

In Figure 1, an example scenario shows us the existing monitoring approaches may not be able to prevent privacy leakage. The privacy vulnerabilities in the scenario is as below:

1) Users and/or administrators have to create their own policies manually. To do so, they must first learn

policy languages, such as P3P, APPEL, XPref, or WS-XACML, prior to online monitoring.

2) An administrator needs to manually insert monitoring code to conduct online monitoring/analytics but the task requires IT knowledge about monitoring code and web programming languages.
3) Both users and e-health applications should have their own privacy policies written in existing policy languages. Due to the lack of a standard data schema to specify health data, however, they are unable to define their privacy using a single schema that is shared by the user and the application to express privacy policies in a consistent manner. (In Figure 1, AP^- and UP^- represent the application policies and the user policies in the absence of a standard health data schema.)
4) User preferences regarding data sharing with third-parties might be ignored during online monitoring.
5) There is no way to detect applications' mistakes or fraud. For example, untrusted policies are represented as AP_f in Figure 1, but it is not possible to enforce the mutually agreed privacy policies systematically. Currently, policy "enforcement" amounts to trusting the application.

B. Requirements

Towards trustworthy and highly usable online monitoring in e-health applications, the following requirements should be satisfied:

1) For strict enforcement of user privacy policies

- Online monitoring services that are aware of user privacy policies rather than application policies.
- Verification methods to ensure that an application complies with policies mutually agreed by providers and users on user side.
- Enforcement methods to protect user privacy on user side in case of privacy violation during online monitoring.

2) For practical use by non-IT users and staff

- User-friendly interfaces to intuitively specify privacy policies and monitoring objects.
- Automatic generation of privacy policies for e-health applications.
- Systematic conversion of existing applications to privacy-preserving applications in which privacy-aware monitoring code is embedded.

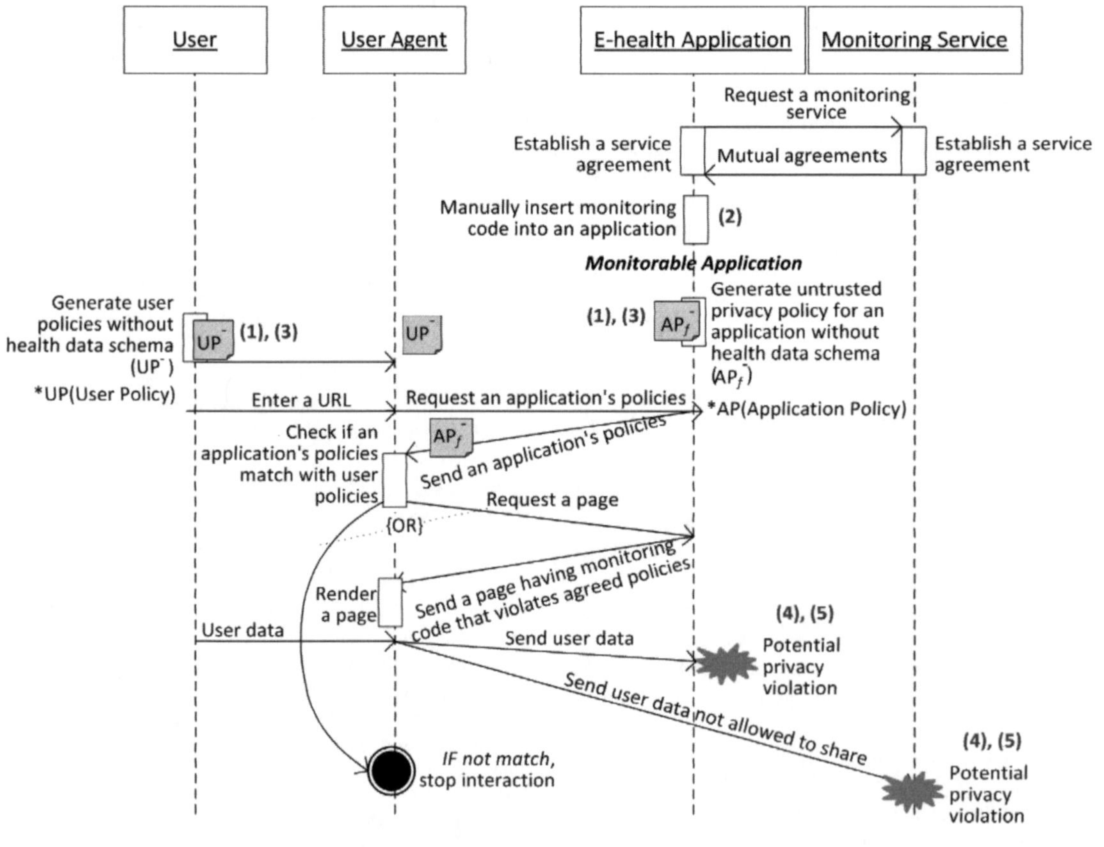

Figure 1. Existing online monitoring approach (User agent is a software agent in a user browser that acts as a client of the application server and captures a user's privacy preferences from a user policy (UP)).

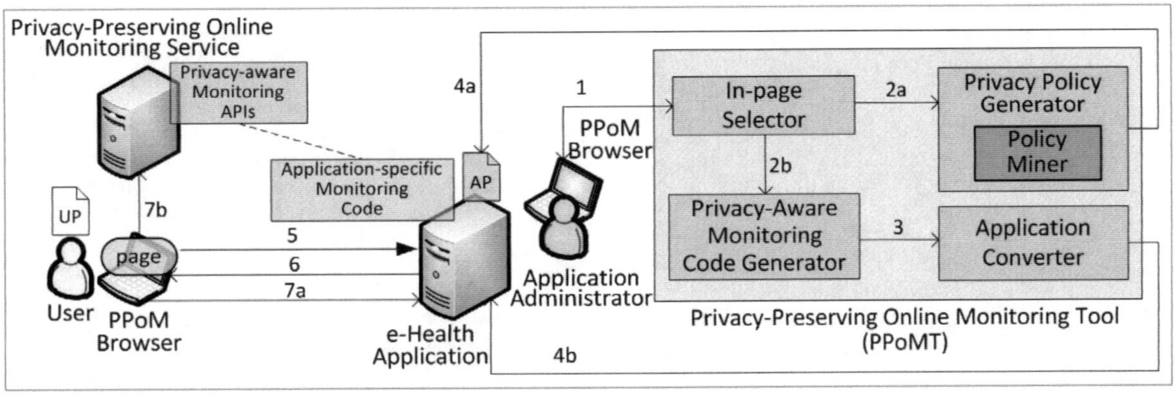

Figure 2. Overall architecture of PPoM framework.

III. Overall Architecture

To fulfill the requirements described above, the idea [1] of the PPoM framework that rigorously protects user privacy by referring user policies and enforcing them on user side was proposed as shown in Figure 2. In this paper, we describe the details of individual services that include the privacy-preserving online service (PPoM service), user browser (PPoM browser), and online monitoring tool (PPoMT).

A. *Privacy-Preserving Online Monitoring Service*

The problems of existing online monitoring services are twofold. First, users are forced to choose between two options: opt-in or opt-out, and not allowed to choose data to be monitored in fine-grained level. Second, user data are collected based on the decision of providers. Once a user gives consent to monitoring (opt-in), all user activities on the web objects in which monitoring code is embedded are captured regardless of users' privacy preferences. None of existing services considers users' privacy preferences during monitoring.

To address the issue of ignorance of user preferences during monitoring, the PPoM service gathers only authorized user/usage data that users allow to monitor. By specifying privacy policies, users determine which data can be monitored. User policies will be then enforced by the PPoM service that understands users' privacy policies and enables selective monitoring based on user preferences. As for other monitoring services, we envision a set of service APIs (Application Programming Interfaces) that can be invoked by code snippets inserted in the original application code in order to monitor events or data of interest. The PPoM service does not capture data that users want to never be monitored. The PPoM service enables the monitoring code snippets inserted into the original application program to interpret user preferences written in the policy language and send only allowed data to the monitoring code on the server side. Unlike the existing approach where user data are collected based on application policies and enforced by the application itself, users will now be able to choose what data can be monitored and the user policies will be enforced by a trustworthy third-party monitoring service.

Using the PPoM service, e-health applications can guarantee effective protection of user privacy by providing a way to enforce user policies in a systematic manner rather than simply providing a written agreement. The concrete enforcement framework assists both monitoring services and e-health applications in building better reputations and accelerates active participation in e-health applications without concerns about privacy leakage. In addition, it can be used to compel and enable e-health applications to obey HIPAA regulations.

B. *Privacy-Preserving User Browser*

If an e-health application is trustworthy and both users and administrators can correctly specify their own policies, then existing approach may be able to guarantee secure online monitoring. However, this protection presumes the application's honesty and the user's ability to specify privacy policies precisely. Let us assume that an e-health application publishes untrusted policies that differ from its actual monitoring behavior. Once user's policies and the application's policies are matched, a user will start interaction with the dishonest application and his browser will send unconsented data to the application.

However, user privacy must be protected even if a user is exposed to untrustworthy parties, such as indiscriminate monitoring/analytics services that conduct online monitoring in violation of user privacy, naïve applications that do not have their own privacy policies or dishonest applications that write and/or enforce their policies dishonestly. To address the above security requirement, the PPoM-enabled web browser (in short, PPoM browser) protects user privacy on the user side. For easy adoption, we implemented a browser extension that enables existing browsers (e.g., Explorer and Chrome) to perform following tasks:

1) It presents all user data being monitored by online monitoring service(s) in web pages: A PPoM browser displays all monitoring data and recipients of data on the screen by analysing source code of web pages.

2) It allows a user to decide which usage and his/her data can be disclosed: A PPoM browser inspects data types, data values, and destinations of all outgoing message according to user policies. If it detects malicious monitoring that violates user policies, it alerts a user to unauthorized monitoring. For example, if the user decides to disclose his/her medical history to a first-party (e.g., an e-health application website), then a PPoM browser only allows outgoing messages to the e-health application and blocks other messages to other entities (such as advertisement companies or social networking sites) even if the application does not have its own privacy policy or inserts monitoring code to collect his/her medical history data.

3) It refines the user's privacy policies based on updated user preferences: If a user' privacy policies are naïve or incorrect, the proposed approach based on user policies cannot protect user privacy successfully. However, It is very difficult for average users to specify precise privacy policies because it is hard to understand the correlation between description of user data in privacy policies and web objects in web pages. To enhance the proposed PPoM framework, a PPoM browser allows users to refine their preferred policies by intuitively selecting what user/usage data can be monitored in web pages.

C. Privacy-Preserving Online Monitoring Tool

As pointed out in Section II.A, it is difficult for non-IT administrators to have professional IT knowledge and skills that necessary for trustworthy online monitoring. The PPoM Tools (PPoMT) helps health professionals by enabling them to specify privacy policies for their healthcare applications, and/or to easily convert their existing applications into privacy-preserving applications that analyze user/usage data without violating user privacy. While motivated by the needs of non-IT administrators, this tool could also improve the efficiency of developers who are required to insert monitoring code into their applications.

To achieve the purpose described above, the PPoMT provides the following services:

1) *User-friendly interfaces for selection of web objects*: To allow application administrators to intuitively select web objects to be monitored and specify privacy policies associated with the objects, the PPoMT provides user-friendly interfaces for selection of web objects to be monitored and corresponding policies.
2) *Semi-automatic conversion of existing e-health applications to privacy-preserving applications*: The biggest obstacle to the use of existing monitoring services is that they require manual insertion of monitoring code, often written in JavaScript, into online applications. Even if a privacy-preserving policy language and monitoring client code are successfully developed, they may still remain too complex for use by users other than IT experts. Existing website optimizers (such as Optimizely, Visual Website Optimizer, Loop11, and ABtests) allow users to improve applications through a visual editor without manual modification of source code. However, those tools are mainly focused on page layout and design tasks, such as changes on size/style/color/position of page elements (e.g., buttons and text). They typically do not have functionalities to modify existing applications in order to conduct online monitoring. To overcome this limitation, the PPoMT allows using online monitoring and analytics services without needing prerequisite IT knowledge or skills.

To do so, it is necessary to analyze an application's source code. There are two types of web pages, static web pages and dynamic web pages. Static web pages, called flat pages or stationary pages, are displayed in user browsers as they are stored in an application server, while dynamic web pages are displayed with dynamically changing web content based on environmental or user context such as time, user location, and browsing history. There are two different approaches to the creation of dynamic web pages: server-side and client-side. In the server-side approach, web contents can be dynamically created by server-side scripting languages such as PHP or ASP. JavaScript code, once downloaded into a browser, can also communicate with a server, enabling web contents to also be changed in a browser by real-time interactions with a server after the downloading of a webpage. In the client-side approach, web contents may be dynamically created by client-side JavaScript. In case of server-based dynamic web pages, an administrator needs to enter URLs for each page to monitor. In case of static pages and client-side dynamic pages, an administrator can either upload HTML source code or simply specify page URLs. Once the source code is obtained, the PPoMT analyzes HTML source code exactly and then add intuitive and friendly user interfaces into web pages in order to enable non-IT administrators to select monitoring objects and describe corresponding policies. To support fine-grained privacy-preserving monitoring in web-object level, each monitoring object must be uniquely identified. If an e-health application does not maintain unique identifiers (IDs) for web objects, the PPoMT must create object IDs before creating monitoring code.

Once the code modifications are made, e-health administrators must deploy the modified source code in their server. In the cases of static web pages or client-based dynamic web pages, administrators can easily deploy the updated source code by overwriting the directory that contained existing source code. In the case of server-based dynamic web pages, they will have to manually insert the newly created monitoring code into a PHP snippet that generates the head part of each HTML page.

3) *Semi-Automatic generation of privacy policies for e-health applications*: Many online applications have published their privacy policies to be regarded as a trustworthy application that cares about protecting user privacy. Because of the sensitivity of data being exchanged, it is especially important for e-health applications to gain users' trust. However, specifying a privacy policy is often beyond the ability of most e-health administrators who may be unfamiliar with online privacy policy languages. To help them publish privacy policies for their own applications, the PPoMT creates the privacy policies for a given application by gathering all monitoring objects and corresponding privacy policies that are selected by administrators.

Each service listed above is implemented as an individual tool within the PPoMT: the In-page Selector, the Monitoring Code Generator, the Privacy Policy Generator, and the Application Converter. The detailed explanations for each tool are following:

1) In-page Selector: It is a server-side software module that is capable of generating modified webpages that have user-friendly interfaces for selection of web objects to be monitored and corresponding policies and delivering user selections to the Privacy Policy Generator and the Monitoring Code Generator. It sends modified webpages to the PPoM browser of an administrator. By clicking on web objects that are displayed in the PPoM browser (e.g., a button, link, text, image, or video), the administrator can select which objects to monitor without any IT knowledge and skills. Additionally, he can declare corresponding privacy policies.

2) Monitoring Code Generator: It generates privacy-aware monitoring code by receiving an administrator's selection on monitoring objects and associated privacy policies from the In-page Selector. The generated code will be varied depending on monitoring services.

3) Privacy Policy Generator: It helps non-IT administrators who may be unfamiliar with online privacy policy languages publish privacy policies for their own applications. Using the Policy Miner, it derives the privacy policies for a given application by analyzing all monitoring objects and corresponding privacy policies that are selected by the administrator.

4) Application Converter: It helps non-IT administrators to update source code of an existing application by assisting them in inserting the privacy-aware monitoring code generated by the Monitoring Code Generator. Once the code modifications are made, an administrator needs to deploy the modified source code in their server.

IV. PRIVACY-PRESERVING MONITORING USING PPOM

For better understanding of how the PPoM framework protects user privacy during online monitoring and supports non-IT administrators and users, we present its operation flows and example use cases in this section.

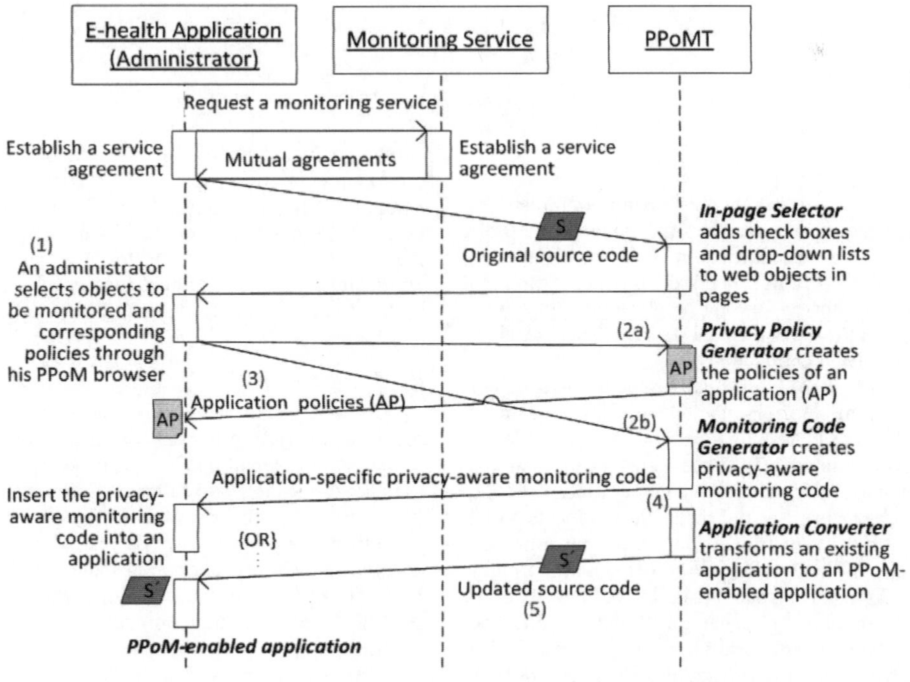

Figure 3. Operation flow of administrators of e-health applications in PPoM.

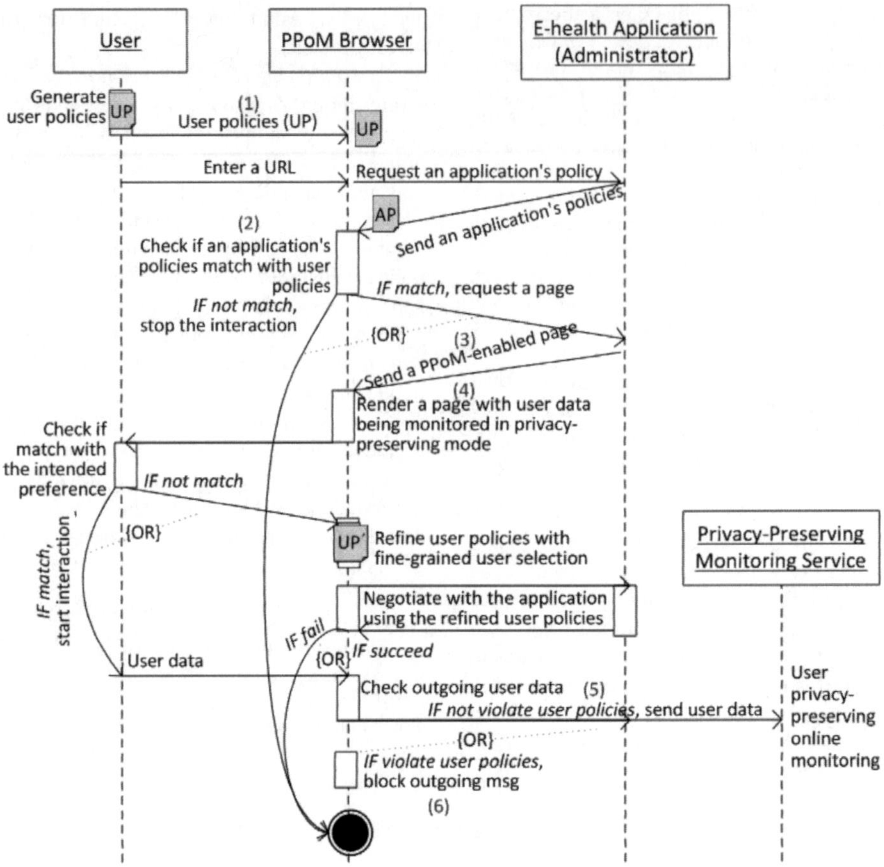

Figure 4. Operation flow of users in PPoM.

A. *For online healthcare providers*

Let us assume that Alice, a medical doctor, administrates an e-health application that assesses the impact to people following exposure to traumatic events. To trace patients' activities and collect health-related data, Alice wants to conduct online monitoring on her applications, but it is impossible for her to use existing online monitoring services due to lack of IT-related knowledge.

If she uses the PPoMT, she can easily transform her application into a monitoring-enabled application without any IT knowledge. For better understanding, let us look at its operational flow in detail. First, she needs to upload the source code of her application or enter the URL(s) of the application's webpage(s). Second, she selects objects to be monitored through the user-friendly interfaces generated by the In-page Selector, and specifies corresponding privacy policies (Figure 2.(1) and Figure 3.(2)). Third, her selections are delivered to the Policy Generator (Figure 2.(2a) and 3.(2a)) and the Monitoring Code Generator (Figure 2.(2b) and Figure 3.(2b)). Forth, the Privacy Policy Generator then creates the applications' policies by analyzing selected monitoring data and policies, while the Application Converter (which enables the application to perform privacy-preserving monitoring by inserting monitoring code generated by the Monitoring Code Generator into the original source code (Figure 2.(3) and Figure 3.(4)) produces the updated source code. Fifth, Alice deploys the created application policies (Figure 2.(4a) and Figure 3.(3)) and the updated source code in the application server (Figure 2.(4b) and Figure 3.(5)).

B. *For Users*

Let us assume that Bob is one of Alice's patients. Alice recommends Bob to use her e-health application every week to assess his mental and physical health but he hesitates to use the application due to privacy concern. If Bob uses the PPoM browser, he may want to use an e-health application without privacy loss. Towards this, Bob is first required to specify his own privacy policies and store his policies in his browser before using an e-health application (Figure 4.(1)). His PPoM browser then compares his privacy policies and application policies when he enters a url of Alice's application (Figure 4.(2)). If they match, the application server sends PPoM-enabled pages, which privacy-aware monitoring code is embed in (Figure 2.(5)(6) and Figure

4.(3)). As he interacts with the application, the PPoM browser displays all user/usage data being monitored to enable users to verify privacy protection during online monitoring (Figure 4.(4)). If there is no privacy violation, the privacy-aware monitoring code collects only authorized user data according to policies of him specified by him for the sake of himself (Figure 2.(7a)(7b) and Figure 4.(5)). To ensure enforcement of user policies, his PPoM browser blocks outgoing messages that violate his privacy policies on the user side (Figure 4.(6)).

V. PROTOTYPE IMPLEMENTATION

As a first step in the development of the PPoM framework, we implemented a prototype of the PPoM browser. It analyzes an e-health application's HTML source code in which PPoM monitoring code is embedded, displays a webpage with information about monitoring status, and prevents privacy leakage by blocking outgaining message containing unconsented user/usage data.

An example use of the prototype of PPoM browser is shown in Figure 5. When a user specifies a URL of an e-health application such as the online weight tracker of *sparkpeople.com*, an interaction is initiated. If a user turns on privacy-preserving mode, a PPoM browser renders a web page that is sent by an server of *Spark People* with information about which data being monitored in that page. The PPoM browser displays which user data is being monitored and who is the recipient of the data by using different-colored check marks. In addition, the types of data that are being monitored such as HTTP, Clientevents, and Cookies types of Dynamic data schema are also shown in the status bar (e.g., Your Goal and An answer to the question, How Active Are you? in this example).

If the monitoring data are in accordance with the user's policy, then the user keeps interacting with the application. Otherwise, the user can change the monitorable user data and its disclosure level using one of the following four options: 1) can be disclosed to all parties (*green check mark*), 2) can be disclosed to only first party (*orange check mark*), 3) prohibited to use/disclose it (*red check mark*), and 4) preference is not described yet (*no check mark*). With the specific user selection in web object level, the browser refines user policies and negotiates with the application using the refined policies. If the negotiation is successful, the interaction continues, otherwise the browser stops the interaction with the application. If online monitoring is acceptable to the user (e.g., based on past usage experience and/or trust), he/she can turn off the privacy-preserving mode. Using the PPoM browser, users can easily know what user data are being monitored and prevent user data from unwanted monitoring by turning on privacy-preserving mode.

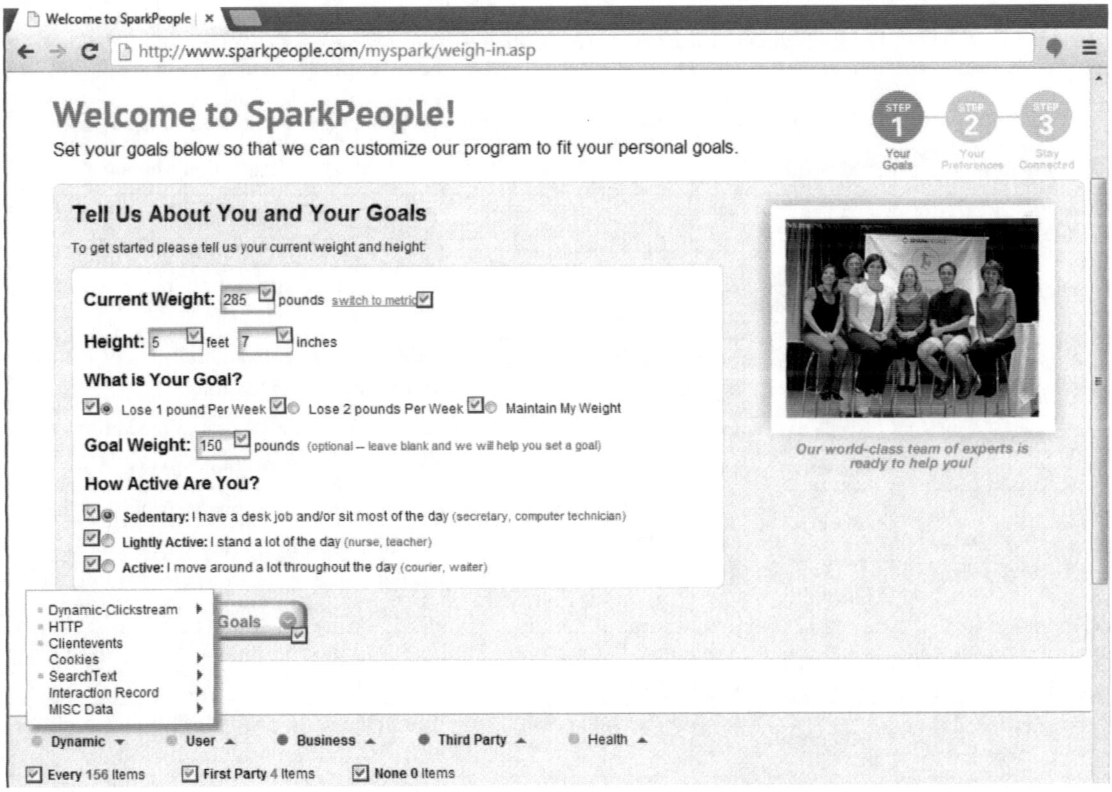

Figure 5. Example use of the prototype of PPoM browser

VI. Evaluation Plan

To evaluate the successful blockade ratio of the number of messages containing unauthorized data ($c+d$ in Table I) to the number of blocked messages among them (c in Table I).

TABLE I. OUTGOING MESSAGES FROM A PPOM BROWSER

	Sent	Blocked
Allowed	a	b
Not allowed	c	d

In addition, the performance of the proposed privacy-preserving monitoring service will be evaluated by calculating the ratio of the number of all monitoring data ($e+g$ in Table II) to the number of unauthorized but collected data (g in Table II).

TABLE II. USER DATA MONITORED BY THE PPOM SERVICE

	Monitored	Not monitored
Allowed	e	f
Not allowed	g	h

To test the developed tool, PPoMT, we plan to compare a user's selections of monitorable data and corresponding privacy preferences with monitoring scripts or source code that generated by PPoMT. The correctness of the generated privacy policies is also required to be verified. To do so, we plan to create a variety of anticipated privacy policies and then compare the anticipated policies with resulting policies that are generated by PPoMT. In Table III, the overall evaluation plans and methods for each components of the proposed PPoM framework are presented. As you can see, we plan to conduct human-subject usability tests in order to test the user-friendliness of the PPoM framework.

VII. Related Work

In this section, we introduce existing work on privacy protection during online monitoring on both the service provider side and the user side.

A. *Provider-side privacy protection for online monitoring*

To guarantee secure user monitoring, a few methods have been proposed by providers of online applications and monitoring services. Some third-party advertising companies have voluntarily begun to insert an 'Adchoices' icon into their ads to increase user awareness of online tracking. However, it has been found that the icon was not very effective at making users aware of tracking occurrences [21]. As a middleware approach, Privad [22][23] is proposed to conceal a user's activities from an advertising network by interposing an anonymizing proxy between the browser and the ad network. However, the adoption of a proxy-based middleware may not be a feasible solution to small-size e-health applications because of its huge overhead requirements. In addition, it is useless if an e-health application requires identifiable user data to analyze performance of applications at the individual level.

TABLE III. OVERALL EVALUATION OF THE PPOM FRAMEWORK

Component		Measure	Control group	Evaluation Method
PPoM Service		Accuracy	All data collected by existing monitoring services	Percentage of data that can be protected by the privacy-aware monitoring service
PPoM Browser		Accuracy	All data collected by existing monitoring services	Percentage of data that can be protected by the PPoM browser extension
			Monitored data collected by a monitoring service	Completeness of monitored data displayed by PPoM browser extension
			Ideal user policies	Comparison of ideal and refined policies
		Usability	None	Usability survey of PPoM browser
PPoMT	In-Page Selector	Accuracy	Webpages displayed by existing browsers	Completeness of webpage display by the In-page Selector
		Usability	None	Usability survey of webpages modified by the In-page Selector
	Application Converter	Accuracy	Monitored data without privacy violations	Privacy violations of monitored data collected by deployed code
	Privacy Policy Generator	Accuracy	Ideal application policies	Correctness of Policy Miner policies

B. User-side privacy protection for online monitoring

There are two major approaches for privacy protection on the user side, the browser-based approach and the policy-based approach.

1) Browser-based approaches: To protect user privacy in user side, browser-based approaches have been proposed. Adnostic [24], a browser extension, is capable of behavioral profiling and targeting in users' browsers to select effective ads while not sending user data to third-party ad companies. RePriv [25] enables browsers to conduct user interest mining and only share the resulting encapsulated interests with third-parties. Both Adnostic and RePriv have only focused on targeted advertising and/or personalization but have not considered online monitoring services. As a simple solution to indiscriminate online monitoring, using opt-out cookies and/or a blocked-application list have been recommended. Opt-out cookies are, however, fragile because they can be easily disabled or deleted by a third party [26][27]. Setting a block list in a browser can effectively block malicious applications but currently this approach blocks any listed application in its entirety and does not support fine-grained blocking at the data level.

2) Policy-based approaches: As a policy-based protection approach, Privacy Bird [28] has been proposed. It is a P3P user agent that reads P3P policies of online applications and lets users know whether the application policies and user preferences are matched. If policies are not matched, a bird icon turns red. A user can get information by clicking on a red bird icon. However, Privacy Bird is only able to check the acceptability of application's P3P policies, so users cannot check all user data monitored at the application's data level. Moreover, it does not support policy refinement. HummingBird [30] is a privacy-preserving online social network system. Unlike Twitter where all tweets are visible to everyone, it restricts accesses to tweets based on user-defined access control list (ACL). In HummingBird, tweeters encrypt their tweets and then the HummingBird server distributes a decryption key to only authorized followers. To do so, the Hummingbird client deals with key management and tweet encryption. However, it does not encrypt other user data such as user profile and also is not concerned with user privacy policies at all.

VIII. CONCLUSION AND FUTURE WORK

Although many e-health applications enable people to access healthcare services in easy and convenient way at the reduced cost, the lack of reliable and effective methods of privacy protection has been the biggest obstacle to the growth of e-Health applications. Without a suitable solution, people keep hesitating to use e-Health applications due to privacy concerns. To address the privacy protection issue above, the PPoM framework was proposed as a preliminary work in order to protect user privacy in both the application side and the user side. By using the PPoM framework, secure online monitoring can be guaranteed in the entire process of online monitoring, and in turn it accelerates practical use of e-health applications. Towards this goal, in this paper, we enhance mechanisms for each component in the PPoM Framework and present a prototype of the PPoM browser as a first step in the development of the PPoM framework but a few challenges still need to be pursued in the future:

- Development of the privacy policies to specify preferences on healthcare data in fine-grained level.
- Development of a privacy-preserving monitoring service that protects user privacy based on user policies.
- Development of the PPoMT for non-IT medical staff.
- Evaluation of individual components and an integrated framework.
- Field test associated with actual clinical trials.
- Development of a threat model for PPoM and security test using a threat model.

REFERENCES

[1] Y. Jung, "Privacy-Preserving Online Monitoring Framework for e-Health Applications," Proc. Fourth International Conference on Global Health Challenges (GlobalHealth), IARIA, Nice, France, 2015, pp. 18-24.

[2] J. R. Mayer and J. C. Mitchell, "Third-Party Web Tracking: Policy and Technology," Proc. IEEE Symp. on Security and Privacy (SP '12), IEEE Press, 2012, pp. 413-427, doi=10.1109/SP.

[3] B. A. Schroeder, "On-line monitoring: A tutorial," IEEE Computer, vol. 28, no. 6, pp. 72-78, 1995.

[4] N. Schmucker, "Web Tracking," SNET2 Seminar Paper (Online), 2011. http://www.snet.tu-berlin.de/fileadmin/fg220/courses/SS11/snet-project/web-tracking_schmuecker.pdf [retrieved: May, 2016].

[5] Google Analytics. http://www.google.com/analytics/index.html [retrieved: May, 2016].

[6] Adobe Analytics. http://www.adobe.com/sea/solutions/digital-analytics.html [retrieved: May, 2016].

[7] M. N. K. Boulos et al., "CAALYX: a new generation of location-based services in healthcare," Int. J. Health. Geogr., vol. 6, no. 1, March 2007, pp. 1-6, doi:10.1186/1476-072X-6-9.

[8] C. Hoelscher and H. Dietrich, "E-Commerce Personalization and Real-Time Site Monitoring," in Designing personalized user experiences in eCommerce, Kluwer Academic Publishers, 2004, pp. 95-117.

[9] C. L. Borgman, S. G. Hirsh, and J. Hiller, "Rethinking online monitoring methods for information retrieval systems: from search product to search process," J. Assoc. Inf. Sci. Technol., vol. 47, no. 7, pp. 568-583, July 1996.

[10] G. Eysenbach, "What is e-health?," J. Med. Internet. Res., vol.3, no. 2, 2001.

[11] J. M. Bernhardt and J. Hubley, "Health education and the Internet: the beginning of a revolution," Health. Educ. Res., vol. 16, no. 6, pp. 643-645, 2001.

[12] E. M. Daley, R. J. McDermott, K. R. B. McCormack, and M. J. Kittleson, "Conducting web-based survey research: a lesson in internet designs," Am. J. Health. Behav., vol. 27, no. 2, pp. 116-24, March-April 2003.

[13] D. F. Duncan, J. B. White, and T. Nicholson, "Using Internet-based Surveys to Reach Hidden Populations: Case of Nonabusive Illicit Drug Users," Am. J. Health. Behav., vol. 27, issue 3, pp. 208-218, May-June 2003.

[14] K. E. Evers, "eHealth promotion: the use of the Internet for health promotion," Am. J. Health. Behav., vol. 20, issue 4, pp. 1-7, March-April 2006.

[15] M. Bilenko and M. Richardson, "Predictive client-side profiles for personalized advertising," Proc. ACM SIGKDD conf. on Knowledge discovery and data mining (KDD '11), ACM New York, 2011, pp. 413-421.

[16] A. McDonald and L. F. Cranor, "Beliefs and Behaviors: Internet Users' Understanding of Behavioral Advertising," TPRC 2010 Social Science Research Network (SSRN), August 16, 2010, pp. 1-31, http://ssrn.com/abstract=1989092.

[17] Wikipedia, "The Honor System," http://en.wikipedia.org/wiki/Honor_system [retrieved: May, 2016].

[18] U.S. Department of Health and Human Services Office for Civil Rights, "HIPAA Administrative Simplification Regulation Text, 45 CFR Parts 160, 162, and 164 amended through March 26," 2013, pp. 1-115, http://www.hhs.gov/sites/default/files/ocr/privacy/hipaa/administrative/combined/hipaa-simplification-201303.pdf.

[19] P. Ohm, "Broken Promises of Privacy: Responding to the Surprising Failure of Anonymization," UCLA Law Review, vol. 57, pp. 1701-2010, August 2009.

[20] P3P 1.1. http://www.w3.org/TR/P3P11/ [retrieved: May, 2016].

[21] WS-XACML 1.0. http://xml.coverpages.org/Anderson-WS-XACML v10.pdf [retrieved: June, 2015].

[22] M. Hastak and M. J. Culnan, "Online Behavioral Advertising "Icon" Study," Future of Privacy Forum (online), 2010, http://www.futureofprivacy.org/2010/02/15/online-behavioral-advertising-icon-study [retrieved: June, 2015].

[23] S. Guha, B. Cheng, and P. Francis, "Privad: practical privacy in online advertising," Proc. USENIX conf. on Networked systems design and implementation (NSDI'11), USA, 2011, pp. 169-182.

[24] A. Reznichenko, S. Guha, and P. Francis, "Auctions in do-not-track compliant internet advertising," Proc. ACM conf. on Computer and communications security (CCS '11), New York, 2011, pp. 667-676, doi=10.1145/2046707.2046782.

[25] V. Toubiana, H. Nissenbaum, A. Narayanan, S. Barocas, and D. Boneh, "Adnostic: Privacy Preserving Targeted Advertising," Proc. Symp. on Network and Distributed System Security, March 2010, pp. 1-21.

[26] M. Fredrikson and B. Livshits, "REPRIV: Re-Envisioning In-Browser Privacy," Proc. IEEE Symp. on Security and Privacy, May 2011, pp. 1-15.

[27] P. Leon et al., "Why Johnny can't opt out: a usability evaluation of tools to limit online behavioral advertising," Proc. ACM Conf. on Human Factors in Computing Systems (CHI '12), USA, 2012, pp. 589-598, doi=10.1145/2207676.2207759.

[28] M. Ayenson, D. J. Wambach, A. Soltani, N. Good, and C. J. Hoofnagle, "Flash cookies and privacy II: Now with HTML5 and etag respawning," Social Science Research Network, July 2011, http://dx.doi.org/10.2139/ssrn.1898390.

[29] Privacy Bird, http://www.privacybird.org [retrieved: May, 2016].

[30] E. D. Cristofaro, C. Soriente, G. Tsudik, and A. Williams., "Hummingbird: Privacy at the Time of Twitter," Proc. of the IEEE Symposium on Security and Privacy (SP'12), IEEE Computer Society, Washington D.C., USA, 2012, pp. 285-299, DOI=10.1109/SP.2012.26

Agile Medical Device Software Development

Introducing Agile Practices into MDevSPICE®

Fergal McCaffery, Marion Lepmets, Kitija Trektere
Regulated Software Research Centre & Lero,
Department of Computing & Mathematics
Dundalk Institute of Technology
Dundalk, Co. Louth, Ireland
e-mail:{fergal.mccaffery, marion.lepmets, kitija.trektere}@dkit.ie

Özden Özcan-Top
Information Systems, Informatics Institute
Middle East Technical University
Ankara, Turkey
e-mail: ozdentop@gmail.com

Minna Pikkarainen
Centre of Health and Technology
University of Oulu, VTT, Technical Research Centre of Finland
Oulu, Finland
e-mail: minna.pikkarainen@oulu.fi

Abstract— **Medical device software is usually embedded within the overall system as one of the sub-systems. It needs to be integrated with other sub-systems such as the electrical and mechanical for a functional medical device to be developed. In order to develop a working medical device system through integrating its sub-systems, the sub-systems' requirements have to be derived from the overall medical device system requirements. The system requirements are continuously collected, analysed and built from the needs of different stakeholders such as patients, health professionals and other companies offering relevant devices, interfaces and software related to the medical device system under development. Various regulatory requirements have to be achieved for a medical device to be allowed market access. We have developed and piloted a medical device software process assessment framework called MDevSPICE® that integrates the regulatory requirements from the relevant medical device software standards. This paper describes how the MDevSPICE® framework has been designed to enable medical device software developers to produce software that will be safe and easily integrated with other sub-systems of the overall medical device. We also describe the lessons learned from piloting MDevSPICE® in the medical device industry and introduce an agile methodology together with its benefits and challenges. This paper outlines how MDevSPICE® can be extended to include agile practices to enable medical device software development to be performed in a more flexible manner.**

Keywords- medical device software; MDevSPICE®; medical device risks; medical device software development; agile methods; agile software development practices.

I. INTRODUCTION

Safety-critical software systems are increasingly affecting our lives and welfare as more and more software is embedded into safety critical systems such as hospital systems, medical devices, cars and airplanes. New approaches and international standards are being developed to ensure the safety of these systems before they are delivered. The integration of software into the complete medical device requires particular attention [1].

In order to market a medical device, the manufacturer has to satisfy a number of regional regulatory requirements commonly achieved by following international standards and guidance issued by international standardizing bodies and regional regulatory authorities. Additionally, in order for the solution to sell, the medical device also needs to fulfil the requirements of patients, health professionals and other medical system interface providers.

To help software companies in the medical device domain reach regulatory compliance, we have developed an integrated framework of medical device software development best practices called MDevSPICE®. This framework integrates generic software development best practices with medical device standards' requirements enabling robust software process assessments to be performed while preparing for a regulatory audit. The "SPICE" in MDevSPICE® reflects its foundation in the ISO/IEC 15504 (SPICE) [2] series of standards for process assessment. In this paper we describe the validation of the MDevSPICE® framework that provides evidence of the importance of traceability between the system and software levels of development. We also explain how the establishment of robust interface requirements for these two levels supports more effective software integration

In Section II, we provide an overview of the regulatory requirements medical device software development companies face before they are able to market their devices. In Section III, we describe the development of the MDevSPICE® framework. Section IV, outlines the lessons learned when validating the framework in expert reviews and in industry through MDevSPICE® pilot assessments. We also discuss the importance of traceability between system and software development processes when developing an embedded medical device software system as it increases the safety and quality of the developed medical device. In Section V, we introduce an agile methodology by outlining its benefits, challenges and its suitability for medical device software development. The paper concludes in Section VI with areas of future research related to agile medical device software development.

II. Medical Device Regulations

A medical device can consist entirely of software or have software as a component of the overall medical device system. In order to be able to market a medical device within a particular region it is necessary to comply with the associated regulatory demands of that region. Two of the largest global bodies responsible for issuing and managing medical device regulation belong to the central governing functions of the US and EU.

In the US, the Food and Drug Administration (FDA) issues the regulation through a series of official channels, including the Code of Federal Regulation (CFR) Title 21, Chapter I, Subchapter H, Part 820 [3]. Under US regulation, there are three medical device safety classifications: Class I, Class II and Class III. The medical device safety classification is based on the clinical safety of the device. Class I devices are not intended to support or sustain human life, and may not present an unreasonable risk of harm. A thermometer is a Class I device. Class II devices could cause damage or harm to humans. An example of a Class II medical device is a powered wheelchair. Class III medical devices are usually those that support or sustain human life, and are of significant importance in the prevention of human health impairment. An example of a Class III device is an implantable pacemaker. All implantable devices are Class III medical devices as the surgery required carries with itself additional high risks from anaesthesia and possible infections that go beyond the safety risks of the medical device.

In the EU, the corresponding regulation is outlined in the general Medical Device Directive (MDD) 93/42/EEC [4], the Active Implantable Medical Device Directive (AIMDD) 90/385/EEC [5], and the *In-vitro* Diagnostic (IVD) Medical Device Directive 98/79/EC [6] - all three of which have been amended by 2007/47/EC [7]. Similarly to the US, the EU device safety is also based on the clinical safety of the device embodying similar classifications and limitations, where Class I in the EU corresponds to Class I in the US, Class IIa and IIb to Class II, and Class III to Class III.

A further safety classification applies to the software in medical devices as outlined in IEC 62304:2006 [8], where the safety classification is determined based on the worst possible consequence in the case of a software failure. In the case of failure of software that is of safety Class A, no injury or damage to the health of a patient can occur. When software of safety class B fails, injury may occur but it is not serious or life-threatening. Class C medical device software is the highest risk and in the case of failure of such software death or serious injury can happen. Depending on the functionality of software within the medical device, the software safety classification may vary from the overall medical device safety class. When software involves critical functionality of the medical device, it will carry the same classification as the device, i.e., Class C software in a Class III device. The safety classification of software may be lower but cannot be higher than the overall medical device safety class, e.g., software of safety Class B, may be embedded in Class III device but there cannot be software of safety Class C, in a Class I or Class II device.

Medical device manufacturers in the US as well as in EU must satisfy quality system requirements to market their developed devices. In the medical device domain, ISO 13485:2003 (ISO 13485 from hereon) [9] outlines the requirements for regulatory purposes from a Quality Management System (QMS) perspective in medical device domain. ISO 13485, which is based on ISO 9001 [10], can be used to assess an organization's ability to meet both customer and regulatory requirements in the medical device domain. ISO 13485 does not, however, include requirements for software development. IEC 62304, which can be used in conjunction with ISO 13485, does offer a framework for the lifecycle processes necessary for the safe design and maintenance of medical device software. As a basic foundation, IEC 62304 assumes that medical device software is developed and maintained within a QMS such as ISO 13485, but does not require an organization to be certified against ISO 13485. Therefore, IEC 62304 can be considered to be a software development specific standard supplement to ISO 13485, similar to ISO 90003 for ISO 9001.

IEC 62304 is based on ISO/IEC 12207:1995 [11], which although a comprehensive standard for software development lifecycle processes, has effectively been decommissioned following the publication of the more extensive ISO/IEC 12207:2008 [12]. Furthermore, other developments in the ISO and IEC communities for software development, such as ISO/IEC 15504 [13], have provided significant additional levels of software process detail to support ISO/IEC 12207:2008. IEC 62304 is a critical standard for medical device software developers as it is the only standard that provides recommendations for medical device software implementations based on the worst consequences in the case the software failure causing hazards. For general medical device risk management, IEC 62304 is used in conjunction with ISO 14971 [14] and IEC 80002-1 [15] that provides guidance on the application of ISO 14971 for software development.

Since IEC 62304 considers a medical device system to consist of software as a sub-system, the system or product level requirements are not included within IEC 62304 but instead within the medical device product standard of IEC 60601-1 [16]. Due to the increasing importance of usability of devices within the medical device industry, organizations should also adhere to the medical device usability engineering process requirements outlined in IEC 62366 [17]. When the Medical Device Directives were amended in 2007 [6], this defined standalone software to be a medical device in its own right. Previously, software had always been seen as a subsystem embedded in a medical device. This amendment revealed a gap in international standards as none of the published standards were addressing the concerns for standalone software as a medical device. Today, IEC CD 82304-1 [18] applies to the safety of healthcare software that is designed to operate on general purpose IT platforms and that is intended to be placed on the market without dedicated hardware, e.g., iPad applications.

All companies planning to market a medical device in the United States need to register their product with the US FDA. Most Class I devices can be self-registered but most Class II devices require a 510(k) submission. For Class III devices, a Pre-Market (PMA) submission is needed. To support manufacturers in addressing the relevant guidance, the FDA has issued an overview of their guidance documents for medical device manufacturers and software developers [19]. The FDA Guidance on Premarket Submissions [20] provides guidance and recommendation for premarket submissions for software devices, including standalone software applications and hardware-based devices that incorporate software. Premarket submission includes requirements for software-related documentation that should be consistent with the intended use of the Software Device and the type of submission. The FDA Guidance on Off-The-Shelf Software Use in Medical Devices [21] was published in 1999 with the purpose of describing the information that should be provided in a medical device application that uses off-the-shelf (OTS) software. Many of the principles outlined in this guidance document may also be helpful to device manufacturers in establishing design controls and validation plans for use of off-the-shelf software in their devices. The FDA General Principles of Software Validation [22] outlines general validation principles that the FDA considers to be applicable to the validation of medical device software or the validation of software used to design, develop, or manufacture medical devices. This guidance describes how certain provisions of the medical device Quality System regulation apply to software. The scope of this guidance is somewhat broader than the scope of validation in the strictest definition of that term to support a final conclusion that software is validated.

The challenge that software development companies in the medical device domain face when they want to market a device is in the adherence to a large number of regulatory requirements specified in various international standards that can often become overwhelming. In order to help these companies better prepare for the demanding and costly regulatory audits, we developed the MDevSPICE® framework. MDevSPICE® includes requirements from all the previously mentioned standards and FDA guidance documents rendering the task of regulatory compliance much less complex. Following is a description of the development of the MDevSPICE® framework that integrates the requirements from various international medical device standards and guidance documents with the generic software development best practices while providing a possibility to assess processes.

III. MDevSPICE® Framework

This section describes the development of the MDevSPICE® process reference model, the MDevSPICE® process assessment model, the support MDevSPICE® provides for software and system integration, and the validation of the MDevSPICE® framework through pilot assessments in medical device industry.

A. *Development of the MDevSPICE® Process Reference Model*

A process reference model (PRM) describes a set of processes in a structured manner through a process name, process purpose and process outcomes where the process outcomes are the normative requirements the process should satisfy to achieve the purpose of the process. In order to develop a PRM that integrates requirements from various standards allowing the processes to be evaluated in terms of their achievement of their purpose statements, we followed the format of the process description illustrated in ISO/IEC 24774 [23]. With that in mind, we first mapped and integrated the requirements from ISO/IEC 12207:2008 and IEC 62304 into what today is called the PRM for IEC 62304 that also reflects the updates to ISO/IEC 12207 from the 1995 to the 2007 version. A systematic approach of memoing and constant comparison, which is based on the principles of Grounded Theory [24] was followed when developing the PRM, further details of which are to be found in [25. The Process Reference Model of IEC 62304 was published in June 2014 as IEC TR 80002-3 [26].

While IEC 62304 describes only the software lifecycle processes, additional processes should be in place for system development in the case where software is not embedded as part of an overall medical device. These additional processes were derived from ISO/IEC 12207:2008. Design and development related requirements from ISO 13485 and ISO 14971 were also added to the MDevSPICE® Process Reference Model. Both ISO 13485 and ISO 14971 are *de facto* standards for medical device software organizations. ISO 13485 requirements are primarily related to system level processes and ISO 14971 is concerned with risk management (and therefore aligned with the Software Risk Management process of the PRM.

The final MDevSPICE® PRM consists of 23 processes of which 10 are system lifecycle processes, 8 are software lifecycle processes and the remaining 5 support both the system and lifecycle processes as can be seen in Figure 1.

The MDevSPICE® PRM was then extended with additional elements to create a process assessment model (PAM). The aim of the MDevSPICE® PAM is to provide a comprehensive model for assessing the software and systems development processes against the widely recognized medical device regulations, standards and guidelines that a software development organization in the medical device domain has to adhere to. The MDevSPICE® PAM, similar to ISO/IEC 15504-5 (SPICE) [26], has two dimensions – a process dimension and a capability dimension. The process dimension lists three groups of processes from various models and standards, i.e., systems lifecycle processes, software lifecycle processes and support processes. Each process is described in terms of a Process Name, Process Purpose, Process Outcomes, Base Practices, Work Products and Work Product Characteristics.

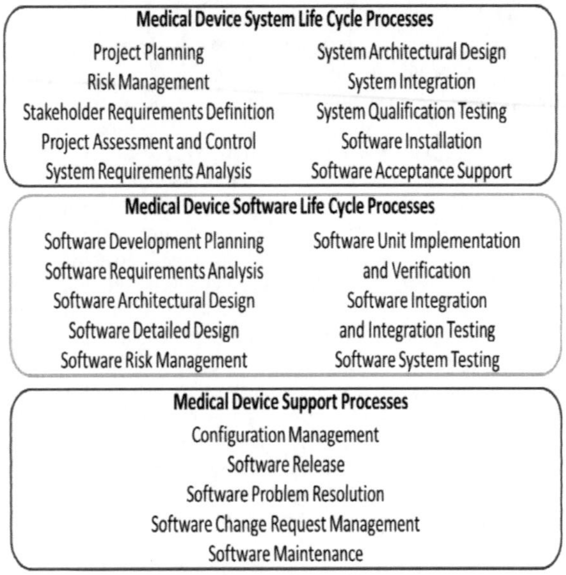

Figure 1. Processes of MDevSPICE® PRM

B. *Development of the MDevSPICE® Process Assessment Model*

The MDevSPICE® PRM is based on IEC 62304, ISO/IEC 12207:2008, ISO 14971 and ISO 13485. The MDevSPICE® PAM then extends this PRM with base practices and work products, some of the latter also being normative as they are described in IEC 62304, ISO 14971 or ISO 13485 as requirements. Where process outcomes are derived from ISO/IEC 12207:2008, their corresponding base practices and work products are derived from ISO/IEC 15504-5. Where process outcomes are derived from ISO 14971, their corresponding base practices are derived from IEC 80002-1. In addition to these sources, FDA guidance on premarket submissions, software validation and off-the-shelf software have been added to the informative base practices where the base practice did not already address the requirements of the corresponding FDA guidance. Product safety requirements have been added to the MDevSPICE® PAM from both IEC 60601-1 and IEC CD 82304-1, while the usability engineering requirements have been incorporated from IEC 62366.

The capability dimension of the MDevSPICE® PAM is derived directly from ISO/IEC 15504 together with the Capability Levels, Process Attributes, Generic Practices, Generic Resources and Generic Work Products.

While integrating processes from different standards and guidance documents for the MDevSPICE® PRM and PAM , a focus on the traceability between and within system and software lifecycle processes was maintained [27]. Both the FDA General Principles of Software Validation [22] and ISO/IEC 12207 [12] incorporate traceability of risks, changes and requirements throughout the development lifecycle. This interaction and traceability of requirements is a key enabler of subsequent integration, and it has a vital role to play in raising the safety of medical device software.

C. *MDevSPICE® Framework's support for integration*

The MDevSPICE® framework contains key facilities for integrating medical device software. Since MDevSPICE® is grounded in IEC 62304, the software sub system decomposition is consistent with the requirements of IEC 62304, meaning that the language of a *software unit*, a *software item* and a *software system* is adopted.

A software system is the integrated collection of software items to accomplish a specific function or set of functions; a software item is any identifiable part of a computer program; and a software unit is a software item that is not subdivided into other items. This software system hierarchy has an important role to play when a software developer wishes to decompose a system into parts of varying software safety classification. A benefit of such decomposition is that those parts of the software subsystem that are vital for safety (and which require additional safety activities when under development) can be isolated until they are later integrated with the other software components. It is also important that when the components are integrated that the safety implications are reflected in test cases that are pre-defined, then tested and the results are checked to ensure that they match the expected results. Otherwise sign-off cannot take place at the various levels – unit tests, integration tests and system tests.

Integration activities in the MDevSPICE® framework start by integrating software units into software items, and thereafter software items are further integrated with each other (and possibly with other units as well) into the software subsystem (which in turn is integrated into the overall medical device system). In other words, there are several levels of integration and they must take into consideration the safety implications at each step. It is further the case that the bi-directional traceability of requirements (including requirements related to safety) from the product level right down to the individual software unit level is supported in MDevSPICE® thus further enhancing medical device software safety at the integration stage and beyond.

D. *Piloting the MDevSPICE® Framework*

The MDevSPICE® framework has been validated in various stages of its development by different parties through both international expert reviews and industrial trials. The foundation of the MDevSPICE® PAM, IEC TR 80002-3 (the development of which was led by the authors), was published after several iterations of development and analysis by the standardization working group responsible for the publication of IEC 62304 (i.e., ISO/IEC Sub-Committee 62A, Joint Working Group 3). An international standard is published only after the national delegates of the standard's working group have agreed on every detail of that standard.

In addition to working with the international medical device standards community, the MDevSPICE® PAM has also been developed together with and analysed by experts of the Working Group 10 of ISO/IEC Joint Technical Committee 1, Sub-Committee 7, responsible for the development and maintenance of the series of process assessment standards. These standards are currently being revised from ISO/IEC 15504 series to ISO/IEC 330xx series of standards. MDevSPICE® framework keeps abreast of these updates as well as with the updates of any other standard and guidance document information, which is contained in the MDevSPICE® framework.

Upon successful completion of international expert reviews, the MDevSPICE® process assessment framework was then validated in the medical device software industry through pilot assessments over the past two years. MDevSPICE® process assessments were conducted in different types of organizations: (1) a small software company wishing to supply software to a large medical device manufacturer who wants them to demonstrate that they are capable of developing safe medical device software and provide the medical device manufacturer with a feeling that they will not jeopardize the safety of their overall medical device or the reputation of their organization; (2) three different assessments (across a 2 year period) were performed in two different international sites of a multinational medical device manufacturer who wants to ensure that they are incorporating best practices within their software development processes to not only achieve regulatory compliance but also reduce the likelihood of recalls through developing better quality and more robust software; (3) a software development company seeking to achieve regulatory compliance against IEC 62304 so that they can become medical device software suppliers; and (4) a large automotive manufacturer experienced in developing safety-critical embedded automotive software now wishing to also develop embedded medical device software.

IV. Lessons Learned from Piloting MDevSPICE®

As a result of the MDevSPICE® pilot assessments we have witnessed different types of needs and challenges that companies face in medical device software development.

Companies that manufacture medical devices as well as develop embedded software for their devices, manage traceability and integration between systems and software lifecycle processes well. This might be due to systems and software engineers working closely together in building a safe medical device where the software developers are aware of the system risks and requirements.

For companies that develop and supply software to large medical device manufacturers it can be very difficult to become aware of the overall system level requirements including the requirements of end users, e.g., patients, health professionals and related interfaces, as well as the risks before the software development project commences. Medical device manufacturers working on innovative devices are sometimes reluctant to provide their software subcontractors with the details of their device design and end user requirements as this could jeopardize device novelty or competitive advantage. Yet, the safety risks related to the performance of medical devices can outweigh such business risks when the medical device manufacturer has a proper legal know-how and proficiency about the market needs. When the system requirements are not provided to the software developers, the traceability engineering and integration of the sub-systems of the medical device will be hindered. Therefore, we would recommend medical device manufacturers to communicate with their software subcontractors more openly in order to best support risk and requirements management throughout their device design – even if this only encompasses those product requirements, which are related to the software requirements (and especially those, which are safety related). Although there is a potential issue in capturing, managing and changing requirements throughout the development of a medical device, the ultimate goal for all device manufacturers is to have a safe medical device on the market and not risk liability or damage of their brand as a result of a recall of a faulty device.

V. Agile for Medical Device Software Development

It is generally believed that technology will automatically improve health care efficiency, quality, safety, and cost, however, few people consider that technologies may also introduce errors and adverse events. Nearly 5,000 types of medical devices are used by millions of health care providers around the world [28]. While this technology holds much promise, the benefits of the technology are not always realized due to poor technology design that does not adhere to human factors, a poor technology interface with the patient or environment and an inadequate plan for implementing a new technology into practice [29].

Future trends indicate that medical software and devices in which clinical decisions will be guided by individual patient preferences, combined personal and medical data as well as specific needs and values [30]. In this case, continuous requirements collection and involvement of different stakeholders such as patients, health professionals and interface providers can be seen as essential for the future success of medical device software development. However, the development of medical devices that target the needs of either the patient or the health professional can be difficult when adopting a traditional, plan-driven software development approach where all system requirements should be known at the beginning of the development process. We believe that agile software development methodologies could provide support in achieving this challenge when delivering medical device software. In the next sub-sections we describe agile software development together with its benefits, the challenges it presents when adopted in the medical device domain and a brief justification as to why agile practices should be integrated into the MDevSPICE® framework.

A. What is Agile?

In recent years agile software development methodologies have gained significant interest in the IT community with proposed solutions to the problems of traditional, plan-driven software development approaches. "It's a framework, attitude, and approach to software delivery that is lean, fast, and pragmatic. It's no silver bullet, but it dramatically increases your chances of success while bringing out the best your team has to offer" [31].

Agile software development is a set of principles and practices used by self-organizing teams to rapidly and frequently deliver customer-valued software. It follows an Incremental and Evolutionary lifecycle, emphasizing close collaboration between the software development team, the customer, and other stakeholders. It is adaptable, emphasizing the need to adjust the principles and practices to fit the context and environment in which the software is being created [32].

B. Benefits of Agile

A priority of an agile methodology is to "satisfy the customer through early and continuous delivery of valuable software" [33]. There are many benefits that an agile methodology promotes, such as, improved quality, sustainable development, continuous attention to technical excellence as well as changing to this sort of approach is welcomed even late in the development. By adopting agile practices, the speed to market is improved, supporting the achievement of competitive advantage in the market.

Agile principles also dictate that "working software is the primary measure of progress" [33]. Unlike the extensive upfront planning and heavyweight processes and bureaucracy required for plan-driven development; agile development focuses on delivering highest business value to the customer through: short time-boxed iterations; receiving and providing fast feedback; collaborating with stakeholders, making use of self-organizing teams, embracing requirements changes, balancing up-front and just-in-time work, and favouring adaptive and exploratory development approaches [33]. A key factor in the agile process of system delivery is the close collaboration between clients (i.e., clinicians, patients and related interface providers) and developers, which assists decision making and optimizes the market value of the developed solution. The client–developer collaboration and the continuous requirements prioritization are also important parts of a typical agile requirements engineering (RE) approach [34].

C. Challenges with Agile Adoption

Despite the abovementioned benefits of agile methods in software development, the experience reports and case studies indicate that there are several challenges to adopting agile methods in the medical device development domain. These challenges can broadly be grouped into the following three groups:

1) Challenges in relation to the perceived unsuitability of agile software development approaches for safety critical domains because of the conflicts with satisfying regulatory requirements;

2) Challenges in relation to the tailoring of agile practices to conform with the regulatory requirements;

3) Challenges in relation to the acceptability of agile adoption when conflicts occur between executives/high level managers and development teams.

In this subsection, we are going to address the problems related to the perception of adopting agile practices for medical software development and we aim to change this perception through providing empirical support. We title each section with a different misconception:

1) Undisciplined nature of agile software development

2) Approaches vs the demands of a highly regulated medical device development domain:

"Discipline" against "agility" was first used by Boehm and Turner in 2003 in their book titled "Balancing Agility and Discipline: A Guide for the Perplexed" [3]. The traditional software development methods and quality standards (SW-CMM at that time) were defined as the disciplined side of the contradiction. They clearly stated that "*agility is the counterpart of discipline. Where discipline ingrains and strengthens, agility releases and invents*" [36]. Ambler, the author of the disciplined agile delivery approach, states that when "properly executed, agile is not an excuse to be undisciplined" [37]. High ceremony procedures of traditional approaches such as formal document reviews or formal document approval are a sign of bureaucracy rather than discipline [37]. The misperception of agile being an undisciplined approach could be due to its empirical nature, self-organizing teams and an emphasis on less documentation. On the contrary, agile software development methods have to focus more on establishing discipline than other approaches to achieve built-in quality and to remove the *cost of non-conformance* in the first instance. Ambler states that discipline in agile projects is what makes the difference between successful and unsuccessful agile adoption [37].

The discipline, which also means consistency is established in agile software development by people applying a set of rules and practices. One of the significant practices of agile software development such as continuous integration brings commitment and discipline to development teams. It was stated by Humble and Farley that when the necessary discipline for this practice is not adopted, the improvements in quality will not be as expected [38]. Continuous integration requires being disciplined in; refactoring, ensuring that the mainline is never broken, coding automated tests, and maintaining acceptance tests over a long term [38]. Having focused people, trusting and respecting each other in a safe environment where there is no hesitation to share ideas or no fear to fail is the start of the disciplined agile environments [37]. Therefore, *the undisciplined nature of an agile software development approach* is an expression of belief, not an expression of fact.

3) Documenting the evidence required by regulatory standards vs little emphasis of agile on documentation:

In medical device development projects, evidence is required in order to prove that the executed process ensures a safe and reliable product. Essential phases of software development process including requirements, architecture, design and test phases are recorded in a traceable way from initiation to release of the product. In the CFR 21:820 Quality System Regulation of FDA, it has been stated that design and development planning, design inputs, design outputs, design reviews, design verification, design validation, process validation, design changes, traceability and much more have to be established, which means "defined, documented (in writing or electronically), and implemented" [3].

One of the four values of the agile manifesto states that *working software is preferred over comprehensive documentation* [33]. This would suggest that following an agile software development method would not support the development of sufficient documentation necessary to achieve regulatory approval or it could be misinterpreted that an agile approach emphasises "no documentation". From either of the perspectives, it would be reasonable to accept that the documentation required for regulatory purposes needs to be developed regardless of the SDLC (Software Development Life Cycle) adopted in the company [32].

It should be noted that plan-driven methods such as the V-model and waterfall model are well-suited to the addressing the documentation needs as the SDLC phases and process outputs are in correlation. Agile software development methods do not undermine the value of necessary documentation. In a user story mapping approach, it is stated that a story-driven process needs lots of documents to work but those documents don't always look like traditional requirements documents [39]. Furthermore, there is also no clear emphasis on traceability of the software development process either in the agile manifesto or in agile principles. Evidence obtained from the literature suggests that documentation and traceability concerns in agile projects are resolved by managing the artefacts with appropriate tailoring and using software tools effectively [40], [41], [42]. For example, to ensure that all the necessary documentation is achieved within a sprint, a person who is responsible for documentation and support was established as a permanent member of an agile project at the QUMAS medical company [41]. This enables both adherence to regulations and standards without slowing down the process. Also, *living traceability* was achieved with the support of the integrated tools of *Atlassian* in the same company, which enabled an accurate snapshot to be provide of the system in real-time.

Rottier and Rodrigues report that a Use Case document can be used for the validation of the medical product in an agile project with a supplement to a Software Requirements Specification (SRS) document, which details non-functional requirements [43]. Obviously, use of use cases instead of traditional requirement specifications makes a difference for *Cochlear* in terms of agility as it was mentioned. At this point, the agility level that was achieved with use cases needs to be evaluated. Manjunath, Jagadeesh and Yogeesh mention that user stories acquired from customers were documented in the SRS in another agile medical project [42]. The founders of a *user story mapping* approach, Patton and Economy, make an important statement for the use of user stories as software requirements: "Stories in agile software development get their name from how they should be used, not what you write down. If you are using stories in development and you are not talking together using words and pictures, you are doing it wrong" [39].

D. Agile in Medical Domain

Based upon the Chaos Report of Standish Group, among 1500 software projects developed between 2011 and 2015, 39% of all the software projects that were developed using agile software development methods were successful [44]. While agile software development projects were 3 times more successful than waterfall projects, the ratio of the "challenged projects" (52% and 60% for agile and waterfall projects, respectively) cannot be underestimated. Those challenged projects refer to projects that were delivered with incomplete functionality, or exceeded the planned budget or schedule. The ratios present that there are challenges in relation to adoption and adaptation of the agile practices, interpretation of the agility principles and mindsets in the IT and medical community.

The regulatory requirements and audits, that safety critical projects are subject to, bring more concerns about the applicability of the agile approaches in the field and increase the challenges. For Class II and Class III type projects and some of Class I type projects, the FDA requires formal approval of most of the steps and items in the SDLC. The reason why traditional approaches like waterfall or V-model are being used in medical device domain could be explained with the rigid predictability and linear flow that the models present. On the other hand, Sutherland states that according to the leading research and analysis firms, such as Gartner, Forrester Research and Standish Group [45], the old style work characterized by command and control and rigid predictability is *obsolete*.

Regulatory issues are not a barrier for the implementation of agile approaches [46]. It is indicated in a mapping study [46] that SCRUM practices could be successfully used in medical device software development. Similarly, Perline [47] recommends agile methods like SCRUM for lightweight and proven framework for managing work in the complex software development projects like those in the safety-critical domains.

The Association for the Advancement of Medical Instrumentation (AAMI) published a guidance for the use of agile practices in the development of medical device software [32]. The report (AAMI TIR45:2012) provides high level guidance of agile practices, which have been found useful and appropriate for medical device software development. The report is a good resource that states major challenges for the agility implementation such as review and verification activities, use of documentation, managing the change, risk and traceability. However, the guidance was kept at an abstract level.

Evidence shows that agile development approaches in medical domain are being widely used with proper adoptions and tailoring [1], [42], [46], [48], [49], [50], [51], [52]. For instance, it has been indicated that user stories can be used as an up-front planning technique; iterative testing and test driven development to assure that all the software will be fully tested before releasing; and configuration management is used to perform necessary traceability between initial requirements and released solutions [46].

E. How can MDevSPICE® be improved by using Agile practices

The sequential flow of development processes in the MDevSPICE® framework might suggest that the V-model could be best suited for medical device software development, making the software development process long and tough on budget, especially when requirements changes are introduced later in the lifecycle. When integrating agile practices into the MDevSPICE® framework, the overall medical device software development process could become more flexible and faster. It will take some time to gather the best agile practices that would be most suitable for MDevSPICE®, but once done – this framework will be a comprehensive guide to all medical device software companies.

MDevSPICE®, similarly to international medical device standards and FDA guidance documents, does not dictate or recommend the use of any specific software development lifecycle approach. MDevSPICE® is an integrated set of regulatory requirements, practices to achieve these requirements and work products that need to be delivered in order to be allowed to sell the software on the market. With the reported and abovementioned benefits of agile methods in safety-critical software development, a medical device software development organization needs to select the most appropriate agile practices that their organization should follow and integrate them into the development lifecycle model applied in their organization [53]. It is important for the medical device software organizations to realize that the key values of the agile manifesto [33] are not contradictory but can be aligned to be complimentary to the development of medical device software, resulting in a quality management system that produces high-quality medical device software [32].

VI. CONCLUSION

Safety-critical domains are characterized by heavy regulatory demands that companies have to adhere to before they can place their devices on the market. Regulatory audits are conducted regularly to evaluate these companies and the safety of their devices. In order to pass these audits, medical device manufacturers have to ensure that all regulatory requirements have been adhered to in the design and development of each of the medical device subsystems.

In this paper, we have explained the medical device regulatory requirements and the related standards and guidance documents. We have described how MDevSPICE® addresses all concerns regarding regulatory requirements in a single medical device software framework. The key to developing this framework was an acknowledgement that the overall medical device requirements have a direct impact on the safety of the device, and it is therefore critical that top level product requirements are fully realized in the software system and its related requirements. This can be especially difficult to achieve in environments where device manufacturers decide to outsource software development without necessarily sharing all top level product requirements with the subcontractors. To address this critical interface, the MDevSPICE® framework incorporates not just software development lifecycle processes but also the system level processes. Hence, system requirements that have an impact on software requirements are identified in MDevSPICE®, and through the implementation of bilateral requirements traceability, decisions taken during the software subsystem development are fed back to the top level system requirements – thus providing a closed loop for requirements management, which can help to increase the overall safety of the device.

In this paper we have argued that agile practices such as iterative development cycles, continuous integration, sprint planning meetings and continuous requirements prioritization should be tested when assuring the development of better technology design and technology interfaces. We have also illustrated the benefits and the challenges of agile practice integration into traditional medical device software development. Through providing empirical support to these challenges we have established the basis for our future research work in which we will decide upon the most appropriate agile practices that will be integrated into the MDevSPICE® framework. We will then integrate the selected agile best practices into the MDevSPICE® framework to shorten the medical device software development lifecycle as well as the time to market for the resulting medical devices.

ACKNOWLEDGMENT

This research is supported by the Science Foundation Ireland under a co-funding initiative by the Irish Government and European Regional Development Fund and by Lero - the Irish Software Research Centre (http://www.lero.ie) grant 10/CE/I1855 & 13/RC/20194. The research is also supported by Digital Health Revolution project and Tekes, Finnish Funding Agency for Innovation.

REFERENCES

[1] F. McCaffery, M. Lepmets, and P. Clarke, "Medical device software as a subsystem of an overall medical device: The MDevSPICE experience," The First International Conference on Fundamentals and Advances in Software Systems Integration FASSI, Aug. 2015, Venice, Italy, doi:10.1002/smr.1731

.[2] ISO/IEC 15504-5. Information technology - process assessment - Part 5: an exemplar process assessment model. 2012. p. 211.

[3] FDA. Chapter I - Food and drug administration, department of health and human services subchapter H - Medical devices, Part 820 - Quality system regulation. Available from: http://www.accessdata.fda.gov/scripts/cdrh/cfdocs/cfcfr/CFRSearch.cfm?CFRPart=820. Last date accessed - 26.05.2016.

[4] Directive 93/42/EEC of the European Parliament and of the Council concerning medical devices. 1993. European Commission, Brussels, Belgium. p. 43.

[5] Council directive 90/385/EEC on active implantable medical devices (AIMDD). 1990. Brussels, Belgium. p. 35.

[6] Directive 98/79/EC of the European Parliament and of the council of 27 October 1998 on in vitro diagnostic medical devices. 1998. Brussels, Belgium. p. 43.

[7] Directive 2007/47/EC of the European Parliament and of the Council concerning medical devices. 2007. EC: Brussels, Belgium. p. 35.

[8] IEC 62304: Medical device software - software life-cycle processes. 2006. IEC: Geneva, Switzerland. p. 151.

[9] ISO 13485: Medical devices - quality management systems - requirements for regulatory purposes. 2003. ISO: Geneva, Switzerland. p. 57.

[10] ISO 9001:2000 - Quality management systems - requirements. 2000. Geneva, Switzerland. p. 27.

[11] ISO/IEC 12207:1995 - Information technology - software life-cycle processes. 1995. ISO/IEC: Geneva, Switzerland. p. 106.

[12] ISO/IEC 12207:2008 - Systems and software engineering - Software life cycle processes. 2008. ISO/IEC: Geneva, Switzerland. p. 138.

[13] ISO/IEC 15504-2:2003, Software engineering - Process assessment - Part 2: Performing an assessment. ISO: Geneva, Switzerland. 2003.

[14] ISO 14971 - Medical devices - Application of risk management to medical devices 2009. ISO: Geneva, Switzerland. p. 82.

[15] IEC TR 80002-1 - Medical device software - Part 1: Guidance on the application of ISO 14971 to medical device software. 2009. IEC: Geneva, Switzerland. p. 58.

[16] IEC 60601-1 - Medical electrical equipment – Part 1: General requirements for basic safety and essential performance 2005. IEC: Geneva, Switzerland. p. 20.

[17] IEC 62366 - Medical devices - Application of usability engineering to medical devices. 2007. IEC: Geneva, Switzerland. p. 104.

[18] IEC 82304-1: Health software -- Part 1: General requirements for product safety. 2012. IEC: Geneva, Switzerland. p. 30.

[19] FDA Guidance documents (medical devices and radiation-emitting products), 2015. FDA: USA.

[20] FDA Guidance for the content of premarket submissions for software contained in medical devices. 2005. FDA: USA. p. 20.

[21] FDA's Guidance for industry, FDA reviewers and compliance on - Off-the-shelf software use in medical devices. 1999. FDA: USA. p. 26.

[22] FDA's General principles of software validation; final guidance for industry and FDA Staff. 2002. FDA: USA. p. 43.

[23] ISO/IEC 24774 - Systems and software engineering - Life cycle management - Guidelines for process description. 2010. Geneva, Switzerland. p. 15.

[24] B. Glaser, and A. Strauss, "The discovery of grounded theory: strategies for qualitative research", Ed: A.d. Gruyter. Hawthorne, NY, USA, 1976.

[25] M. Lepmets,P. Clarke, F. McCaffery, A. Finnegan, and A. Dorling, "Development of a process assessment model for medical device software development," in Industrial Proceedings of the 21st EuroSPI Conference, 2014, Luxembourg, pp. 2.25-2.35.

[26] IEC TR 80002-3: Medical device software -- Part 3: Process reference model of medical device software life cycle processes (IEC 62304). 2014. IEC: Geneva, Switzerland. p. 23.

[27] G. Regan, M. Biro, F. McCaffery, K. McDaid, and D. Flood, "A traceability process assessment model for the medical device domain, " in EuroSPI, Luxembourg, 2014, pp. 206-216, doi: 10.1007/978-3-662-43896-1_18.

[28] J. Gaev,"Technology in healthcare. in: clinical engineering handbook, " Ed: Dyro, Joseph F., p.342, 2004.

[29] G. Powell-Cope, A. Nelson, and E. Patterson, "Patient safety and quality: an evidence-based handbook for nurses," Chapter 50. Hughes RG, Ed. Rockville (MD): AORN Journalvol.90(4), pp.601-602, 2008.

[30] M. Dulin, C. Hugh-Jones, M. Pitts, and G. Hughes, "Applying data to improve patient-centric and personalized medicine," Conclusions Paper, 10th Annual Health Care & Life Sciences Executive Conference, SAS, 2013.

[31] J. Rasmusson, The Agile samurai: how agile masters deliver great software, Pragmatic Bookshelf, 2010.

[32] AAMI, AAMI TIR45:2012 -- "Guidance on the use of agile practices in the development of medical device software," 2012.

[33] (2001). Agile Manifesto. Available from: www.agilemanifesto.org. Last date accessed: 26.05.2016.

[34] M. Daneva, E. van der Veen. C. Amrit, S. Ghaisas, K. Sikkel, R. Kumar, N. Ajmeri, U. Ramteerthkar, and R.Wieringa, "Agile requirements prioritization in large-scale outsourced system projects: An empirical study, " The Journal of Systems and Software, vol. 86(5) , pp.1333– 1353, 2013.

[35] B. Boehm and R Turner. "Balancing agility and discipline: A guide for the perplexed," Addison-Wesley Professional, 2003.

[36] B. Boehm and R. Turner. "Balancing agility and discipline: Evaluating and integrating agile and plan-driven methods," 26th International Conference in Software Engineering (ICSE), 2004, pp. 718-719.

[37] A. Scott W. and M. Lines, Disciplined agile delivery: A practitioner's guide to agile software delivery in the enterprise, IBM Press, 2012.

[38] J. Humble and D. Farley, Continuous delivery: reliable software releases through build, test, and deployment automation, Pearson Education, 2010.

[39] J. Patton and P. Economy, User story mapping: Discover the whole story, build the right product, O'Reilly Media, Inc., 2014.

[40] U. Upender, "Staying agile in government software projects," in Agile Conference Proceedings, July 2005, pp. 153-159, doi: 10.1109/ADC.2005.41.

[41] B. Fitzgerald, K.-J.Stol, R. O'Sullivan, and D. O'Brien, "Scaling agile methods to regulated environments: An industry case study," 35th International Conference in Software Engineering (ICSE), May 2013, pp. 863-872.

[42] K.N. Manjunath, J. Jagadeesh, and M. Yogeesh, "Achieving quality product in a long term software product development in healthcare application using Lean and Agile principles: Software engineering and software development," International Multi-Conference on Automation, Computing, Communication, Control and Compressed Sensing (iMac4s), March 2013, pp. 26-34, doi: 10.1109/iMac4s.2013.6526379.

[43] P.A. Rottier, and V. Rodrigues "Agile Development in a Medical Device Company," in Agile Conference Proceedings, pp.218-223, Aug. 2008, doi: 10.1109/Agile.2008.52

[44] Chaos Report, Standish Group, based upon Jeff Sutherland Scaling Agile course notes, 2015, Boston.

[45] J. Sutherland, Scrum: The Art of Doing Twice the Work in Half the Time, Crown Business, 2014.

[46] M. McHugh, F. McCaffery, and V. Casey, "Barriers to adopting agile practices when developing medical device software," in Software Process Improvement and Capability Determination Proceedings, vol.290, pp. 141-147, 2012, doi: 10.1007/978-3-642-30439-2_13.

[47] J. Pelrine, "On understanding software agility - A social complexity point of view," E:CO. vol.13(1-2), pp. 26-37, 2011

[48] Z.R. Stephenson, J.A. McDermid, and A.G Ward, "Health modelling for agility in safety-critical systems development," in the 1st Institution of Engineering and Technology International Conference Proceedings, June 2006, pp. 260-265, doi: 10.1049/cp:20060225.

[49] J. Sutherland."Future of scrum: parallel pipelining of sprints in complex projects," in Agile Conference Proceedings, July 2005, pp. 90-99, 24-29.

[50] M. Sumrell, "From waterfall to agile - how does a qa team transition?," in Agile Conference Proceedings, Aug. 2007, pp. 291-295, doi: 10.1109/AGILE.2007.29.

[51] J. W. Spence "There has to be a better way! [software development]," in Agile Conference Proceedings, July 2005, pp. 272-278, doi: 10.1109/ADC.2005.47.

[52] K. Weyrauch, "What are we arguing about? A framework for defining agile in our organization," in Agile Conference Proceedings, July 2006, pp. 213-220, doi: 10.1109/AGILE.2006.62.

[53] M. McHugh, F. McCaffery, and V. Casey. "Adopting agile practices when developing software for use in the medical domain," Journal of Software: Evolution and Process, vol.26(5), pp. 504-512, 2014, doi: 10.1002/smr.1608.

The MedITNet Assessment Method – Development and Validation using Action Design Research

Self Assessment against IEC 80001-1

Silvana Togneri MacMahon, Fergal McCaffery, Frank Keenan
Department of Computing & Mathematics
Dundalk Institute of Technology
Co. Louth, Ireland
{e-mail: silvana.macmahon, fergal.mccaffery, frank.keenan}@dkit.ie

Abstract— **The provision of care to patients has moved away from episodic acute care due to the increase in chronic diseases such as diabetes. This has changed the relationship between the patient and the care team. The management of chronic disease requires the use of information technology including networked medical devices to facilitate the establishment of an ongoing relationship between the patient and care team. The use of networked medical devices can provide benefits to patients such as reduced cost of care, reductions in adverse events and improved care through the provision of accurate and up-to-date information. However, the placement of a medical device onto an IT network can lead to risks to the device. These risks may lead to incorrect or degraded performance of the device impacting patient care and negating the potential benefits of using the device. While, IEC 80001-1 was developed to assist Healthcare Delivery Organisations in addressing these risks, HDOs may struggle in implementing the requirements of the standard. This paper discusses the development of an Assessment Method that forms part of MedITNet, an assessment framework that can be used by Healthcare Delivery Organisationss to assist them in implementing the requirements of the standard by providing a flexible, consistent and repeatable approach to assessing the capability of their risk management processes relating to networked medical devices. The assessment highlights weaknesses in the process and can be used as a foundation to improve these processes. This paper also discusses the development and validation of the Assessment Method using Action Design Research.**

Keywords- Risk Management; Medical IT Networks; IEC 80001-1; MedITNet; Assessment Framework; Assessment Method.

I. INTRODUCTION

This paper extends the discussion of the development of the MedITNet assessment method in [1] by extending the discussion of the pilot implementation of the assessment method and examining the recommendations that were implemented as a result of the implementation. This paper also discusses the expert review of the overall assessment framework.

The recent downturn in the global economy has led to an increased focus on ensuring that a high standard of care is provided to the patient while reducing the cost of care. Interoperability of medical devices has been recognised for its potential to achieve this goal [2-4]. Such is the potential that governments have provided incentives to promote the meaningful use of interoperable medical devices and Health Information Technology (HIT), such as Electronic Health Records (EHRs) [5-7]. The use of interoperable medical devices has resulted from the increased prevalence of chronic conditions such as diabetes, which has resulted in a move away from acute episodic care. The management of chronic disease requires the establishment of an ongoing relationship between the patient and their care team facilitated by carefully designed care processes and requiring the support of information technology [8-11] As a result of this change, the number of networked medical devices in use continues to increase [12-14].

A number of benefits of the use of networked medical are recognised. These include reducing the instances of adverse events improving patient safety, reducing the time spent by clinicians manually entering information, reducing redundant testing due to inaccessible information, improving patient care, reducing healthcare costs and ensuring comprehensive and secure management of health information [15, 16]. These benefits have resulted in medical IT networks becoming a critical, integral component of the medical system [17]. However, as medical devices increasingly interface with other equipment and hospital information systems the integration complexity of the systems is increased and this presents additional operational risks [14, 18-20]. Proprietary networks were traditionally used when a device was placed onto a network. However, these are being used less with medical devices being designed to be placed onto the hospitals general IT network. This means that medical device manufacturers no longer exercise control over the configuration of the network [21]. This lack of control can lead to risks potentially resulting in unintended consequences outside the control of the medical device manufacturer. The placement of the device onto the hospital network creates a new system in which the device has not been validated [22]. These risks can result in the incorrect and degraded performance of the medical device [23, 24] compromising patient safety, effectiveness and the security of the IT network [25-27].

IEC 80001-1: *Application of risk management for IT-networks incorporating medical devices* [28] was published in 2010 to address the risks associated with the incorporation of a medical device into an IT network. However, Healthcare Delivery Organisations (HDOs) face challenges when implementing the requirements of this standard [29]. HDOs vary in size and in terms of the capability of their risk management processes [17, 30] and the regulatory requirements of the region in which they provide care differ meaning that the implementation of the requirements of the standard will vary depending on the relevant regulatory requirements. The effective performance of risk management activities requires interaction between different stakeholder groups. An understanding of the context of the HDO is also required in order to manage the identified risks [18, 31]. In addition, organisational changes are required to facilitate the necessary level of interaction among stakeholders and HDOs may be unprepared for this [14] due to the fact that departments within the HDO typically operate in silos [8]. These challenges make the requirements of the standard confusing and difficult to implement.

These difficulties in implementing the requirements of the standard highlighted the need to provide HDOs with assistance. This research has focused on the development of an assessment framework which provides HDOs with a flexible approach to assessing the capability of their current risk management processes relating to medical IT networks. The use of the assessment framework enables communication among stakeholders groups allowing HDOs to implement the requirements of the standard.

The rest of this paper is organized as follows. Section II describes the development of the Assessment Method component of the MedITNet assessment framework while Section III described the stages of the Assessment while the validation of the resultant Assessment Method is discussed in Section IV. The conclusions are presented in Section V.

II. Development of the Assessment Method

The Assessment Method described in this paper is one of three components that make up the MedITNet assessment framework [32, 33]. In addition to the Assessment Method, MedITNet contains a Process Reference Model (PRM) and Process Assessment Model (PAM). The PRM provides a description of 14 processes, which address the requirements of IEC 80001-1. The processes within the PRM are described in terms of the purpose of the process and the outcomes achieved as a result of performing the process. The PAM extends the description of the processes by including a description of the base practices or activities performed during the process and the work products used or produced as a result of performing the process. The PAM also introduces the concept of a measurement framework or scale on which the capability of the process can be measured. The presence of the PRM and PAM within the MedITNet framework mean that the framework can be used regardless of the context of the HDO, including the regulatory environment in which the HDO provides care.

The Assessment Method provides a consistent approach to assessing the capability of the processes in the PAM using questions related to each of the base practices. The Assessment Method can be used as presented in the technical report or can be tailored for use based on the context in which the HDO provides care. In order to tailor the Assessment Method, the HDO can rephrase the questions that are being asked in order to address specific aspects of the context in which they provide care. For example, there may be additional regulatory requirements for risk management that apply to HDOs due to the geographical location in which they provide care. The HDO can either rephrase the questions to take into account the regulation or may choose to add additional questions during the performance of the assessment. Any alterations to questions or additional questions that are added, must be reviewed against the relevant base practices in the PAM. The HDO must ensure that the questions continue to address the assessment of the performance of these base practices before making any amendments or additions. This ability to tailor the Assessment Method addresses the issue of HDOs providing care within differeing regulatory environments. In addition, the ability to tailor the Assessment Method allows HDOs to take into account the size of the HDO and also the capability of the HDO in terms of the risk management of medical IT networks. For HDOs operating at a lower level of maturity, the PRM and PAM can be used to identify processes and practices that need to be implemented to achieve a higher level of maturity. These HDOs may wish to perform an initial assessment of the capability of risk management processes using the Assessment Method, which will highlight areas for improvement. Based on the assessment, the HDO can then refer to the PRM and PAM to assist in the definition and implementation of processes at a higher capability level. The HDO can then perform a follow-up assessment at a later date to ensure that the identified improvements have been implementated and that the target capability level has been achieved.

A. Development Approach

The approach to the development of the Assessment Method combines the learnings from a literature review with knowledge of risk management practices in a HDO.

In order to understand the risk management practices within the HDO, focus groups sessions were conducted with risk management stakeholders within a HDO. These sessions were performed during the Practice-Inspired Research phase of the Action Design Research (ADR) process [34] used in the development of the Assessment Method and also in the development of the MedITNet Assessment framework. The focus of the Practice Inspired Research phase of the ADR process is to validate the findings of the literature review, which is this case focused on an examination of the risk management of medical IT networks and the challenges experienced by HDOs in the implementation of risk management processes, and confirm that these challenges are experienced in practice within HDOs. This phase of the research process is useful in ensuring that the solution, in this case the assessment framework, will be suited for use in the context in which it will be used.

B. Literature Review

The Assessment Method was developed following the development of the PRM and PAM components of the MedITNet Assessment Framework. During the development of the PRM and PAM, a literature review in the area of process assessment, focusing on process assessment standards was conducted. This literature review was extended in order to develop the Assessment Method.

In order to inform the development of the Assessment Method, a review of Assessment Methods for similar standards was completed. This review focused on ISO/IEC 15504-3 [35] and Appraisal Requirements for CMMI [36] Domain specific including Rapid Assessment for Process Improvement in Software Development (RAPID) [37], Express process appraisal (EPA) [38], Adept [39], Med-Adept [40] and Tudor IT Service Management Process Assessment (TIPA) [41] were also reviewed. While this review informed the development of the Assessment Method, the results of the review were not sufficient in themselves to develop the Assessment Method. In order to develop the Assessment Method, the results of the literature review were combined with the knowledge gained during the Practice-Inspired Research conducted as part of this study. This approach allowed the researcher to take into account the concerns that HDOs express in relation to the implementation of the IEC 80001-1 standard.

The literature review provided an understanding of the challenges that HDOs encounter when incorporating a medical device into an IT network. Each of the identified challenges was considered when developing the requirements for the Assessment Method, using a similar approach to that used by Mc Caffery and Coleman [42] using criteria for Assessment Methods as outlined by Anacleto et al. [43]. The criteria were adapted to take into account the domain in which the Assessment Method will be used, that is, within the HDO rather than in the context of software development. The development of the requirements for the Assessment Method also took into account the challenges related to the management of risk associated with the incorporation of a medical device into an IT network which were highlighted as part of the Literature Review and Practice-Inspired Research. The requirements for the Assessment Method were defined as follows:

- Due to the constraints on resources within HDOs, the Assessment Method should be lightweight in its approach and facilitate self-assessment;
- The Assessment Method should be based on the processes described in the MedITNet PAM;
- Guidance should be provided for tailoring the Assessment Method for use in various scales of HDOs and in different geographical contexts. The Assessment Method should also facilitate assessments based on conformance with the standard as well as those seeking to assess the capability level with which risk management processes are being performed;
- The Assessment Method should support the identification of risks and improvement opportunities;
- The Assessment Method should not assume any previous knowledge of process assessment on the part of those conducting the assessment;
- The Assessment Method should facilitate the development of tool support in the future;
- The Assessment Method should be publicly available;
- The Assessment Method should encourage a culture of communication among various multidisciplinary risk management stakeholders including those within and external to the HDO;
- The Assessment Method should be validated for use within the HDO context.

In addition to the literature review and, to augment the Practice-Inspired Research, members of the Clinical Engineering team (CE) and the Clinical Informatics team in a HDO were consulted throughout the development of the questions for the Assessment Method. This was an iterative process, which is described in the following section.

C. Question Development

The involvement of HDO risk management stakeholders in the development of the Assessment Method was considered to be vital. HDOs may use the Assessment Method in its form within the technical report and without reference to the PRM and PAM. This means that the process for conducting the assessment oulined in the Assessment Method and the questions that are used during the assessment must be understandable to a range of risk management stakeholders.

The Assessment Method assesses against ISO/IEC 15504-2 compliant models, i.e., the MedITNet PRM and PAM. These models describe processes at the level of the *process purpose, outcomes, practices and work products*. This approach to the development of the Assessment Method ensures its applicability beyond the HDO assisting with its development, across varying geographical and regulatory contexts. The use of ADR also ensures that all components of the framework are developed initially based on a combination of the results of the literature review combined with Practice- Inspired Research. The resultant components are then validated by both practitioners in the field and end users. This ensures that the components are both suited to use in a particular context and suited for use across a range of contexts. The development of the assessment questions, which form part of the Assessment Method, was completed in two phases.

a) Question Development – Phase 1

During phase 1 of the question development process, a meeting was held in the HDO with the Principal Physicist and a Physicist/Clinical Engineer. Both had taken part in the initial phase of the Practice-Inspired Research and were already familiar with the provisions of the standard and the proposed MedITNet framework.

During the previous discussions on the current risk management practices within the HDO, it was agreed that the *Risk Analysis and Evaluation Process* was the main process relating to the identification and classification of risks. It was noted during the previous focus groups session that discussion of the Risk Analysis and Evaluation process lead to discussion of other aspects of risk management outside the scope of that process. This was due to the fact that the discussion of the Risk Analysis and Evaluation process led to a discussion of the overall HDO risk management policy and also to a discussion on the evaluation and subsequent application of risk control measures. The discussion also revealed how risk was documented in the HDO. Therefore, it was decided that questions should be developed for this process first.

The development of these questions would inform the development of the assessment questions for the remaining processes. In order to develop the questions for the Risk Analysis and Evaluation process, a number of steps were followed [44]. Firstly, each of the base practices was reviewed and the participants were asked to formulate a question that could be used to assess the base practice being described. The base practices in the PAM describe the activies that must be performed in order to bring about the process outcomes and achieve the overall purpose of the process. To facilitate gaining an understanding of each of the base practices, each base practice was discussed in the context of the standard with the relevant section of the standard being consulted and reviewed if required. This was useful for the participants as it provided an understanding of how the requirements of the standard were expressed in the PAM in terms of activities to be performed.

Once all participants were clear on the meaning of the base practice, the participants from the clinical engineering team were encouraged to think of a "real" scenario where the relevant base practice had been implemented in the past. The discussion of the scenario would focus on how the base practice was implemented in the context and any constraints that may have affected the implementation of the base practice. This assisted the participants in identifying how the requirements of the standard were and could be implemented in the specific context of their HDO.

Once the practice had been discussed in context, the participants were encouraged to formulate questions that could be used to assess the degree to which the base practice had been implemented during the proposed scenario. All questions that were formulated by the participants were recorded and the participants were encouraged to rephrase the questions in order to decrease the number of questions used to assess each base practice. This was an interative process and resulted in discussions around how the questions should be phrased. This discussion was useful as it allowed participants to examine and understand the terms used in the standard and ensure that a common understanding of the concepts related to risk management was established. The approach outlined in this section was also noted by participants as being a useful way in which to gain a better understanding of the standard and the context in which the HDO provides care. Participants also suggested that this approach would also be useful in the tailoring of questions to a specific context as the questions could be reviewed to see where amendments should be made to take into account the context of the HDO in which the assessment is being performed.

The Risk Analysis and Evaluation Process contains five base practices against which 14 questions were eventually formulated. This draft of questions was used in the validation focus group within HDO A conducted as part of the ADR process. However, the set of questions (presented in Table I) does not represent the final set of questions which were developed to be used in the assessment of this process.

b) Question Development – Phase 2

During the second phase of the development of the questions, the questions for the remaining 13 processes were developed. These questions were developed with the assistance of the Clinical Informatics Manager (CIM) of the HDO. The CIM is a former nurse who oversees the systems administration tasks of the Clinical Information System within the Intensive Care Unit. The CIM was briefed on the research being carried out on the development of the Assessment Method and was given the PRM and PAM to review and was briefed on the requirements of the IEC 80001-1 standard. Following the development of the assessment questions for the remaining 13 processes, the CIM was also shown the questions developed during phase 1 for the Risk Analysis and Evaluation Process. The CIM was asked to review and reformulate the questions, as required, for this process based on their experience of development of the questions for the remaining processes. When reviewing the questions related to this process, the CIM and the researcher rephrased some of the questions to ensure that they were more closely based on the base practices of the process. The original set of questions was determined to be too specific to the context of the HDO in which the question development had taken place. In addition, some questions were, on review, considered to be unnecessary, again being too context specific and were removed accordingly.

TABLE I. SAMPLE ASSESSMENT QUESTIONS

Base Practice Summary:	***Question Number:***	***Question:***
BP.1 - Identify likely hazards.	BP.1 Q.1	How do you identify likely safety hazards for individual devices?
	BP.1 Q.2	How do you analyse the system as a whole to identify likely safety hazards?
	BP.1 Q.3	How do you consider the impact of the device on the environment?
	BP.1 Q.4	How do you consider the impact of the device in terms of effectiveness?
	BP.1 Q.5	How do you consider the impact of the device in terms of data and system security?
BP.2 - Estimate associated risks.	BP.2 Q.1	Do you have a procedure for estimating risk?
	BP.2 Q.2	What approach do you use to estimate the risk associated with each source of harm?
	BP.2 Q.3	What information sources do you use to estimate the risks associated with each source of harm?
	BP.2 Q.4	Are risks reviewed throughout the life cycle?
BP.3 - List possible consequences of harm.	BP.3 Q.1	How do you identify possible consequences of harm?
BP.4 - Record results of Risk Analysis and Evaluation activities.	BP.4 Q.1	How are risk management activities recorded?
	BP.4 Q.2	Are instances where risk estimate is so low that risk reduction is not required recorded?
BP.5 - Implement Risk Control Measures.	BP.5 Q.1	How are risk control measures implemented?
	BP.5 Q.2	Are risk control measures implemented in line with risk management policy?

In general, one question was related to each of the base practices. However, the assessment of some base practices required more than one question. The CIM was asked to participate in the development of the questions in order to ensure that the questions were phrased in a way that could be understood by various risk management stakeholders within the HDO. The questions were developed using the same steps as those outlined in section C sub-section a). The questions were also developed based closely on the base practices defined within the PAM to ensure that the questions could be applied across multiple HDO contexts and were not specific to the HDO in which the research was being carried out.

III. STAGES OF THE ASSESSMENT METHOD

The stages of the assessment process are illustrated in Figure 1 and discussed in the remainder of this section.

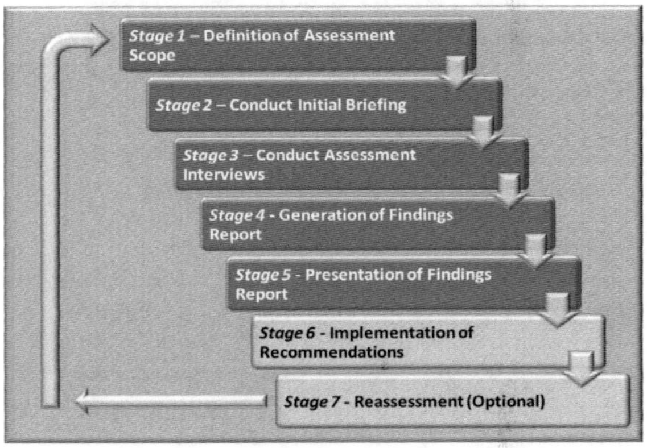

Figure 1. Stages of the Assessment Process

Participants in the assessment process include the lead assessor, a risk management stakeholder from within the HDO, who will manage the assessment on behalf of the Top Management (TM) of the HDO. Focus group interviews are used during the assessment to ensure communication among risk management stakeholders. An additional Assessor (A) may be required to assist the LA. In addition to sponsoring the assessment, TM will ensure that Risk Management Stakeholders (RMS) are available to participate in the assessment. The RMS will be drawn from a multi-disciplinary team from within the HDO and will include members of the IT, CE and Clinical Teams and any other relevant RMS as required. The RMS may also include participants who are external to the HDO such as MDMs. The inclusion of participants external to the HDO is more typical during the procurement phase of a new system or devices. However, it should be noted that the IEC 80001-1 standard notes the importance of the participation of external risk management stakeholders throughout the life of the medical IT network. The participation of relevant internal and risk management stakeholders is necessary to perform the successful risk management of the medical IT network.

It should be noted that Stages 1 to 5 above complete the assessment activities. Stage 6 involves the implementation of recommendations made during the assessment. Where a follow-up assessment is required, stage 7 is performed. A reassessment can be used to confirm that the recommendations for improvements to the risk management process have improved risk management processes as envisaged.

a) Stage 1

The lead assessor meets with Top Management and the scope of the assessment is discussed. The system, which is to be the focus of the assessment, is defined and the context of the system is understood. At this time, the availability of relevant risk management stakeholders to participate in the assessment is confirmed.

b) Stage 2

The lead assessor meets with relevant risk management stakeholders who will be taking part in the assessment to explain the Assessment Method and give details of what their participation will involve.

c) Stage 3

The lead assessor conducts focus group interviews based on the scripted questions with the relevant risk management participants and evaluates the responses. The assessor makes notes on the interviews and additional questions are asked if clarification is required. The resonses to these questions will highlight areas of weakness in the risk management process. The identification of these weaknesses forms the basis for the findings report that will be generated during the the next stage of the assessment. Relevant work products are reviewed at this stage to highlight areas where risk management documentation may be missing or incomplete.

d) Stage 4

A findings report is prepared based on the data gathered and the weaknesses identified at stage 3. Each process is reviewed in turn and where relevant particular strengths and weaknesses are identified based on the evaluation and interview notes. Suggested recommendations are made for actions to address these issues and to facilitate process improvement are outlined and discussed.

e) Stage 5

The findings report is presented. The lead assessor presents the findings of the assessment. The finding will generally be reported to Top Management within the HDO and to relevant risk management stakeholders. It is important that the findings report is thoroughly reviewed and tha t recommendations are carefully considered. The findings report may be refered to during follow-up assessments and may also be used as a source of information for the identification of risks on future projects.

f) Stage 6

Having allowed time for the contents of the report to be considered, the findings are discussed and a plan for improvement of the processes with specific improvement objectives is agreed. At this stage participants may schedule a reassement to be conducted at a later date.

g) Stage 7

The HDO having implemented the agreed improvements have the option of performing a reassessment to ensure that improvements have been implemented and that risk management processes have improved accordingly.

It should be noted that the interviews conducted during Stage 3 are conducted as focus group interviews. The focus group interviews are conducted with risk management stakeholders. Prior to the commencement of an assessment, HDOs should ensure that all relevant risk management stakeholders are identified and are available to participate in the focus group interviews. Participation of relevant risk management stakeholders in the focus group interviews ensures: that a shared understanding of the concepts related to the risk management of medical IT networks are understood; that risk management stakeholders, through the assessment process and discussion of risks, gain a greater understanding of the IEC 80001-1 standard, greater level of communication are established between risk management stakeholder groups as the assessment process requires that these groups operate outside of their "silos".

IV. VALIDATION OF THE ASSESSMENT METHOD

The Assessment Method was validated from the perspective of its utility in a specific HDO context. The validation of the Assessment Method was performed in two stages. The first involved a pilot assessment performed in a HDO, while the second stage involved the validation of the assessment method by the standards community. Both of these stages of the validation of the Assessment Method, which were conducted using ADR, are discussed in this section.

a) Stage 1- Validation – Pilot Assessment

The first stage of validation consisted of performing an assessment of current risk management practices within a HDO context using the Assessment Method. This phase consisted of a pilot implementation of the Assessment Method by performing an assessment of the Risk Analysis and Evaluation process using the questions from the Assessment Method.

A focus group session took place in the HDO with participants from various risk management stakeholder groups taking part. The assessment allowed for areas of weakness in the current risk management processes related to medical IT networks to be highlighted and addressed. A findings report was provided to the HDO and a summary of the recommendations is provided in Table II. This phase of the validation ensured that the developed questions could be understood by risk management stakeholders and were suited for use for the performance of an assessment in the specific HDO context.

TABLE II. SAMPLE ASSESSMENT RESULTS SUMMARY

BP.1 - Identify likely hazards
Develop a standardised process for the identification of hazards, including the identification of hazards during the tendering process
Maintain the same level of documentation in the recording of identified hazards, regardless of when in the lifecycle the hazard is identified
Store information related to risk management in a manner which can be accessed as an information source for the estimation of future risks
BP.2 - Estimate associated risks
Establish a policy detailing risk acceptability criteria
Formalize and document a procedure for the estimation of risk which stipulates which risk management stakeholders should be involved
BP.3 - List possible consequences of harm
Consider consequences of harm based on the risk acceptability criteria
Consider consequences of harm based on the risk management policy
BP.4 - Record the results of Risk Analysis and Evaluation activities
Record Risk Analysis and evaluation activities in the risk management file
Ensure accessibility of emails containing information on Risk Analysis and Evaluation activities
BP.5 - Implement Risk Control Measures
Establish a process for risk control
Ensure that risk control measures are implemented in line with the risk control process
Document risks which have been considered so low as not to require additional risk control measures

A follow-up focus groups session took place nine months later to review which recommendations had been implemented. Not all of the recommendations made during the assessment were implemeted by the HDO [44]. However, the performance of the assessment resulted in improvement to not only the risk analysis and evaluation process within the HDO, but participants also reported improvements in the overall risk management of medical IT networks within the HDO. Participants also confirmed that the recommendations, which were made in the findings report, were considered to be appropriate. Where recommendations had not been implemented, this was due to the constraints on resources within the HDO. Recommendations which had not been implemented at the time of the follow-up focus group session were scheduled for implementation at a later date. Participants had also highlighted the importance of the implementation of the requirements of the IEC 80001-1 standard in future medical IT network projects.

At the time of the follow-up session, the CE team had secured agreement from Top Management that a Medical IT Network Risk Manager would be recruited for an upcoming medical IT network project. The responsibilities of the medical IT network risk manager role was to be defined based on those as outlined in IEC 80001-1. The agreement to recruit for this position and to base the reponsibilities of the role on IEC 80001-1 requirements was agreed with Top Management based on the results of the pilot assessment. The CE team identified this as a major improvement in risk management processes as prior to this they felt that the skills required to perform effective risk management of the network were not currently present in the HDO. The CE team noted that the performance of the assessment was instrumental in gaining Top Management engagement in the promotion and adoption of the standard. This sentiment was repeated during an expert review of the overall MedITNet framework where experts contended that without this type of assessment instrument, adoption of the standard may follow a shallow trajectory [44].

The performance of this stage of the validation:

- confirmed the utility of the Assessment Method in a specific HDO context
- confirmed that the questions used in the assessment were understandable to various risk management stakeholders
- confirmed that the Assessment Method could be tailored for use in various HDO contexts.
- confirmed that the Assessment Method could be used to provide appropriate recommendations for the improvement of the risk management process
- confirmed that the use of the Assessment Method improved communication among risk management stakeholders
- confirmed that the use of the Assessment Method may be useful in promoting Top Management engagement with and promotion of the adoption of the standard

This stage of the validation process was conducted as part of the ADR process as part of the "Build, Intervene and Evaluate" stage of the ADR process [34]. During this stage of the process "end-users" of the developed artifact, in this case the Assessment Method, trial the artifact in the context in which it will be used, in this case in a HDO setting. The focus of this phase is to ensure the utility of the developed artifact in a specific context.

b) Stage 2 Validation – Standards Community

In order to confirm the generalisability of the Assessment Method across a range of HDO contexts, the Assessment Method was also validated through expert review by members of the standards community from the International Electrotechnical Commission (IEC) Sub-Committee 62A and the International Organization for

Standardization (ISO) Technical Committee 215 Joint Working Group 7 (JWG7). Members of this group are drawn from risk management stakeholders within HDOs, medical device manufacturers and providers of other IT technology. They are recognised as experts in their field and represent their country in this capacity. The focus of this stage of the validation is to ensure that the Assessment Method can be used across multiple HDO contexts, regardless of the regulatory environment in which the HDO operates. During this phase of the validation the Assessment Method was circulated to members of JWG7 for review. The Assessment Method was circulated with the MedITNet PRM and PAM and members were invited to make comments on any aspect of these components of MedITNet. The review by members of this group resulted in a number of changes to the Assessment Method including the provision of sample templates that could be used by HDOs during the performance of an assessment and in the preparation of the findings report for circulation to Top Management of the HDO.

Table III presents the approach adopted for reviewing the comments. Each of the comments was assigned to one of the four categories listed in Table III and addressed accordingly. All comments were discussed during the comment resolution meetings and resolution was based on the expertise of the JWG7 group which included representatives from HDOs, medical device manufacturers and providers of other information technology. As the Assessment Method had been previously validated in a trial assessment, the comments received from JWG7 were largely editorial in nature and did not result in changes to the questions within the Assessment Method. In total, 298 comments were received related to the Assessment Method. A large number of duplicate comments were received. This was due to one reviewer who raised a comment for each instance of a particular issue, leading to a large number of comments being duplicated.

During the comment resolution period, a total of 298 comments related to the Assessment Method were received from members of JWG7. During an initial review of comments 202 comments were found to be duplicate comments and an additional four comments were deemed to be not applicable. Therefore, those 206 comments required no changes to be made to the Assessment Method and have not been included in the following analysis of comments. An initial review of the remaining 92 comments was completed.

While a large number of the comments received on the Assessment Method were small wording changes, some of the comments required changes to the overall structure of the technical report. These changes included the following:

- Assessment stages were listed before a description of each stage was provided
- Assessment questions were removed from the description of the stages of the Assessment Method and placed in the annex of the Assessment Method.
- Templates for conducting the assessment including a sample question template and a findings report template were also developed and placed in the annex.

These changes were suggested to improve the usability of the Assessment Method and facilitate performance of both conformance and capability assessments and as such were made to the Assessment Method. The Assessment Method developed as part of this research along with the MedITNet PRM and PAM were published as ISO TR 80001-2-7, a Technical Report in the IEC 80001-1 family of standards [45].

TABLE III. COMMENT REVIEW APPROACH

Comment Category:	***Review Approach:***
Duplicate	Multiple comments received related to each instance of a specific issue. Comments are addressed based on the decision relating to first instance of the comment
Editorial Comments	Editorial comments are those that address the structure and flow of the technical report. Editorial comments are accommodated when they improve the structure, understanding and usability of the document and do not impact IEC 80001-1 requirements. Agreement is by consensus
Wording Comments	Wording comments relate to the wording or terms used within the technical report and include grammatical and typographical errors. Wording comments are accommodated when they improve the structure, understanding and usability of the document and do not impact IEC 80001-1 requirements. Agreement is by consensus
Not Applicable	Comments which are received that do not require any update to the Assessment Method. Examples of these comments include a statement of abstention or approval of the

The performance of this stage of the validation:

- confirmed the utility of the Assessment Method in a range off HDO contexts. This was possible due to the composition of JWG7 with members being drawn international experts representing a range of risk management stakeholders
- confirmed that the questions used in the Assessment Method were understandable to various risk management stakeholders
- confirmed that the structure of the Assessment Method was appropriate for use across a range of HDO contexts.
- confirmed that the questions used in the Assessment Method are suited for use or can be tailored for use across a range of contexts.

This stage of the validation process was conducted as part of the ADR process, again as part of the "Build, Intervene and Evaluate" stage of the ADR process [34]. During this stage of the process "pratitioners" in the field of the developed artifact, review the artifact in terms of its ability to be generalised and used across a range of contexts, in this within differing regultory environments in which the HDOs provide care. Practitioners from JWG7 reviewed the Assessment Method, as well as the PRM and PAM.The focus of this phase is to ensure the utility of the developed artifact(s) across a range of contexts.

In addition to the review by members of JWG7, a focus group session was conducted with a selection of experts from the group. These experts were asked to comment on various aspects of the overall MedITNet framework. This again was conducted using a "Practitioner Review" approach as part of the ADR process [44]. The reviewers were asked to comment on:

1. The utility of the assessment framework
2. The usability of the assessment framework for self-assessment of risk management processes within a Healthcare Delivery Organisation
3. The scalability and generalisability of the assessment framework
4. The coverage of the requirements of IEC 80001-1 by the MedITNet framework
5. Suggestions for improvements to the assessment framework

While the comments discussed the review focused on the overall MedITNet framework, a number of comments related to the Assessment Method specifically.

During this session experts reported that the use of the Assessment Method and specifically the assessment questions resulted in risk management stakeholders having a greater understanding of the requirements of the IEC 80001-1 standard. The expert noted that being asked questions related to specific requirements of the standard gave participants in an assessment a greater understanding of the requirements than they would have gained by reading the standard alone [44]. It was also noted that, by having a means to assess the capability of the risk management processes, Top Management understood the weaknesses in the current processes and had a better understanding of why adoption of the standard was important.

The performance of an assessment and the subsequent improvement of risk management processes also provides Top Management with a means to ensure that the benefits, which were intended to be provided to patients through the use of networked medical devices, were realised as expected. The experts also noted that the definition of the requirements of the standard at the level of processes in the PAM enabled the assessment questions to be tailored to take into account of the context in which the HDOs provide care. Experts further noted that the Assessment Method questions are beneficial as a starting point but noted that most HDOs would need to tailor the questions, not only based on the regulatory environment in which they provide care, but also based on the maturity level of the HDO in which the assessment is being performed. One expert taking part had been involved in a trial assessment in a different HDO to the one in which the trial assessment was performed. The expert noted that, while the questions in the Assessment Method are directly related to the base practice that is being assessed, these were rephrased during the assessment to use more open ended questions which were more appropriate to the context of the HDO being assessed. The rephrased questions did not focus as directly on assessing whether the base practices had been implemented but rather were phrased in a more open way that prompted a more general discussion of overall risk management processes before targeting the base practice in question.

Experts also noted that the requirements of the IEC 80001-1 standard had been covered in the MedITNet framework and also noted that the approach taken in the framework was consistent with the approach taken in IEC 80001-1. The experts noted that an improvement may be made to the framework following more trial implementations. These implementations may be able to provide guidance on how the framework could be tailored based on the maturity of the HDO. Experts also suggested the inclusion of a document map within the framework and suggested that a mapping from the Assessment Method questions back to the requirements of the standard may be helpful.

The performance of this stage of the validation:

- confirmed the utility of the MediITnet Framework including the Assessment Method in a range off HDO contexts. This review served as a final "expert" and "enduser" review of the final version of the MedITNet Framework
- confirmed that the usability of the framework across a range of HDO contexts and maturity levels
- confirmed that therequirements of IEC 80001-1 had been covered in the MedITNet framework
- gathered suggestions from improvements to the MedITNet framework

Each of these validation phases was performed iteratively as part of the ADR process and changes suggested by each phase of the validation were incorporated into the next version of the Assessment Method and the overall MedITNet framework.

V. Conclusions

While IEC 80001-1 takes steps to address the risks associated with the placement of a medical device onto an IT network, HDOs may face challenges in understanding and implementing the requirements of the standard. The MedITNet framework has been developed using Action Design Research in order to assist HDOs in addressing these challenges. The use of ADR ensures that the MedITNet Assessment Framework, including the Assessment Method, provides a consistent, repeatable and tailorable approach to the assessment of the capability of risk management processes related to the management of medical IT networks. An assessment of these processes can highlight weaknesses therein and can be used as a foundation for an improvement of risk management processes. The use of ADR ensures that the framework that was developed can be used in a specific context but is also suited for use across a range of HDO contexts. Effective risk management of medical IT networks ensures that the potential benefits of networked medical devices are realised while ensuring the safety of the patient is protected, the effectiveness of the device is assured and the security of the data and system are preserved.

Acknowledgment

This research is supported by the Science Foundation Ireland Principal Investigator Programme, grant number 08/IN.1/I2030 (the funding of this project was awarded by Science Foundation Ireland under a co-funding initiative by the Irish Government and European Regional Development Fund),and by Lero - the Irish Software Research Centre (http://www.lero.ie) grant 10/CE/I1855 & 13/RC/20194.

References

[1] S. T. MacMahon, F. McCaffery, and F. Keenan, "Development of the MedITNet Assessment Method - Enabling Healthcare Delivery Organisation Self Assessment against IEC 80001-1," in *First International Conference on Fundamentals and Advances in Software Systems Integration (FASSI 2015)*, Venice, Italy, 2015, pp. 1-7.

[2] West Health Institute, "The Value of Medical Device Interoperability - Improving patient care with more than $30 billion in annual health care savings," 2013.

[3] A. Hamilton, R. Nau, R. Burke, S. Weinstein, C. K. B. Dlatt, S. Fiore, *et al.*, "Summary of the August 2011 Symposium on the Role and Future of Health Information Technology in an Era of Health Care Transformation," The George Washington University, 2011.

[4] I. Lee, G. J. Pappas, R. Cleaveland, J. Hatcliff, B. H. Krogh, P. Lee, *et al.*, "High-confidence medical device software and systems," *Computer,* vol. 39, pp. 33-38, 2006.

[5] N. Milenkovich. (March 15, 2013). *OCR issues new HITECH regulations* Available: http://drugtopics.modernmedicine.com/drug-topics/news/drug-topics/health-system-news/ocr-issues-new-hitech-regulations. [Accessed: 23-May - 2016]

[6] Centers for Medicare & Medicaid Services, "42 CFR Parts 412, 413, 422 et al. Medicare and Medicaid Programs; Electronic Health Record Incentive Program; Final Rule", Health and Human Services Ed., ed, 2010.

[7] Centers for Medicare & Medicaid Services. (10/04/2013). *EHR Incentive Programs*. Available: http://www.cms.gov/Regulations-and-Guidance/Legislation/EHRIncentivePrograms/index.html?redirect=/ehrincentiveprograms. [Accessed: 23-May - 2016]

[8] Institute of Medicine. (2001). *Crossing the Quality Chasm: A New Health System for the 21st Century*. Available: https://download.nap.edu/catalog.php?record_id=10027

[9] E. H. Wagner, "The role of patient care teams in chronic disease management," *BMJ: British medical journal,* vol. 320, p. 569, 2000.

[10] E. H. Wagner, B. T. Austin, C. Davis, M. Hindmarsh, J. Schaefer, and A. Bonomi, "Improving chronic illness care: translating evidence into action," *Health affairs,* vol. 20, pp. 64-78, 2001.

[11] C. Hoffman and D. Rice, "Chronic care in America: A 21st century challenge," *Princeton, NJ: The Robert Wood Johnson Foundation,* 1996.

[12] J. Comstock. (2013, 23/01/2014). *14M networked medical devices to ship by 2018*. Available: http://mobihealthnews.com/28295/14m-networked-medical-devices-to-ship-by-2018/. [Accessed: 23-May - 2016]

[13] Agency for Healthcare Research and Quality (AHRQ), "Health IT for Improved Chronic Disease Management," Department of Health and Human Services, Ed., ed, 2013.

[14] M. Castañeda, "Connecting devices and data on the healthcare network," *Biomedical Instrumentation & Technology,* vol. 44, pp. 18-25, 2010.

[15] J. Goldman and S. Whitehead, "Advancing the Adoption of Medical Device "Plug-and-Play" Interoperability to Improve Patient Safety and Healthcare Efficiency," 2010.

[16] K. K. Venkatasubramanian, S. K. S. Gupta, R. P. Jetley, and P. L. Jones, "Interoperable Medical Devices - Communication Security Issues," *IEEE Pulse,* vol. Sept/Oct, 2010.

[17] R. Hampton and R. Schrenker, "What Does IEC 80001-1 Mean to You?," *24x7 - Technology and Service Solutions for Biomeds,* 2011.

[18] S. R. Rakitin, "Networked Medical Devices: Essential Collaboration for Improved Safety," *AAMI.org,* 2009.

[19] S. Loughlin and J. S. Williams, "The top 10 medical device challenges," *Biomedical Instrumentation & Technology,* vol. 45, pp. 98-104, 2011.

[20] T. Mehta and C. Mah, "Auto-Provisioning of Biomedical Devices on a Converged IP Network," *Biomedical Instrumentation & Technology,* vol. 43, pp. 463-467, 2009.

[21] T. Gee. (2008, 27/1/2012). *Medical Device Networks Trouble Industry*. Available: http://medicalconnectivity.com/2008/12/18/medical-device-networks-trouble-industry/ [Accessed: 23-May - 2016]

[22] S. Eagles, "An Introduction to IEC 80001: Aiming for Patient Safety in the Networked Healthcare Environment," *IT Horizons,* vol. 2008, 2008.

[23] National Cybersecurity and Communications Integration Center, "Attack Surface: Healthcare and Public Health Sector," ed, 2012.

[24] D. Talbot. (2012, Computer Viruses Are "Rampant" on Medical Devices in Hospitals. *MIT Technology Review*. Available: http://www.technologyreview.com/news/429616/computer-viruses-are-rampant-on-medical-devices-in-hospitals/. [Accessed: 23-May - 2016]

[25] J. Graham and C. Dizikes, "Baby's death spotlights safety risks linked to computerized systems," in *Chicago Tribune*, ed, 2011.

[26] J. Shuren, "Health Information Technology (HIT) Policy Committee Adoption/Certification Workgroup - Testimony of Jeffrey Shuren, Director of FDA's Centre for Devices and Radiological Health," ONC, Ed., ed, 2010.

[27] S. Eagles, 2008. An Introduction to IEC 80001. *IT Horizons*, pp.15-19.

[28] IEC, "IEC 80001-1 - Application of Risk Management for IT-Networks incorporating Medical Devices - Part 1: Roles, responsibilities and activities," ed. Geneva, Switzerland: International Electrotechnical Commission, 2010.

[29] F. J. Hegarty, S. T. MacMahon, P. Byrne, and F. McCaffery, "Assessing a Hospital's Medical IT Network Risk Management Practice with 80001-1," *Biomedical Instrumentation & Technology,* vol. 48, pp. 64-71, 2014.

[30] T. Cooper and K. Fuchs, "The Wireless Challenge - Technology Risk Assessment In Healthcare Facilities," *Biomedical Instruments and Technology,* vol. May/June 2013, 2013.

[31] M. Janssen and R. Schrenker, "Guidelines From 80001: Maintaining a Medical IT Network," *Biomedical Instrumentation & Technology,* vol. 45, pp. 295-299, 2011/07/01 2011.

[32] S. T. MacMahon, F. McCaffery, S. Eagles, F. Keenan, M. Lepmets, and A. Renault, "Development of a Process Assessment Model for assessing Medical IT Networks against IEC 80001-1," presented at the Software Process Improvement and Capability Determination (SPICE) 2012, Mallorca, Spain, 2012.

[33] S. T. MacMahon, F. McCaffery, and F. Keenan, "Transforming Requirements of IEC 80001-1 into an ISO/IEC 15504-2 Compliant Process Reference Model and Process Assessment Model," presented at the European Systems and Software Process Improvement and Innovation Conference, Dundalk, Co Louth, Ireland, 2013.

[34] M. Sein, O. Henfridsson, S. Purao, M. Rossi, and R. Lindgren, "Action design research," 2011.

[35] ISO/IEC, "ISO/IEC 15504-3:2004 Information technology -- Process assessment -- Part 3: Guidance on performing an assessment," ed. Geneva, Switzerland, 2004.

[36] M. Busby, P. Buttles-Valdez, P. Byrnes, W. Hayes, R. Khetan, D. Kirkham, *et al.*, "Appraisal Requirements for CMMI (Registered Trademark) Version 1.3 (ARC, V1. 3)," DTIC Document, 2011.

[37] T. P. Rout, A. Tuffley, B. Cahill, and B. Hodgen, "The rapid assessment of software process capability," in *First International Conference on Software Process Improvement and Capability Determination*, 2000, pp. 47-56.

[38] F. Wilkie, D. McFall, and F. Mc Caffery, "The Express Process Appraisal Method," 2005.

[39] F. Mc Caffery, P. S. Taylor, and G. Coleman, "Adept: A unified assessment method for small software companies," *Software, IEEE,* vol. 24, pp. 24-31, 2007.

[40] F. McCaffery and V. Casey, "Med-Adept: A Lightweight Assessment Method for the Irish Medical Device Software Industry," presented at the EuroSPI, Grenoble France, 2010.

[41] B. Barafort, V. Betry, S. Cortina, M. Picard, M. St Jean, A. Renault, *et al.*, *ITSM Process Assessment Supporting ITIL : Using TIPA to Assess and Improve your Processes with ISO 15504 and Prepare for ISO 20000 Certification* vol. 217. Zaltbommel, Netherlands: Van Haren, 2009.

[42] F. Mc Caffery and G. Coleman, "The development of a low-overhead assessment method for Irish software SMEs," *Journal of Information Technology Management (JITM),* 2007.

[43] A. Anacleto, C. G. von Wangenheim, C. F. Salviano, and R. Savi, "Experiences gained from applying ISO/IEC 15504 to small software companies in Brazil," in *4th International SPICE Conference on Process Assessment and Improvement, Lisbon, Portugal*, 2004, pp. 33-37.

[44] S. T. MacMahon, F. Mc Caffery, and F. Keenan, "Development and Validation of the MedITNet Assessment Framework: Improving Risk Management of Medical IT Networks," presented at the International Conference on Software and System Process (ICSSP), Tallinn, Estonia, 2015.

[45] ISO, "ISO/TR 80001-2-7: 2015 Application of risk mangement for IT-networks incorporating medical devices -- Application guidance -- Part 2-7: Guidance for healthcare delivery organisations (HDOs) on how to self-assess their conformance with IEC 80001-1," ed. Geneva, Switzerland: International Organisation for Standardisation, 2015.

Construction of the emotional environment and media

Treton Cécile and Bourret Christian
DICEN IDF/ – UPEM (Paris East University)
Paris, France
e-mail: ceciletreton@gmail.com
e-mail: christian.bourret@u-pem.fr

***Abstract* – The article addresses the society problem of the perception of individual loneliness in the cohort of elderly people. The researchers conducted empirical research by recorded interviewing a cohort of female and male people between the age of 65 and 99. The researchers postulate that the feeling of solitude is an opportunity to reveal the relational needs of individuals. They question the opportunity of a technical mediation in an emotional situation. Is it possible to consider approaches that are not limited to functional aspects? This work is integrated in the field of anthropology communication as defined by F. Martin-Juchat and B. Galinon- Mélénec. These approaches consider the body as a medium. The studies focus on the singular message that it leaves in its environment. The researchers highlight the unique aspects of communicative experience. The research method examines the interactions including the use of social networks on the web and the significations. The researchers postulate that relational developments correlate to certain life events, e.g., war, death, retirement. Their investigative work focuses on the elderly with the collection of—autobiographical testimonies. They have formalized an interpretative grid from an extended approach of the "situational and interactionist semiotics" developed by A. Mucchielli. This grid was used as a tool of analyze. The relationship to the other people brings a sense of material and emotional security. They have shown that the construction of the relational environment includes a strategic and creative dimension. They concentrated their investigations on the mediation of relational creativity. They rely on works on the biography of C. Delory Momberger and they established a model of mediation. They propose a model called "relational biography". This approach highlights the way the person builds its relational environment. It highlights the emotional and bibliographic part of the action. It takes into account the experience lived in its fictional dimension. This research highlights a relational mode with the object which contributes to the inscription of the person in the reality and participle in the relational feeling of satisfaction. In the situation studied by ageing, it allows to establish the necessity of taking into account the creative dimension of the relation as the esthetics of the everyday life. The old age is marked by breaks in the route of life that accompany intense phases of identity reorganization. The researchers suggest developing a bibliographic approach from the object as a media of evocation. The fictional object favors the conditions of the reorganization of the relational environment. The media like digital networks is integrated in the model like an extension of human being and relational space. It is a research in-progress.**

Keywords - loneliness; strategy of relations; emotional social network; age; friendly.

I. BACKGROUND AND PURPOSE

A. *Felling lonely and infrastructure*

The article takes back the communication "Loneliness and Relational Biography-Affective communication" [1].

The medico-social sector is subject to numerous changes and problems. In this context, there are two aspects that we wish to explore and connect, even though they seem un-related at first. On one hand, the health sector, led by the goals of streamlining services to the population, explores the development of individualized and modeled benefits, provided remotely through digital technologies in the context of the e-health with the approach of patients' pathways. On the other hand, the social sector whose demand is increasing in terms of compliance with standards of hygiene, safety and care, is moved by the increasing isolation situations that mainly affect marginalized people, because of their health status, their age, or their social status. It appears that the infrastructure that offers many opportunities for communication and security cannot resolve the development of a feeling of loneliness for a growing part of the population. The paradoxical aspect of this situation raises issues pertaining to the humanities. It seems conducive to questions that fall within the field of social innovation in terms of communication sciences. The suffering caused by loneliness disturbs. It emerges in a society characterized by the diversity of its technological possibilities of communication: mobile phones, digital social networking, mails, Web 2.0, etc. and by its individualism, according to P. Flichy and the "connected individualism" [2].

It challenges the institutions. Indeed, E. Durkheim [3] demonstrated the consequences of the transition from a traditional to a modern society. In a traditional society, the family ensures the protection of individual. In modern society, the concept of family disappears and the State may support or not the vulnerable individual's protection according to common standards. According to S. Paugam [4], the institutions contribute in reality to the emergence of this loneliness. They use intervention models that tend to dehumanize the relational practices perceived only as "services".

B. Socialisation and relational tensions

The question of the loneliness is for the society the opportunity to wonder about the relational needs for the individuals and the possibilities of answering it through approaches that are not limited to functional and pragmatic aspects. These questioning require to take into account the components from the relation to the other one in its active dimension "as experience of the communication" to resume the terminology of Louis Quéré [5], by integrating his emotional and emotional forms. Norbert Elias [6], demonstrates the interdependence between the individual and the community. He underlines the necessity for the man to communicate to assure his survival through the collective. He highlights the physiological and biological potentialities of the man such as the word that allow him to strengthen his relation with others and his survival. Through the relation of the man with the other one is formed a report of membership in the collective underlined by A. Honneth [7]. The other one is perceived as an object of the social. He offers the opportunity to the individual to experiment its relation to the collective and to negotiate the report between his identity peculiarity and the cultural standards of the community to be integrated. The construction of the relation can be considered as a social knowledge. This knowledge is the object of a learning of psychic nature that according to D.W. Winnicott [8] develops from the childhood in a natural process of detachment between the baby and her mother. The tension led by the distance allows the child to become aware of his intimate being and to consider in the same movement her mother as a subject and not an object at his disposal. D.W. Winnicott underlines the importance of solitary times that offer to the child then the adult the opportunity of return on himself and the desire of new meetings. The emotional sharing is an essential component of the relation. Its biological nature dresses a vital importance for A. Damasio [9]. He describes a process of emotional regulation that allows the individual to react to his environment. In the description of a behavioral adjustment of the individual in real situation, A. Damasio adds a description of the internal processes of regulation that participate in the self-awareness. It emerges from the report between the emotion and the feeling. It ends in an awareness that develops through a double movement of identity reorganization. It pulls an identity transformation that allows to answer the biological and psychic necessity of a preservation of one. The relation in the other one generates a psychic energy that finds its origins in the biological body and pass by a process of awareness of the feelings. These feelings depend on the sociocultural environment of the person and on its history. The constructed dimension from the relation to the other one fluctuates between need and desire. It is established in a duality between autonomy and dependence that expresses itself for P. Ricoeur [10] and P. Malrieu [11] through the narrative of its life. Through the narrative account of oneself, the person highlights its questioning in the way she invests the reality. C. Delory Momberger [12] observes the increase of the narrative of its life in the modern society in particular by means of the digital social networks. She considers that this behavior testifies of the solitude of the people confronted only with multiple social questions. Her approach is shared by B. Spiegler [13] and M. Maffesoli [14]. They hypothesize that a new sociability is made visible by networks. The use of the digital social networks compensate for the absence of recognition of the emotion in the modern society.

C. Communication and relational engagment

In the field of the sciences of the information and the communication, the question is a matter of the ethics. According to E. Illouz [15], the modern society considers communications tools as opportunities of economic development and leans on the mass media as the media of promotion. Nevertheless, the digital social networks let appear behavior of exchanges that give great emphasis to the personal expression. A. Casilli [16] observes that the social networks develop social learnings and that their use is considered as a "continuum" of the everyday life. He specifies that the digital social use of networks does not seem either to increase the feeling of solitude or to reduce it. The limits of networks live for M. Doueihi [17] in the modelling of the exchanges. For M. Doueilhi, the socialization of networks is restricted to the social roles. It joins in a logic of capitalization of the resources that does not take into account the emotional part of the relations. G Simmel [18] demonstrated by means of the structural sociology that the composition of the relational environments reveals distinctive forms among that the relations for two, "dyades", are next to community circles. For C. Bidart [19], these forms testify of the importance of the emotional relation. The individuals make choices and categories in their relations. They constitute a distinction between the relations of group and affective relations as the friendship. M. Mauss [20] considers the friendly relation as a "complete social relationship". She is freely chosen and mixes social gratitude and emotional commitment. O. Renault [21] underlines that Aristote observes from the Antiquity, the moral value of the friendship. According to P. Ricoeur, it allows man to recompose his understanding of his environment through a phenomenology of desire. P. Ricoeur distinguishes a permanent form of personality, "selfhood", around that are developed changing identities. P. Ricoeur emphasizes man's anxiety in front of these continual identity changes. He needs to reassure himself by searching for clues of permanence. According to P. Ricoeur, friendship is "a promise in time of a self-preservation."

The relational theories show the necessity for the individual to compose a relational environment. It develops in a report of desire and need that involves identical and emotional aspects. It tells a search for safety and for assertion of one in the relation to the other one. It is built on the basis of distinctions. They build themselves gradually and end in an emotional commitment. It rests on the confidence and passes by modes of emotional exchanges. The relation develops little by little through transformations of life and the identities in evolution underlined by C. Bidart. Within the framework of our reflection, we wish to envisage approaches that can take into account the emotional part from the relation to the other one. It is a factor of satisfaction, in the sense of the "flow" described by M. Csikszenmihalyi [22]. He considers it as a psychic energy that results from a balance between the current action and the potentialities of the actor. We are particularly interested in the phases of transformations that reveal tensions and relational imbalance. Our objective is to rest the question of the mediation such as she is invested by the sciences of information and communication under the angle of the intervention of a technical device to envisage a model adapted to our observations.

II. Methodological Framework

In the field of Information and Communication Sciences, our approach is in the course of the anthropology of communication and in particular with reference to the emotional communication as defined by F. Martin-Juchat [23]. She perceives the emotional body as "moved" by the emotion. F. Martin-Juchat highlights the lack of studies on this emotional dimension, in the field of science of information and communication that is shared between two conceptual approaches. The first concerns the question of the interpersonal relationship and considers the body in terms of signs verbal and nonverbal. This approach refers to the work of F. Saussure and of the Invisible College of Palo Alto. The second highlights the manipulative attempts of mass media. In both cases, the receiver is not considered in its ability to act as if he had not feelings. According F. Martin-Juchat, the receiver's action must be considered as a media. She proposes to put the "emotional body" in the heart of the communication device like a media.

The current of the anthropology of communication highlights the paradigm of "man-trace that B. Galinon-Mélénec [24] defines as follows: "the human is an anthropologically man-trace in the sense that it is both a trace builder and a trace producer. " She questions herself on action of communication highlighted by Habermas (1987). She considers the verbal expression (speech, writing) as an indicator of the "meaning" of the communication action in its dialectical dimension between internalization and externalization. The central issue of the approach revolves around the objectification of human thought and body. The body is perceived in the community. It is designed sensorily. The separation between the body and the reality is an artifact because the body lives in its reality and his reality is constituted by the body. B. Galinon-Mélénec writes: "When the body lives, he smells, he sees, he moves even when it is stationary. The flesh constantly is tested by the interactions of man with his environment. The separations between inside and out, emotional and cognitive, are artifacts. "The interpretation of the "meaning" of the footprints or traces, verbal or non-verbal left by man's actions in reality gives him the opportunity to access the understanding of his action. The individual is seen in his etymological dimension (from the Latin "individuum", "what is indivisible") of uniqueness. The report to another is established in a process of objectification and the confrontation with the other builds the identity of the individual. The unity of the subject is "still on trial" through interaction and involves regular work of reconstruction and transformation. The body is seen as "a way to exist," it is studied by searcher from his footsteps. These traces are left in real or virtual spaces. Sung do Kim [24] highlights the mediation dimension of places that organize the flow of material and immaterial exchanges. The methodological issue stressed by B. Galinon- Mélénec involves taking into account the analog dimension of the interpretative work of traces. Our approach takes as reference the paradigm of "engaging communication" to the meaning of F. Bernard [25]. The concept of commitment is used to demonstrate the link between the action and the meaning given to action. The commitment depends on the situation. The "engaging communication" is based on the action as a mean of change. The identification is integrated in the processing of change through the action.

These theoretical approaches orientate our research method. We consider communication actions as singular productions of the person in reality. They become relational objects used as expressive media. They reveal the human inscription modes into reality. The media in this sense is seen as an additional space that reveals the communication actions. Mediation is used as awareness means and supports the exploratory process.

III. Relational Investigations

The researchers highlighted the existential challenge of relating to others. They want to study how the person asserts his existence through the relationship to another. They established three questions. With the first question they have attempted to theoretically clarify how individuals build their relational environment. Their second question concerns the meaning that the person gives to its communication activities. Indeed, they consider the person as the driving element of the relation that is established. Their third question deals with the transformation and is particularly relevant with regard to their subject. The question is to identify the impact of the changes in the relationship and the uses of reorganization of the relational environment.

From those questions, they studied the relational dynamics of the elderly in order to highlight their mode of action, the uniqueness of their subjective experience and confirm temporal and spatial changes. The ageing is a stage of life that seems to reactivate interrogations about the relationship to another.

A. Relational dynamics characteristics in old age

Building on the contributions of sociologists and gerontologists, the researchers find that the old age - estimated at the retirement age - is a period marked by many transitions and changes. V. Caradec [26] and P. Pitaud [27] observe that the elderly person is faced with a multiplicity of events and breaks. These changes are due to recurrent affective losses, changes in material and economic living conditions, a decline in physical abilities, often affecting motor skills and the initial conditions of life. The old age is a particularly intense period of identity reorganization. The life changes impact the modes of socialization of the person. C. Bidart emphasizes socialization change with age with a tendency to build proximity links and to keep more distance in the relation. She notes the need to more moments of intimacy. For gerontologist M. Billé [28], these times of intimacy and relational distancing correspond to the need for "interior narrative work." The elderly need to remember the past to ensure a temporal continuity at their life and preserve a consistent and positive image of their life. The collected items highlight the intensity for the elderly from an internal contradiction between autonomy claimed as singularity and social integration as assimilation and dependency. The researchers understand through the transformations described that during the retirement are developing events. Those events can be felt as an accumulation of loss and grief. The activity of the elderly is not an economical necessity normalized by the job. It depends on a determination that seems related to the desire to assert its autonomy, to assert them as active and alive. The aim is to remove the scary specter of the final dependence.

B. Method of collecting and interpreting data

The researchers chose to proceed by biographical interviews. The biographical interview allows the collection of data revealing the subjectivity of relational experience. It can be used as a means to train the person to take action. The approach that they consider analytic allows us to focus on the speech of the person. As stated by C. Delory-Momberger, the biography enables a work on the self images that precede action. The researchers agree with this approach that considers the individual as an agent of his own socialization by the action. This method of data collection takes into account the "emotional body". C. Delory-Momberger notes that it is the place of biographical investment. The researchers have collected 15 hours of recording stories of friendship of fifteen women and men, aged 65 to 99 years. As researchers, their goal was to highlight the words collected. Didier Demazière and Claude Dubar [29] observe that the empirical research work that uses the interview must use a method of analysis. It is necessary to establish comparative tables and explain its translation procedures. Thus the word becomes knowledge. The emerging theory should not be a simple data formatting, but must be the result of the comparison of these data. The aim is to highlight a process common to several behavioral phenomena and accompany them with information. This approach encourages reflection through a data aggregation and translation through explicated categories. The categories are "a structured symbolic world" that tells the speech. The analysis involves exceeding these categories and concepts by theoretical tools described by the researcher. Thus, analytical posture passes successively by several categories, natural, emerging, conceptual and abstract. To analyze the biographies, the researchers used categorization method proposed by A. Mucchielli [30] through situational and interactionist semiotics. A. Mucchielli raises the question of social identification. He writes: "Identify the other is a judgement for define him in a specific context. Identify others is a means to give a meaning to my "being" situated also in a context." This method allowed us to establish a frame of reference from a categorization that we have defined. These categories form "interpretive frameworks' of intentions and needs of the actor, its reference standards, its positioning in relation to other actors, quality of relationships in a historic, temporal and sensory setting. The meaning is defined in a constructivist perspective and shows a schematic representation of the operation of the phenomena studied.

C. Results

The following table shows an extract of the interpretative grid. The researchers highlight three aspects common to all the interviews. They find that relational environment evolves according to life events. Relationships are always built on the same pattern. This pattern is established from emotional factors related to values and beliefs. It shows the specific needs of each person and his way of composing relations through the choice of communication spaces and rhythms of interactions. The researchers established a grid from distinctions between the components of the relationship and the relational modes of action that we have identified. The grid looks like the model below.

BOARD I: EXTRACT INTERPRETATIVE GRID

	Categorizations		
	Biographical frame	***characterization of the relationship***	***Methods activation***
D.	Boarding school Work Marriage Death	Compensatory mode and selective strongly linked to the stages of change and emotional traumas	favors the communal group activity
A.	Childhood Studies Work Travels Marriage Death	Adaptive mode and links marked by complicity siituational Links with family friends Friendship is a family value	friend and ritualized moments : New Year's Day / birthday / holiday
N.	War/Childhood (holocaust)	Selective mode	search for help, support,

	Categorizations		
	Biographical frame	***characterization of the relationship***	***Methods activation***
	Marriage Accident (vision loss)	marked by mistrust, emotional distancing, sharing the difficulty, Classification of friends by period	taking account of disability

Using this method, the researchers have defined the criteria for analysis. They compare the action patterns of people and their interpretation. The action patterns are defined by how they use the resources of their environment, and the spaces in that they organize their time. The study reveals identity and biographical aspects.

D. *Determinations of the congruence of factors in the composition of relations to others*

All the evidence confirms a match between the emotional state and the manner that the persons build their relationships. Loneliness appears in stages of life related to an upheaval. Some have experienced affective breaks (or professional breaks, but most have experienced a feeling of loneliness from events related to advancing age, retirement, deaths of spouses and health problems. The reorganization of the social environment is common to all. It comes along with a strong determination expressed in a very voluntary way. Most of the time, the collective brings protection and validation. The contribution to the collective is imperative. The elective friendships often stemming from these collectives do not exceed 2 or 3 people. The meeting becomes established in an empathic way. The particular attraction for a person is expressed through personality's elements dependent on qualities such as the beauty, the kindness, the intelligence, etc. The relation is strengthened through exchanges based on the friendly listening, the shared gaiety, the common activities. They are perceived as gifts. The notion of gift seems different between the sexes. The men express the values that they bring to the other one: transmit, allow other one to develop, to protect. The women express an expectation of mutual protection or conceive the exchange from shared values. The sharing of centers of common interests or common values is determining. Finally, the anchoring in the territory is essential. He expresses himself in a recurring way. The relation in the territory is identity, bibliographic and relational. The district is particularly quoted by the Parisian. Age is quoted most of the time. He expresses himself by a consciousness of time and of the life story and the anticipation of the losses of physical abilities.

E. *Relational strategy and relational creativity*

This work shows two components of the experience of communication that it can be distinguishing in terms of relational strategy and relational creativity. Those distinctions design a model of mediation adapted to the relational needs for the individuals. The researches observe that the person builds its relation through plans of actions that pass by an operational effectiveness in the reality and are common to all. These actions obey a relational strategy. The relational strategy leans on real or virtual spaces of mediation. They define the relational strategy as a construction of relations leaning on the resources of the reality. The relational creativity depends on the sense that the person gives to its actions. It transfigures the reality. These elements of creativity depend on a personal relational history that gives a coherence to the current relation. The relational creativity is of the interpretation by the person of its relations. It has a nature of identification and idealized image of one according to values. It integrates a symbolic perception of the reality. The reality establishes breaks that put in tension the person. The reorganizations are lived in an identity way. They participate in an emotional regulation and generate a feeling of satisfaction. The researchers perceive dependence between the events of life and the appearance of new relationships. The person builds its relation in time. The events impact on his relational environment. Therefore, there is a relational biography. Each phase change appears to activate a clean relationship strategy to each person and builds on perennial patterns

The relational strategy and the relational creativity are closely imbricated. The person makes a classification and tells for example: "my friends from before" according to his report with his history and to the way he perceives his old age: it concerns the disappearance of friendships: "of my generation has disappeared ...", "we were ten friends... we are no more than two". It expresses itself by the regret of the bonds of the past by comparison with those present: "Yes I have had really good friends, it was great ... today, it is different." It indicates a feeling of loneliness evoked on the mode of a battle to fight, "for not to be alone, you have to go out! ". Classifications depend on elements of identification established on the basis of emotional elements and of elements of social standardization through values such as the value placed on friendship, beliefs and expectations. They direct the sympathies or antipathies. They provide data on the image that the person has of itself in an idealized form. The relational choices appear homogeneous, standardized, selective based on distinguishing elements of social status, age, etc. The interactions are located in social contexts and spaces that have a symbolic value as: schools, companies, unions, associations of hikers, etc. They reinforce the value of belonging.

The relational commitment is linked to an idealized image of one but it is established according to the immediate needs for each. The emotional factors determine the level of commitment in the relationship ex. "When they suffer, I suffer." They compose an imaginary of the relationship: "I wanted to have news from him because it was my first love". The person creates, in this case, an ideal relational environment that is considered like a game. This is particularly noticeable when the person uses digital communication media. "I am in relation to a community of artists". The exchanges are organized according to temporality of pragmatic order, as the definition of a date of meeting, but also symbolic. It takes the form of ritualized interactions whose rhythm varies among individuals: annual exchange of holidays wishes, weekly meal, evening conversations on Skype, etc. They take singular forms appropriate to the needs of the person such as sharing of activities or friendly moments, or its values: "to give is my life ..." etc.

The information collected shows activation of affects in the implementation of the social network and the importance of biographical and identity work that is developed. The experience through the implementation of the relationship strategy own to everyone and the relational creativity appears as the essential element of identity's reorganization.

The relation in the other one does not have a harmless nature. It coincides with a stage of life and takes sense for the person through an event. The relation in the other one is lived as an effective registration of one in the reality. It is about a reality perceived through the subjectivity. It consists of objects, spaces and temporality to that the person gives sense through its personal history. Other one acquires a status of object in this composition of the reality. The relation is a creative expression of one in the reality through the otherness. The search underlines that it is not so much the way we communicate with the other who matters but the way the other one confirms our existence. That is why the hypothesis is to grant a singular place to the relational creativity and to envisage the opportunity of a mediation.of it. The objective would be to facilitate the imaginative report towards the other one.

IV. THE NOTION OF FICTIONAL OBJECT

The difficulty of the questioning is to determine a way of highlighting the subjective part from the relation to the other one. Is it possible to show what belongs to the domain of the close friend? How to envisage a mediation of the imaginary part of the relation?

It is important for the researchers to define the imaginative process that is established in the relation with the other one. The put in perspective of this process appears through the elaboration of a relation that they described the characteristics and that they identified through an interpretative bar.

A. The fictional créativity in the relation to the other one

They lean on surrounding areas of G. Bachelard [31], J. Schaeffer [32] and G. Simondon [33] to study the relation in its report of sensitive immediacy with the object.

G. Simondon postulates that a person can be considered as an object. The object in its materiality and corporeality interests us as object of exploration and object of dialogue with the reality. According to G. Simondon, the object is the opportunity for the person of an imaginary and driving exploration that is of the playful and generates of the inventiveness. This exploration answers a need to solve a problem posed by the reality.

The person finds in this action the opportunity to reveal and strengthen its feeling of one, to assert its desire and to express an intuition as a life force. The imagination allows the individual, as J. Schaeffer demonstrates it, to negotiate its report in the reality to transform it as one pleases. The imagination is a skill.

G. Bachelard highlights the epistemic value of the imagination and J. Schaeffer considers it as a fictional skill. G Bachelard describes a relational experience elaborated through the interiorized image revealed in the contact of the object.

G. Simondon and J. M. Schaeffer show that the imaginative function is borrow of myths transmitted socially. They facilitate the dumping in the reality. G Simondon introduces a distinction into the process of elaboration with the objective reality between the symbol, the image and the imago. It considers the imago as an intermediate stage that allows the individual to act in the reality to transform it in a creative way.

The relational creativity bases on the constitution of an object that possesses in him a fictional dimension because it is perceived through the imagination of the person. This imagination is fed by the history of the person. The momentariness of the relation with the object of election is possible through symbols and feelings. The otherness adds to this subjective projection the division of common feelings.

The relation develops in the time and corresponds to the perception of the friendship evoked by F. Guattari [34] as one " a weaving " that he qualifies as "Third World". The relation passes by a singular approach of co-elaboration and joins in a history built for two. The fiction is fed in a joint way by the reality and its interpretation and develops by a temporal and spatial process that strengthens it. We hold two aspects of this composition, the dumping and the exploration.

On one hand, the object calls out. It generates an internal dialogue made up of images and symbols stemming from the imagination and nourishing of cultural references. The person seems friendly through common affinities that are like "idem".

On the other hand, the person is perceived in a way distanced as an object of the reality with that a communication makes a commitment in an exploratory way on the basis of a subjectivity mixture of images and intimate and socialized symbols. In this relation, the reality is transformed.

The relation takes sense in its biographic dimension. This story builds itself from two movements. These mental movements correspond to two times. We observe a movement of exploration and a movement of immediate apprehension. First engages a work of transformation, the second activates a sensation.

The following board distinguishes these two aspects:

BOARD II : FICTIONAL COMPOSITION

OBJECT	
Exploration (transformation of the reality and construction of the relationship) shared image	Immersion (questioning at the time of the meeting and the dialogue) symbol one/other
SUBJECT	

The fictional composition leans on a strategy of anchoring in the reality but rests on a creativity that amplifies the relational experience lived between two people. The relation in the other one acquires a density through the shared fiction. This fictional dimension develops around three complementary elements, the historicity, the imagination and the inventiveness. The imagination feeds on shared symbolism, on identification, on idealized image, on identical evolutions. It assures an office of one and seals the relation. The feeling of solitude appears when the relation loses its creative meaning. The object of election does not fill any more its role of emotional questioning and the relation does not have the means any more to be invested in an exploratory way. The fiction diminishes and loses its capacity of "tonic effect".

B. The relation in the other one as the experience of the body

The fictional dimension from the relation to the other who develops in a dialogue between subjectivity and reality is of a process of immediacy. In the field of the communication, O. Galibert [35] uses the term "relational anthropology" to indicate the relation to the other one in subjectivity. The notion of identity is studied through the dialogue that develops on the basis of a reciprocity. He suggests using the term of "person" as the one who appears in the relation to the other one. The self-awareness does not precede the communicative activity, it is the result. The relational anthropology does not distinguish the subjectivity, the otherness and the relation but considers them as a whole. In reference to E. Levinas [36], O. Galibert considers that the experience of others passes by the body and writes: " The presence in the other one calls out to me ". Other one is arrested not in a reflexive way but as experience. This experience calls an "ethical" responsibility because contrary to the relation in the thing, the vulnerability of other one moves me in "native way". Our work of bibliographic collection highlights a dialectic of the relation. In a functionalist way, J. Caune [37] describes this dialectic as a "sensitive mediation". It passes by the institution of a "captivating" word fed by shared references and by common experiences. J. Caune defines the relation as a "breach" that is as an interstitial space that gives way to the perception subjective as experience of one and sensitive experience for the understanding of the social standards. He distinguishes this space, of the "contact" the stake in that is the balance between the closeness and the distance and of the "link" as the participation in the community by material, symbolic and imaginary links. We consider that the link testifies in its material dimension of the relation. C. Delory Momberger observes in the narratives of life the emergence of the "faces of one" in "the action" within spaces defined by their materiality and invest by the interpretation of the subject.

C. Materiality and relational creativity: the object of dialogue

In its fictional dimension, the presence of other one can pass in transit by a realized object. G. Simondon tells that the intermediate space allows explorations through concrete objects that D.W. Winnicott considers as "transitional objects" universal. In this notion of object of exploration, G. Simondon adds "the intermediate object" perceived through the memories. G. Bachelard describes the intimate perception of the object in the place of residence. He underlines that the imagination generates a feeling of happiness from the harmless. The process is short-lived and orders. It develops through the subjectivities fed by the history of the subject. It is about a story in movement that implies transformations and fixations revealed by G. Bachelard. He underlines the historic density concentrated in the contact of the object. We observe that the elderly continue to maintain strong relations with people died in a silent dialogue and through objects. The object establishes a particularly perceptible shared dialogue through the work of art. J. Schaeffer demonstrates the implicit understanding of emotional and symbolic nature that develops between the creator and the receiver by means of the object of art. The pleasure is into a symbolic and poetic complicity shared by common cultural references or by feelings. N Alter [38] underlines that the object is a symbolic gift that consolidates the relation. J.M. Schaeffer distinguishes three aspects of the fiction, the "mimesis", the "representation of volitional states" and the creativity. He defines the mimesis as "a learning by modelling " that allows the individual to shape his behavior to that of the group for a better cohesion. The representation of volitional states corresponds to the way the individual is going to adapt himself to the reality while the inventiveness and the creation, on the contrary, adapts the world from an interiorized representation. In the process of communication that develops by means of the work of art, the creator gives a shape to the object on the mode of the mimesis while the receiver perceives the object in a creative way in an immediate echo. R Barthes [39] underlines that without the imaginative contribution, the photography could have that a morbid effect in its temporal fixedness. According to him, the photo testifies of a reality but through a painful fixedness of last moment. The photo livens up only through the emotional projections of the receiver.

This movement of perception corresponds to the immersion. The person is invaded by an emotion activated by the object. This emotion depends of its personal story of life. The object is perceived as a "bait" in the sense of G. Bachelard, that generates and amplifies, according to J.M. Schaeffer, our "emotional skills."

D. Mediation of the relational biography

The consideration of the singular relational creativity pulls a modification of the mediation. The objective of the mediation would be to participate in the highlighting of the fictional object. G Simondon underlines that the project always proceeds of a simplification of the initial imaginative moose. He considers that the intervention of an object of dialogue between the transmitter and the receiver is a relational structure "tertiary sector". The intervention of the third as mediator always reduces the imaginative dimension. As such, C. Delory Momberger considers that the support by the biography proceeds of a "shaping" by the word of the reality lived by the narrator and the "direction" of this reality told by the practitioner. For G. Simondon the relational inventiveness is already a mediation in itself and she can thus lean on an "instrumental mediation". The tool is perceived in its anthropological dimension such as defined by A. Leroi Gourhan [40] as a display of the human capacities. B. Spiegler underlines that he participates in a reification of moment. He names: "tertiary retention" this status of object of the media that keeps the registration of a gesture. In this context, the realization ensures the continuity of the idea. The technique the limits of that we identified by the modelling possesses intrinsically this arrangement in the repetition and in the memory if only in his conception. The project spreads through the possibilities of the technical system. This registration is perceived as a track that can give rise to the interpretation in an understandable way. Within the framework of paradigm of the man-track, the track of the relational gesture passes by an object or by a space. For B. Galinon-Mélénec, the object considered as a track acquires a value of indication. For Sung do Kim [24], the track is a way for the individual "to mark" its space and to establish a membership. He underlines new practices of writings "situated and located" that he considers as private and public scriptural tracks. They express a new relation to the memory because each is a "cultural producer". According to him, all these tracks make up one "ambient commons" or "ambient consciousness" that participates in the constitution of an increased "reality". The space is increased by the imaginary interpretations of each through the flows of information and by exchanges. P. Lévy [41] underlines that spaces are invested in a way abstracted through the expression of movements. The space is not a decoration but it is established through the human experience. According to A. Cassili, the interpersonal media make visible a set of dialogues and of relations perceived as a "continuum" of the everyday life. The interpersonal media compromise for P. Levy the "fourth space" of exploration and exchanges that train a "cosmopédie". P Lévy considers that the cyberspace exists only through the interactions. The cyberspace is topographic and kinetic. This space, according to P. Lévy, tells a singular creativity that questions the model of the mediation through the expert. For P. Lévy, the relations in this space are an immanence because each is responsible for it. P. Lévy considers that the human sciences have to take into account the multiplicity of the processes of interpretation generated by these interactions.

The bibliographic approach perceives the track in its projective dimension. The moment is registered in the unique and singular route. The biography that we name "relational" takes for track "rundowns" of the gesture relational as anchor point in the reality and the subscriber in a temporal continuity. The relational biography is interested in the emotional echo of the track in the "real-life" body. It strengthens the relational creativity through an interpretation of the action in the reality perceived through the fictional and subjective dimension. It has for objective to offer the conditions of an elaboration of a relational environment piloted by the subject. The goal is to be capable of taking into account the creative dimension of the elaboration from the relation to the other one in its emotional dimension. It would be a question then of developing the objective outlines of a dynamic scenario of the relation as the "history in movement".

V. CONCLUSION AND PERSPECTIVES

This research concerns loneliness in the relationship to another. The researchers question the appropriateness of technical mediation in situations of loneliness. They search highlighted reveal two crucial needs for the human being related to safety and self-awareness. These needs are expressed through communication with the external environment composed of other people. This constitutes the means for a person to validate their existence through recognition and to strengthen their feeling of protection. The relation that takes shape in the form of mutual identical projections is transformed into attachment when the reliable feeling evolves and strengthens. This reliable feeling comes along with the certainty of a common cultural belonging that expresses itself through shared activities and symbolic exchanges. This process consolidates the commitment in the relationship.

The experience of communication is perceived as a construction. It develops on the basis of choice and of classifications that the structural sociology highlighted. Every relation corresponds to a level of commitment. It is defined according to identical stakes on that depends the self-awareness. The strong links contributes to strengthen a positive self-awareness based on the mutual identification. The aspects of identity are expressed through emotions and feelings. The self-awareness answers a desire of self-idealization through values that direct the emotions.

The feelings are the mainspring of the composition of the relational environment because they play beforehand a role in the choice of selected people.

The theoretical contributions of the sociology and the psychology, allowed us to adapt to our subject of search our interpretative railing. The study of the bibliographic testimonies of the elderly highlighted the way the person recomposes its relational environment in a phase of transformation further to a break in its life. The researchers emphasized the friendly relation considered as a symbolic relation of the report between the peculiarity and the standard. They observe that the relational composition depends on actions in the reality. They have a strategic dimension and a creative dimension. The relationship strategy composed of the elements highlighted in our study shows the importance of the building of the relationship as a sensitive experience level. The relational creativity develops in the meeting with the other one perceived as an object of stimulation. The history of the relation participates in this creativity. The relation evolves during the relational process between the people. It constitutes a story made up of events of that the meeting with the person is a part, and is characterized by the importance of the moments and shared activities. From this point of view, the story of the relation can be the object of a biography characterized by the ascendancy of its emotional contents. The experience of the communication reveals a fictional dimension. The fictional dimension depends on the imagination. The imagination allows the individual to accommodate the reality to its expectations. It contributes to the inventiveness. The communication stimulates the creativity. Other one is perceived as an object with that to enter into a relationship generates a tension. The relational balance requires the appeal to the personal imagination. A. Damasio demonstrated how the perception of the object pulls a process of internal consciousness likened to a social knowledge. The fiction is objectified in the reality. She can move on an object perceived as transitional support or as object of dialogue. Within the framework of a question about opportunity of a mediation, the researchers are confronted with the questioning of B. Spiegler and G. Simondon on the simplifying dimension of the media as the "tertiary retention" in its transposition of a polymorphic reality. They observe that the object through the feeling that it inspires to the person contributes to a validation of its existence in the present reality. So, the capacity of reification of the present moment of the technical mediation observed by B. Spiegler can contribute to underline this capacity of the man to register his intimate history in the present real-life experience. From the functional approaches but also in reference to the phenomenological approaches, the researchers propose a model. They qualified it as "relational biography". The relational biography consists of a work of specific singular interpretation of the actions and the meanings in the context from the relation to the other one. The media is perceived as a space of mediation. It is used in its dimension of memorization of the track perceived as relational gesture and the shaping of the relational environment. The track and the shape are "baits" of the experience of the communication. The mediation appears as a way to strengthen the experience of communication as assertion of one and registration in an "increased" reality that participates in the "history in movement".

REFERENCES

[1] C.Treton, C. Bourret, "Loneliness and Relational Biography – Affective Communication", Proceedings / The First International Conference on Human and Social Analytics, HUSO 2015 – IARIA, Saint Julians, Malta – pp. 45-48, October 11-16, 2015.

[2] P. Flichy, "L'individualisme connecté : entre la technique numérique et la société, Nouvelles réflexions sur l'internet," *Réseaux*, vol. 22, n° 124, pp. 17-52, 2004.

[3] E. Durkheim, "The division and labour in society," PUF 1930, Paris, p. 124, new collection « Quadrige », 2012.

[4] S. Paugam, "The social bond," PUF, Paris, pp. 7-103, 2013.

[5] Louis Quéré, "Des Miroirs équivoques, Aux origines de la Communication Moderne," Aubier Montaigne, coll. Res Babel, Paris, 1982.

[6] N. Elias, "Society of individuals," Fayard, Paris, pp. 208-301, 1991.

[7] A. Honneth, "The struggle for recognition," Folio essais (n° 576), Gallimard, Paris, pp. 114-235, 2013.

[8] D. W. Winnicott, "Capacity to be alone," Payot-Rivages, Paris, pp. 46-108, 2012.

[9] A. Damasio, "The feeling of what happens, body, emotions, conscience," édition Odile Jacob, Paris, pp. 110-299, 2002.

[10] P. Ricoeur, "Oneself as Another,"édition du Seuil, Paris, 1990.

[11] P. Malrieu, "The process of personnalization," in ERES, «Pocket - Society», Toulouse, p.280, 2003.

[12] C. Delory-Momberger, "The biographical condition. Essai on self-narrative in the advanced modernity," Teraèdre, Paris, pp. 61-75, 2010.

[13] B. Stiegler, "Le bien le plus précieux à l'époque des sociotechnologies - Réseaux sociaux," Editions FYP, Paris, p.235, 2012, Cit. G. Simondon G. (1997), L'individu et sa genèse hysico-biologique, Paris, Edition Jérôme Milliam, p.156.

[14] M. Maffesoli, "Le Réenchantement du monde. Morales, éthiques, déontologies," éd. Table Ronde, Paris, 2007.

[15] E. Illouz, "Les sentiments du capitalisme," Seuil, Paris, 2006.

[16] A. Casilli, "Les Liaisons numériques : vers une nouvelle sociabilité ? " Editions du Seuil, Paris, 2010.

[17] M. Doueihi, "Pour un humanisme numérique," Seuil, Paris, 2011.

[18] G. Simmel, "Studies on forms of socialization," PUF, trad., 1999.

[19] C. Bidart, "Studying networks, contributions and prospects for social sciences," Informations sociales n° 147, pp. 100-259, 2008.

[20] M.Mauss, "Essai sur le don. Forme et raison de l'échange dans les sociétés archaiques " in sociologie et anthropologie, PUF, Paris,1923.

[21] O. Renaut, "Platon, la médiation des émotions, L'éducation du thymos dans les dialogues," Vrin, Paris, 2014.

[22] M. Csikszenmihalyi, "Finding Flow," in Editions Basic Books, New York, p103-125, 1997.

[23] F. Martin-Juchat, "Think the emotional body as a media, anthropology of affective communication," Revue Le corps n°4, Paris, pp. 85-92, 2008.

[24] B. Galinon Mélénec, F. Liènard et S. Zlitni, S. do Kim " L'homme Trace inscriptions corporelles et techniques," CNRS Edition, Suresnes, 2015.

[25] F. Bernard and R. V. Joule, "Link, meaning and action : Towards an engaging communication," Communication and organization, online, URL : http://communicationorganisation.revues.org/2918, mars 2012 [retrieved : August, 2015].

[26] V. Caradec, "Sociology of the elderly and aging," Armand Colin, Paris, pp. 87-103, 2012.

[27] P. Pitaud, "Loneliness and isolation of the elderly," Editions Erès, Toulouse, pp. 25-103, 2010.

[28] M. Billé and J. Pollard, "Manifesto for age and life, reenchant old age," Editions Erès, Paris, pp. 63-112, 2012.

[29] D. Demazière, C. Dubar, "Analyser les entretiens biographiques," les Presses de l'Université de Laval, 2009.

[30] A. Mucchielli, "Situation and Communication," Ed. Ovadia : Nice, 2010.

[31] G. Bachelard, "La poétique de l'espace," PUF, Paris,1957.

[32] J. M. Schaeffer, "Pourquoi la fiction ? " Le Seuil, Paris, 1999.

[33] G. Simondon, "Imagination et invention," Les éditions de la transparence, réédition, Paris, pp. 237-268,2008.

[34] F. Guattari, "Chaosmose, Penser avec Félix Guattari," Revue Chimères, N°77, éditions Erès, pp. 17-107, 2012.

[35] O. Galibert, "L'incommunicabilité dans le cadre de la pratique anthropologique," Socio-anthropologie, pp.23-241 2009.

[36] E. Levinas, "Altérité et trasncendance," Edition Poche, Paris, 2006.

[37] J. Caune, "Pour une éthique de la médiation. Le sens des pratiques culturelles,"Presses universitaires de Grenoble, 1999.

[38] N. Alter, "Donner et prendre," Editions La découverte, Paris, 2010.

[39] R.Barthes, "La chambre claire, note sur la photographie," Editions de l'Etoile Gallimard Seuil , 1980.

[40] A. Leroi Gourhan , " Les racines du monde", Edition Belfond, Paris, p.290, 1982.

[41] P. Lévy, " L'intelligence collective, pour une anthropologie du cyberespace," Editions La découverte, Paris, 1997.

The Mental Organization of Air Traffic and its Implications to an Emotion Sensitive Assistance System

Linda Pfeiffer, Georg Valtin, Nicholas Hugo Müller and Paul Rosenthal
Technische Universität Chemnitz, Visual Computing Laboratory and Institute for Media Research
Chemnitz, Germany
Email: {georg.valtin, nicholas.mueller}@phil.tu-chemnitz.de,
{linda.pfeiffer, paul.rosenthal}@cs.tu-chemnitz.de

***Abstract*—The StayCentered Project at Technische Universität Chemnitz aims for assisting air traffic controllers in stressful traffic situations. Therefore, we are seeking to comprehend air traffic controllers' principles of operation within the dyadic team structure. First exploratory research revealed insights into air traffic controllers' practices, their information processing (mental models), potential stressors, and related emotional effects. This paper discusses the results and the implications for air traffic controllers' work in general and the StayCentered project in particular.**

***Keywords*–Air traffic control; HCI; Decision Support; Mental Models.**

I. INTRODUCTION

In our paper "'Aircraft in Your Head: How Air Traffic Controllers Mentally Organize Air Traffic"' [1] we presented first results on air traffic controllers' practices, their information processing (mental models), potential stressors, and related emotional effects. These were collected during semi-structured interviews and observations at the facilities of the German air traffic service provider DFS in Munich and Langen. The observations were limited to experiences with the P1/ATCAS air traffic management system. As there is a second air traffic management system (VAFORIT) in use at DFS Deutsche Flugsicherung we decided to conduct further observations. This article discusses the expanded results and their implications to the StayCentered project. In the first sections we will introduce the air traffic controllers work, outline related work in research, describe the interfaces currently in use at Deutsche Flugsicherung and introduce the StayCentered project and our goals. Section IV outlines the methods used to gain the findings that will be discussed in Section V. The following sections relate findings to the project context in terms of the future mental and emotional model as well as future interfaces. Section VIII summarizes and concludes the paper.

II. THE WORK OF AN AIR TRAFFIC CONTROLLER DYAD AS A POTENTIAL STRESSOR

German airspace is divided into sectors of differing size and form. By traveling on an airway an aircraft passes several sectors. In the state of normal operation, a dyad of two air traffic controllers is responsible for such a sector. Both have access to task relevant information, such as radar data, weather reports, and flight schedules. Within the dyad, the air traffic controllers take different roles: one (executive) is responsible for the communication with the pilots using spoken traffic commands over the radio, while the other one (planner) is coordinating the acceptance or handover of flights from or to other sectors. This is necessary, since each sector has its individual operation of flight-levels and is generally only accepting flights within a certain flight-level threshold in order to keep a smooth vertical alignment between adjacent flights. While arranging the handovers, the planner is also responsible to verify the communication between the executive and the pilots and to intervene, if necessary. Therefore, the division of responsibilities is not just depending on team coordination regulations but on a good internal communication and a transparent work situation. Expediting and maintaining orderly traffic flows, without infringing separation minima, can be characterized as the main goal of air traffic controllers' work. However, the adherence to strict separation standards for safety reasons sets nonnegotiable rules that act as constraints [2, p. 341]. The combination of these two characteristics results in a demanding work, especially because air traffic controllers have to make most of their decisions in a narrow time frame [3] [4]. Due to the characteristics of their work and the general limitations of the human ability to process information, air traffic controllers often experience time pressure [2, p. 339] that can lead to a stress response. A stress response is the activation of several physiological systems on the affective, cognitive, neural, endocrinal, and muscular level [5] when individuals are facing a stress inducing stimulus (stressor). However, stress is not per se a negative state, since the evaluation of the stressor depends on the interplay of the situational demands and the abilities of the individual to cope with the situation [6]. Since time pressure is a situational characteristic in the daily work of air traffic controllers, the occurrence of negative stress and its emotional and psychological consequences (short term: anxiety, despondence, anger, cognitive impairments; long term: fatigue, health issues, depression) is likely (see for instance [7] [8]). Therefore, the reduction or rather avoidance of stress inducing situations is an important goal in the daily work of air traffic controllers.

III. STATE OF THE ART

Within this section we will have a brief look on related research on modelling human emotions as well as on worload and emotion sensitive user interfaces. We will consider the state of the art in practice and, thus, describing the current workspace of a German air traffic controller.

A. Modeling Emotion

Albeit our own research within the domain of modeling emotions via the simulation of the real world scenarios [9] [10] [11] [12], there are numerous other attempts of calculating human emotions. The range of applications in this field varies greatly. For example, the correct pronunciation of text to speech software depends on the software being able to carry affective states [13], and the ability to correctly identify human affective states might even increase social amicability in human-robotic-co-existence [14]. In order to compute affective states, emotions are viewed as values within the dimensions of either valence, being positive (good emotions, e.g., joy) or negative (bad emotions, e.g., fear, disgust) as well as the energy provided to fuel an emotion by stating a factor for arousal (high or low, the subject being either excited or calm) [15]. Analysis of emotional states in human subjects lead to the conclusion that there is a correlation of self-reported arousal and the activation of the sweat glands, leading to a higher level of electronic conductance (galvanic skin response - GSR). Since there is no correlation regarding the direction of the valence, the GSR is a valid indication of the arousal level and thus can be used as an objective measure [15] [16] [17]. The objective measurement of a valence, however, is right now only possible by implementing the Facial Action Coding System as proposed by Ekman [18], since other measurements like fMRI or fNIRS studies yielded varying results [19] [20]. Main points of criticism about the modeling of human emotions, however, are the missing experimental validation of models and the sheer complexity of different psychophysiological processes involved in establishing a certain emotional affect [21]. As Marinier and Laird point out, established computational emotion models either rely on appraisal theories and the step-by-step analysis and predictions of reacting to an event to compute a corresponding emotion. Since these events can be compared to self-reports and objective measurements like the aforementioned skin conductance [22] they are exceptionally well suited for a valid implementation. Neural Networks, on the other hand, attempt to model brain activities related to the emergence of emotional affections which can be linked to actual observable processes as they happen inside the brain [23] [24] and could be evaluated using brain imaging techniques.

B. Emotion and Workload Sensitive User Interfaces

The detection and simulation of the humans emotional and cognitive state is quiet useless without an adequate system reaction. Emotion and workload sensitive user interfaces are key issue of a whole research area (affective computing [25]) and are present in a multitude of areas of application. They are widely considered within intelligent tutoring systems. Ranging from piano teaching systems, increasing difficulty,when the students workload is low, [26] to emotion sensitive systems that support learning of the Japanese language by adapting the difficulty of tasks to the emotional state of the student. The effects of emotion and motivation on learning performance are a well known relationship within educational research. [27] So, within the last decade there was plenty of research done on emotion sensitive tutoring systems. An overview can be found in [28] and [29]. Even in application areas crucial to safety there is a trend to affective system support: a minimal invasive surgery robot controlled through a speech interface, initializing safety feedback loops, when negative emotion is detected in the surgeons command [30]; a mobile robot on exploration tour with an astronaut, that reacts on the astronauts anxiety level by assisting with hints or by triggering an alarm and hurrying to the astronaut [31]; or a simple application that redirects incoming calls to the mailbox, during high workload in driving situations [32]. Emotion and workload sensitive interfaces in air traffic control are mainly part of the so called adaptive automation research. Within adaptive automation the tasks are dynamically allocated to the system and the human controller. Thus, it is expected to counteract the out-of-the loop performance problem and the loss of situational awareness [33]. Parasuraman et. al compared a workload adaptive Automation with an non-adaptive and a reverse workload automation. They found an improved performance within the workload adaptive situation [34]. According to Langan-Fox et. al there is a huge corpus of research on measuring workload and situational awareness in the area of air traffic control [35]. In projects like NINA - Neurometrics Indicators for ATM [36] workload measurement and the development of workload adaptive interfaces are/were investigated. So far, emotions were rarely considered for triggering adaption within the air traffic control context. We are seeking an adaptive user interface sensitive to workload as well as to emotional and air traffic situation aspects.

C. The Layout of an Air Traffic Controller's Workspace

We had the chance to observe controllers at three control centers of the German air traffic service provider DFS. The systems and the available visualizations and tools in use differ between control centers. Working spaces in Langen and Munich (Fig. 1a) were comparable using the P1/ATCAS system, whereas in Karlsruhe the VAFORIT system is used (Fig. 1b). This section outlines obvious similarities and differences between the controllers workplaces.

What tools have these work spaces in common? Every controller is working with a radar screen, that is showing a top-down view on the sector. Each aircraft is shown as a little icon (square) followed by some dots, indicating the aircraft's former positions. Each target is accompanied by a label, showing the most important information (an example is shown in Fig. 2). (Semi-)Static sector characteristics as beacons, airways or restricted airspaces may be shown or hidden. Tools for distance measurement on radar screen are available and direction vectors of aircraft (a function of time adopting constant ground speed and direction) are displayable. Additional screens show on demand additional data, as aircraft properties, meteorological data, or a status display. The status display is a tool for communicating the own stress level to other controllers and the supervisor. For further communication

Figure 1. The air traffic controllers workspace at the air traffic control centers in Munich (P1/ATCAS system) and Karlsruhe (VAFORIT system). The major differences regard display polarity and the presence/absence of digital flight progress strips. (Sources: DFS Deutsche Flugsicherung GmbH).

with other controllers and pilots each workplace includes a phone and a radio communication equipment.

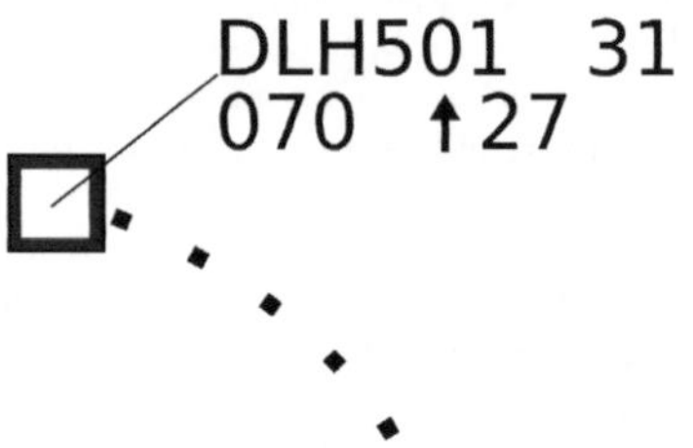

Figure 2. An aircraft target on a radar screen. The label shows callsign, groundspeed, flight level and rate of climb/descent. The history indicates the speed and the aircraft's tendency.

There obviously are differences in workplace design. The radar display of the VAFORIT system has positive display polarity. This means, in contrast to a negative display polarity, that dark symbols are shown on a light background (Fig. 3). With positive display polarity, there are less changes in pupil dilatation, because the secondary screens have nevertheless positive display polarity and surroundings are also rather light. With a work place design, that needs less adaptions between light an dark, there is less eyestrain [37]. Furthermore, a positive display polarity provides better legibility of small letters [38].

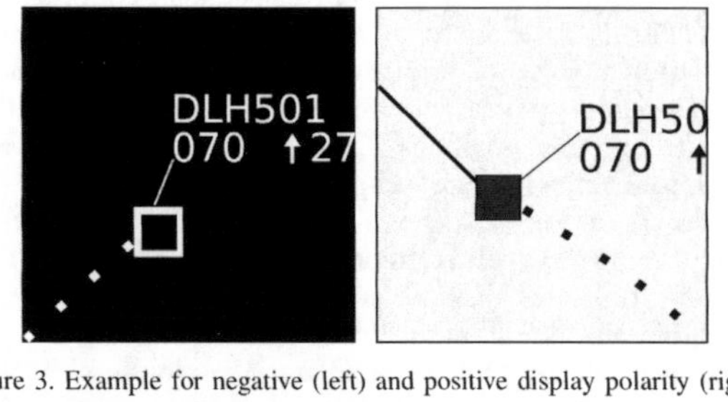

Figure 3. Example for negative (left) and positive display polarity (right).

The main difference between the systems is the use of flight progress strips for displaying flight plan data. The advantages of paper flight progress strips, digital flight progress strips and systems without strips were discussed since the 1990s. Vortac and Edwards et al. recommended, after analysis on flight progress strip activities, an automation of these, in order to decrease controllers effort [39] [40]. Albright et al. investigated the contollers operation methods, while paper flight strips are missed [41]. They found that there was a gain in time due to the loss of strip marking, but that the presentation of flight plan data on every flight at a time was perceived as more informative. Scince then, systems with digital flight progress strips have been developed [42] [43] as well as systems without any flight progress strips. In contrast, MacKay advises against replacing paper strips [44]. They take advantage of visual and tactile memory, they are flexible, reliable and support cooperative work due to their physical presence and visual forms of interaction. The advantages of automation bring the necessity of electronic flight strips about, while the reliability and the whole functionality of paper flight strips should be kept [45]. A possible back door would be the use of augmented flight progress strips. Hurter et al. combined in their system Strip'TIC the advantages of the digital and the physical world [46] [47]. In Munich and Langen air traffic controllers are facing digital strips shown on a screen lying on the desk and edited by a digitizer pen. (Fig. 4). The air traffic controller

Figure 4. Digital flight progress strips of the P1/ATCAS system at DFS Deutsche Flugsicherung. Each aircraft is referenced by its callsign. Data as flight plan and cleared flight level is provided by texts. The arrows at the upper left corner indicate the overall direction of the aircraft's route in 3 dimensions. (Source: DFS Deutsche Flugsicherung GmbH)

in Karlsruhe has the possibility to access flight plan data via extended labels on the radar screen. Thus, making the use of flight progress strips (nearly) redundant. This difference is resulting in slightly different routines in the air traffic controllers ways to reach their goals. Thus, making different aspects of their work observable. There are further quite small differences in interfaces, e.g., a different visualization of the status display, but a complete analysis of them goes beyond this section.

D. The StayCentered Project

Typically, the work of an air traffic controller involves managing various flight routes, aircraft, and altitude as well as air speed differences. Additionally, meteorological data, technical maintenance activities or, in rare circumstances, emergencies can occur at any given moment and require swift and correct reactions by the controller. As air traffic controllers often work in dyads, in order to have an inherent corrective at all time and to provide redundancies, the StayCentered project at Technische Universität Chemnitz aims for enhancing the already high security standards of air traffic controllers, and for identifying as well as for offering assistance within cognitive stressful flight situations. Therefore, the dyadic team structure has to be analyzed comprehensively: both their voiced interactions between themselves and with the pilots within their controlled airspace. The goal is to be able to identify human error potential in voicing commands, interpreting visual data representations and to identify limits in cognitive processing capabilities. The resulting model of a working controller dyad is then used to simulate the emotional and cognitive state of the dyad in regards to upcoming air traffic some hours in advance. For example, planned but delayed flights (e.g., a sandstorm in Dubai and a thunderstorm in Moscow) will lead to an increased number of aircraft in their destination sector. Flight control management would then be able to split sectors and to call in additional controllers in order to keep the workload at a comfortable level. In addition, the controller stations themselves already offer the possibility for the controllers to signal an increased workload. However, the implementation of the projects biophysiological measurements would allow for an objective and immediate feedback to the controllers about their current cognitive state and troubleshooting capabilities [48], as well as for a workload regulation [49]. Therefore, the galvanic skin response, facial action coding, body posture, vocal properties, eye movements and pupil dilation are recorded and used to infer an emotion valence, arousal level, and cognitive load [50]. The paper at hand presents a set of initial exploratory studies. The identification of parameters to the emotional model as well as issues for an emotion sensitive user interface in this context were possible.

IV. Methodology

To assess whether or not an air traffic controller experiences stress and the associated negative emotions, it is necessary to fully understand how the controller is receiving and processing the crucial information and how this is converted into practical actions. Since it is not possible to gain insight into the information processing objectively from the outside, it is necessary that the air traffic controllers verbalize their cognitive processes. For this purpose, we used semi-structured interviews outside the work situation to gather general information about how air traffic controllers experience work-related stress and how they cope with it. Among others, we let them describe exceptional situations, which were especially demanding, how they solved them, and how they felt afterwards. Furthermore, we used the thinking-aloud approach in interviews to get a basic understanding on how air traffic controllers process information. We confronted them with a typical radar screen printout. The sector and scenario were unknown to the participants. There was a situation containing 8 aircraft described, a mid term conflict of two aircraft with same heading and differing speed and a lateral conflict of two aircraft with opposite heading, but vertically divided. We asked them to evaluate the given flight situation regarding the salience of important information as well as the order, in which critical data are perceived and processed. Additionally, we observed the air traffic controllers during their work at the level of moderate participation, allowing us to ask specific questions. Here, we also used the thinking-aloud approach to get information and explanations about certain actions and events. The observation under real working conditions is especially important since cognitive and emotional reactions are known to be a combination of person and situation, and thus only the inclusion of the given situational characteristics allows for a meaningful interpretation of the data gathered in the interviews. We decided to use this combination of methods in an exploratory approach in order to get the information of the air traffic controllers as authentic and natural as possible. Expressing thoughts, ideas and considerations in their own words in an actual work situation as well as in the reflecting, meta-cognitive form of an interview appears to be the adequate methodical approach for this kind of research problem. The data was collected between February and April 2015 at the facilities of the German air traffic service provider DFS in Langen and Munich and in August 2015 at the center in Karlsruhe. To assure a sufficient variability in the data, we interviewed and observed experienced and novice air traffic controllers likewise. Altogether, we collected data of N=37 air traffic controllers (age: 18 to 57). 21 of them are used to the ATCAS system and 16 to the VAFORIT system. Since the evaluation of the air traffic controllers' work requires a basic level of expertise regarding the work station, work processes and air traffic, all researchers received an introduction to the air traffic controller's work by an expert of the DFS before data collection. Since recording audiovisual material is problematic due to security reasons, all interviews and observations were recorded by pen and paper. For the purpose of the analysis, all data was coded and categorized. Due to the exploratory nature of the research, we did not follow a standardized coding scheme. Instead, we tried to identify all relevant factors regarding the cognitive and emotional constitution and experiences of the air traffic controllers in relation so the given work situation.

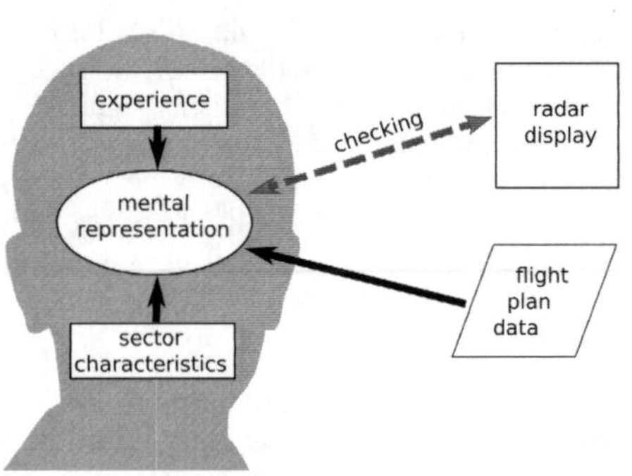

Figure 5. Schematic diagram of the air traffic controllers mental situation representation and its sources. Access to flight plan data depends on interface design.

V. Findings and Discussion

By fulfilling their daily tasks, air traffic controllers face highly demanding situations. They need to process plenty pieces of information simultaneously that are arriving on multimodal channels (primarily auditive and visual). Based on this information, controllers have to make quick and reliable decisions to ensure safety of the aircraft, and thus people, under their control. We were able to identify procedural, communicative and emotional aspects that form the controller's course of action.

A. Procedural Aspects

Procedural aspects deal with the mental organization of air traffic by a controller. This includes their work organization, their information processing and their internal representation of the flight situation. In order to ensure save and fluid traffic flows the air traffic controller has to have a good overview of the air traffic situation. Thus, he constructs and continuously updates a mental representation that is mainly based on flight plan data, the controllers experiences and internalized knowledge on sector characteristics. This representation includes more than just the current situation but also a prediction future air traffic situations. The controllers stated that their prediction reaches about 3 minutes into future in order to proactively control the traffic. The current situation, as shown by the radar screen, is continuously cross-checked against the mental representation (Fig. 5). In the whole process of building and updating the internal representation and cross-checking the role of the radar screen differs between systems. This depends on the place where flight plan data is presented. In a system with digital or paper flight progress strips the radar screen is rather a secondary tool that is used for cross-checking. But the inclusion of flight plan data into radar screen makes it the most important tool for the air traffic controllers work. We also asked the controllers to describe the structure of their mental representation. While in literature mental representations of air traffic situations are often described as somewhat three dimensional models [51] [52], our controllers explicitly stated that they do not build up a three dimensional model of the situation. They described it as two dimensional, similar to the radar display that is expanded by a variable indicating vertical layers. Other studies revealed that air traffic controllers [53] [54] and already controller students [55] do not necessarily build up a three dimensional mental model. Each of them develops an individual mental structure to represent the three dimensional data over time. In situations where the controller first constructs his representation (e.g., during a handover) he has to consider flight plans as well as current positions. The first variable controllers focus on, while scanning the radar display, is altitude information and whether an aircraft is climbing or descending. After this, aicrafts heading and position are considered and lastly, ground speed gives a hint on the existence of a potential conflict. Likewise, Rantanen et al. identified in their experiments [2] altitude as the information that is processed first for conflict detection. Mogford et al. emphasized altitude and heading as the most important information for air traffic controllers situation awareness [51]. In standard situations, when a flight strip appears that represents an airplane that is about entering the sector soon, the controller is first looking for the route the aircraft is tending to take and at which flight level. For first conflict detection, the controller is checking overflight times at the fixes. If overflight times are overlapping ten minutes to the ones of another flight he marks a potential conflict. When the involved aircraft appears on the radar screen, the controller is checking a second time for the conflict and then improving gradually the quality of his prediction about a potential problem. First, he is estimating vertical and horizontal separations according to his experience (rule of thumb). He can use distance measuring tools provided by the system on the radar screen, but he is also able to do exact calculations using mental arithmetics, if necessary.

The observation of the air traffic controllers working these differing systems revealed some interesting aspects about the controller routines. Although the VAFORIT system can be operated without any flight progress strips, there is still a digital representation of them shown in an additional screen. All controllers stated that they do not need these strips. Nevertheless, they used them. The functionality of flagging a strip was still used as a reminder for oneself or the team partner. By sorting the strips by the cleared flight level the controllers gained an overview of the vertical situation in airspace. This fact shows the importance of the altitude information once again. There are small arrows on the flight progress strips in Karlsruhe, that indicate from which roughly direction the aircraft is arriving and where it is going (in 3 dimensions). These are also used by the controllers for an overview of movements. Furthermore, the number of flight strips is also an indicator for the amount of work coming up in the next 10 to 20 minutes. This is often influencing the controllers decision on allowing directs or the like. At some point during the air traffic controllers work (e.g., while considering a request) it is necessary to check the current position of an aircraft, that isn't visible in the current map excerpt, shown by the radar screen. In Munich and Langen it was observed, that controllers started zooming and moving the map while scanning the map with the eyes for this aircraft. In Karlsruhe the controllers used the functionality of showing the aircraft's intended route and thus, they found the target more rapidly. Controllers also stated that there are situations where the information is needed whether the handover to a following sector has passed successfully. This is also an information that was accessed through the flight progress strips in both of the systems. Another clearly observable point in between systems is the importance of handwritten notes. While in the P1/ATCAS system free strip marking can be used for communication with the team partner, there is no corresponding functionality in the VAFORIT system provided. But some controllers used pen and paper for writing notes, as a reminder for themselves or their team partner.

B. Communicative Aspects

The air traffic controllers work is a highly cooperative work. There is long term (e.g., with the team partner) as well as short term (e.g., with pilots) cooperation necessary to ensure save air traffic. Our communicative findings take the communication partners and the communication channels, as well as the controllers perception of cooperative groupings, into account. In order to figure out with whom the air traffic controllers affiliate themselves we asked them for the term 'team'. When we initially used the word team within the context of air traffic controllers, we had the air traffic controller dyad in mind. However, the air traffic controllers understanding of team covers more than initially assumed. On the one hand, they used the term when speaking about all the air traffic controllers responsible for German airspace and adjacent sectors. When recognizing a potential conflict situation that would happen in the neighboring sector, but could be prevented or already solved within their own sector, they would do so. When recognizing a conflict situation within an other sector, they would warn the responsible controller. When recognizing that controllers responsible for an adjacent sector have high traffic load and they are stressed, they try to keep further traffic away from that sector or try to avoid more stress for their colleagues by organizing the flights in their own sector in a way that makes them easy to handle in the next one. This understanding of a team is also supported by the fact that air traffic controllers are on a first-name basis with each other. On the other hand, the term team was used while talking about the air traffic controllers organizational entity. In German air traffic control centers, there are groups of air traffic controllers that are responsible for several neighboring sectors. These sectors share borders and in times of low traffic load they can be combined. Each controller out of this group has the admission to work on every position within these sectors. Thus, each of the controllers will at any time constitute a dyad with any controller out of this group. Beyond these affiliations there are further cooperation necessary. We identified the following cooperation partners and the communication channels in between. An overview is shown in Figure 6. Short-term collaboration with pilots consists of speech over radio using predefined terms and routines in order to minimize the number of misunderstandings. If suitable equipped, there is also the possibility to send text messages to the aircraft. This is not yet common, but controllers like the decrease of misunderstandings in numbers about it. The supervisor communicates to the controllers through the display

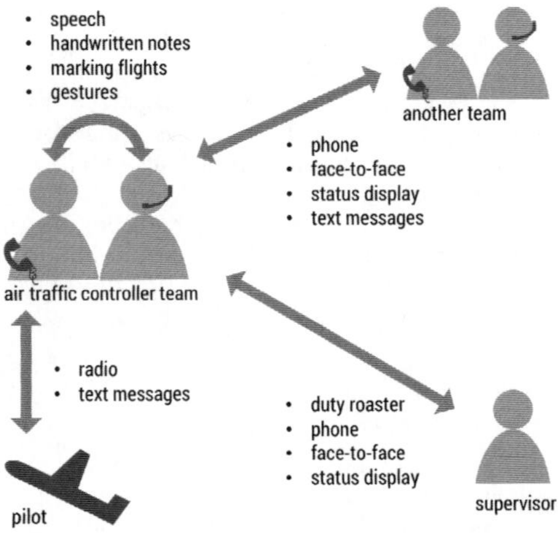

Figure 6. Overview of communication partners and channels. Beside cooperation within the dyadic team, the executive gives instructions to pilots and the planner coordinates handovers with the other sectors planners. The supervisor is responsible for the duty roaster and sector splitting.

of a duty roster. Change requests, rapid updates, and the like are communicated via telephone connection. But it has been observed that both sides are often leaving their place, going to and talking to the other one about these concerns. Most of the communication is happening within the dyad responsible for a sector. Both, the executive and the planner have to build up a shared mental picture of the situation and keep it updated. In order to do so and to solve potential conflicts, they communicate using gestures (pointing gestures to guide the others attention onto the screen, sometimes they are also using the other ones mouse), the flagging of flight strips (to highlight potential conflicts), and handwritten notes (either on digital flight strips or an extra sheet of paper), but speech remains the dominant communication channel. Nevertheless, both verbal and nonverbal communication within the dyadic team are crucial to good cooperation. The importance of nonverbal communication (strip marking and observation of the partners actions) additional to speech was also highlighted by Soraji et al. [56]. In times of high traffic load, controllers are sitting up straight, speaking concisely about traffic concerns. In times of low traffic load they are more relaxed and they are chatting with each other and the surrounding controllers. For coordination in between sectors, the planner is talking to other dyads via telephone, except for the ones sitting spatially near to him, they are addressed directly. Similar to the communication with pilots, this cooperation is guided by regulations and routines. Additionally, in Karlsruhe controllers have the possibility to submit their requests via text messages. They like about it, that they do not disturb the other controllers line of thought at a specific moment. Another tool for controller to controller communications is the status display also called Geneva traffic light. This display shows a color (green, yellow, red) for each sector with green being the default color for normal traffic load. By setting this color, controllers can communicate their current workload to other controllers and to the supervisor.

The status display was introduced due to controllers wishes to communicate their stress level. The visual presentation differs between systems. However, the status display is rarely used. Some controllers do not even recognize it as a tool for communicating their stress level, but as a tool for getting an overview of the sectors that are opened or combined. The status display is experienced as neglected. Most air traffic controllers have no clue when to switch the color to yellow and if they do they feel as if the other controllers ignore their stressed state. However, in the case that they are really stressed then there is often no time left for remembering to switch their status to red.

C. Emotional Aspects

Regarding the emotional aspects of an air traffic controller and their impact on work results, there are numerous factors involved. Since we want to assist the controllers in stressfull situations in order to prevent negative emotional and psychological consequences, one focus was to identify the stressors in the air traffic controllers work. During interviews the air traffic controllers mentioned three main stressors:

High Traffic Load
: The crucial factor for traffic load is the number of aircraft under control. However, the resulting workload goes beyond the sheer number. The structure of the airspace and standard routes as well as directions of the aircraft have an impact on perceived complexity. Plenty of vertical movements, as in approach sectors and sectors in the lower airspace, and lots of crossing trajectories increase the probability for potential conflicts.

Unexpected Events in the Airspace
: Since air traffic controllers tend to have a detailed picture of upcoming events, unconsidered events may cause additional load, since they often require a swift reaction while simultaneously adding a unknown variable to their calculations. Usually, these are events that are neither listed in nor logical consequences of flight plan data. Initially, we considered emergency flights as unexpected things causing stress because air traffic controllers have to clear the way for them. However, most of the emergencies will already be marked in the actualized flight plan by the pilots. Thus, they can be regarded as expected traffic with a higher priority, making them just another variable in the air traffic controller's mental model. Even closures of single airports are not surprising, because every flight has an alternative destination stated in its flight plan. However, an unplanned aircraft calling in or flights within their sector boundaries, which are not under their control, are stress inducing factors. Hence, a pilot who forgot next sectors frequency, just asking for it once again, may cause more confusion than emergencies, because the controller already deleted the associated flight strip and thus also removed the flight and callsign already from his mental model.

Malfunction of Equipment
: Generally, the air traffic controller is dependent on his equipment. Without radar display the controller has to rely on the pilots following his instructions without any misapprehensions. Without flight plan data, the controller would loose the ability to proactively regulate air traffic. Still, air traffic controllers emphasized especially malfunction of the radio as problematic. Without the ability to communicate with the pilots the air traffic controllers are completely incapable of action. They do not know about pilots' plans and are not able to forewarn them of an upcoming danger. For these reasons there are for each system independent fallback systems in use at DFS Deutsche Flugsicherung.

Other Things Indirectly Being Relevant
: For efficiently building their mental picture, air traffic controllers rely mostly on their experiences and internalized information, such as standard routes and sector borders. If controllers are returning after a period of absence (e.g., illness or holidays), they perceive their work as more demanding, due to changes in standard routes, sector boundaries, or agreements. Also, other impact factors like general well-being, mood, private problems, etc. were mentioned by air traffic controllers to influence the work performance. Therefore, personal factors often change the perceived demands. According to the air traffic controllers' experience, the same taskload can be experienced differently.

These results on potential stressors align with the five most stressful items found by Brink [57]. South African air traffic controllers rated the number of aircraft, extraneous traffic, unforeseeable events, peek hour traffic and limitations, and reliability of the equipment to be most stressful factors out of a questionnaire with 20 items.

Emotional aspects within the air traffic controller's work include awareness of own sentiments and awareness of the emotional state of others. Generally, controllers stated that there are no crucial emotional situations. Sometimes private problems cause the controller to "concentrate a little more" but usually they know how to act out of them. After a critical situation at work they are in need of someone to talk to. Often they prefer talking to their colleagues about it. Except for the controllers wish to do so, they have to visit Critical Incident Stress Management sessions by regulations. During follow-up discussions some other situations were identified. Air traffic controllers said they are feeling proud, after managing a tricky situation smoothly. They have a sense of delight, when pilots thank them for satisfying their wishes (e.g., a direct). During long periods of low traffic the predominant sense is boredom. The most important indicator for the others emotional state is the sound of their voice and their choice of words, especially during communication using telephone or radio. Succinct answers indicate elevated concentration. During communication with their spatial neighbors, gestures and poses can be accessed additionally for emotional awareness.

VI. Implications to the Model

From a psychological point of view, it is not surprising that the mental and emotional states of air traffic controllers are influenced by personal as well as situational characteristics. However, without a detailed analysis of the air controllers work, it is impossible to specify the relevant variables and their parameter values. Based on the collected data we are now able to consider precise variables in our model. Regarding the situational aspects, the number of aircraft as well as their flight characteristics are the main aspects for potential workload, and of course the available time is also relevant. Further research is necessary to identify the concrete relationship between those variables. However, it is already clear that there is limit on how many interactions can take place between the air traffic controller and pilots, because every interaction takes several seconds. Considering the well-known relationship between arousal and performance on difficult tasks [58], such as the work of an air traffic controller, we assume that the optimal efficiency lies far below the physical limit of interactions. It has still to be determined how to express the comfort zone of air controllers by an index. One potential solution is the indication of interactions per minute with the option to weight interactions depending on the situation's complexity. The model requires two kinds of critical values for the index that signals a possible overload: One is relating to situational peaks, which can be understood as episodes of high workload in a rather short time frame. The other one is applying to longer periods of time with an increased workload that is higher than the optimum but lower than the situational peak. Both kinds of overload can result in mistakes, incorrect decisions or just slower reactions and must be prevented. Even though German air traffic controllers can be considered a homogenous group of specialists who are able to work under pressure, the critical values must be personalized due to differences in personality related factors. Our data suggest that many typical personality variables affect the work of air traffic controllers, such as mood, alertness, work experience, private problems, absence due to vacation, or sickness, etc. The main problem for the consideration of those variables is their problematic measurement: Many of them are only available to the air controllers themselves, and even they are not always able to fully specify all factors that might influence their performance or to quantify them. Furthermore, many of those variables are changing on a daily basis, even though they should not fluctuate that much. Therefore, the personality factors can be used to improve the index based on the situational variables. Simply put: The critical values can be adjusted depending on how an air traffic controller feels - if this information is available - or based on objective information like the absence of a controller for several weeks, which lets him experience the work as more demanding during the first days of his work. For a short-term evaluation of the air controllers state, additional diagnostics will further improve the determination of the personality variables influence. Additionally, a cross-validation and combination, respectively, with psychophysiological parameters, eye-tracking, voice characteristics, facial emotion expression as well as poses and gestures will also help to classify flight situation regarding their complexity. For instance, a more complex problem will result in longer times of fixation on the involved flights, an increased skin conductance, shorter voice-commands, a straighter body position and a stern facial expression. Therefore, our model must take many variables into account, some global and some situational. Things become even more complicated, since the air traffic controllers are usually working together as a dyad. The model has to take into consideration not only the individual parameters, but also the specifics of the team. The same flight situation in a sector might result in excessive demands for one dyad but present an acceptable challenge to a team of veterans. This additional set of team related variables complicated the model, since questions about the structures and relationships between all the variables contained in the model are not fully answered yet.

VII. Implications to the Interfaces

One of the main goals of the StayCentered project is to identify and to offer assistance within cognitive stressful air traffic situations. Current interfaces have to be rethought in order to give access to the identified and simulated mental and emotional states. The StayCentered interfaces will be designed to give decision support to the supervisor, to facilitate cooperation, and to adapt with respect to the controller's current state and overall supporting the controllers routines and mental models.

At the moment the supervisor's decision upon splitting up a sector is done by consulting workload predictions, mainly based on the expected number of aircraft, and on the air traffic controllers demands. The StayCentered supervisor interface will present the simulation's forecasts over time. Anticipated stressful or tedious situations should be visible at a glance and supporting decision making on resolving these situations.

As described, cooperation and communication are crucial elements of the air traffic controllers daily work. These communicative situations shall be supported by the interfaces. It is still an open point why controllers tend to leave their place for consulting their supervisor in situations of low traffic. This is a potential risky situation. There should be adequate ways of communication. The controllers workspace should be designed in such a way that the actions of one controller are clearly visible to his partner in the dyad. Thus, we are expecting to support the creation of a shared mental model and enhancing communication. The most obvious advantage of the StayCentered system is that the status display can change its color automatically and providing an objective measurement on the controllers stress level. However, also the displays presentation could be enhanced. Currently, each sector is represented by a colored button (green-yellow-red) on a secondary screen. This part may also be hidden by some other information. Additional short textual remarks for the sectors in stressful situations are available. A permanent visible graphical presentation of this information (possibly integrated into radar screen) would make it accessible at a glance. Another important feature of nowadays interfaces is the possibility to generate reminders, either through flagging of flight strips or through handwritten notes. This feature of communication shall be kept in some way. Additionally, notes can also be used for personal work organization and reminders.

The interface adaption with respect to the controller's state applies to the interfaces at the controllers workspace. The information presentation is independent of the controllers emotional and communicative state, his workload, and the complexity of the actual flight situation. However, the importance of information objects differs from situation to situation. We are currently conducting a survey to identify these differences. A positive effect of an automation degree adapting on the air traffic controllers workload was already presented by related work, e.g., [34] and [59]. Also, visual adaption of the interfaces has been investigated [60]. But an adaption on emotional state is still an open point.

StayCentered controller interfaces will consider the identified state of the controller as an indicator for the chosen representation. We want to pay special attention to situations of low traffic load, because in these situations controllers often feel bored. Since boredom has a negative impact on their attention, we want to consider these situations within the design of the adaptive interface as well. Good user interfaces support the user's mental models and their procedures. Recent research on air traffic controllers' interfaces often considers three-dimensional and even stereoscopic radar displays (e.g., [61] [62] [63] [64]). According to our data, the controller's mental model of a flight situation is not necessarily three-dimensional. Therefore, we would prefer a two-dimensional representation with an implicit coding of altitude information. The interface should allow for stepwise adaption of conflict prediction to the required accuracy. Thus, it should also support distance measurement methods at a different granularity. The controllers need a good overview of vertical and horizontal movements in their sector. Furthermore, they need quick access to data about aircraft, not yet visible on their current radar view. Allover the air traffic controllers would be glad to have access to their own workload predictions. In order to have a more precise prediction that they can use for doing decisions upon pilots wishes. Thus, this information should be included into the controllers interfaces.

VIII. Summary and Conclusion

Within this paper we described the StayCentered project at Technische Universität Chemnitz that aims for assisting air traffic controllers' work by identifying and simulating the air traffic controller dyad's mental and emotional states. Within this context we presented the results of our preliminary study and discussed its implications for the mental and emotional models as well as for the user interfaces.

We identified high traffic load with plenty of vertical movements, unexpected events and a malfunction of the equipment as the most relevant stressors in air traffic control. Furthermore, stress level is influenced by personal factors. The controllers stated not to create a three-dimensional mental representation of flight situations. The information used to create the mental representation consists of internal knowledge about the sectors characteristics and standard routes, their experience and flight plan data. For checking the current situation, information is processed in the following order: altitude, climb/descent, horizontal position, heading, and speed on ground.

The order of information processing should be reflected within the user interfaces, as well as the structure of the air controllers mental model. Identified forms of communication should be supported, especially visibility of the partners actions. Workarounds, which are currently used by the controllers, should be included into future interfaces. The automatic recognition of the air traffic controller's workload and emotional state allows for further improvement in the workflow.

Our findings suggest that sufficiently modeling the cognitive and emotional states of air traffic controllers requires the inclusion of many variables regarding the individual controllers as well as the dyad and the current workload. The next steps in the process of model building are the identification of other relevant variables and generally their measurement and further processing. Even though we already know that cognitive and emotional states can be recognized using our multidimensional approach, the relationships between the variables still needs further research. Possible methodological approaches include the recording of actual or simulated work sessions in combination with post-hoc interviews in order to identify critical or demanding situations. By comparing the measurement data with the information given by the controller, we can identify typical patterns that signal stressful episodes, which can be used in our model.

Acknowledgments

This work was partially supported by the Germany Federal Ministry of Education and Research in the project StayCentered - Methodenbasis eines Assistenzsystems für Centerlotsen (MACeLot). Furthermore, we thank the German air navigation service provider DFS Deutsche Flugsicherung GmbH for their support.

References

[1] L. Pfeiffer, G. Valtin, N. H. Müller, and P. Rosenthal, "Aircraft in your head: How air traffic controllers mentally organize air traffic," in HUSO 2015 - The First International Conference on Human and Social Analytics. IARIA, October 2015, pp. 19 – 24.

[2] E. M. Rantanen and A. Nunes, "Hierarchical conflict detection in air traffic control." International Journal of Aviation Psychology, vol. 15, no. 4, 2005, pp. 339 – 362.

[3] V. D. Hopkin, Human Factors in air traffic control. Taylor & Francis, 1995.

[4] R. J. Roske-Hofstrand and E. D. Murphy, "Human information processing in air traffic control," in Human Factors in Air Traffic Control, M. W. Smolensky and E. S. Stein, Eds. Academic Press, 1998, ch. Human information processing in air traffic control, pp. 65–114.

[5] J. Siegrist, "Stress am arbeitsplatz [stress at the workplace]," in Gesundheitspsychologie [Health Psychology], R. Schwarzer, Ed. Göttingen: Hogrefe, 2006, pp. 303–318.

[6] R. S. Lazarus, Emotion and Adaptation. Oxford University Press, 1991.

[7] B. McEwen and J. Morrison, "The brain on stress: Vulnerability and plasticity of the prefrontal cortex over the life course," Neuron, vol. 79, no. 1, 2013, pp. 16 – 29.

[8] D. N. Khansari, A. J. Murgo, and R. E. Faith, "Effects of stress on the immune system," Immunology Today, vol. 11, 1990, pp. 170 – 175.

[9] L. Pfeiffer, N. H. Müller, G. Valtin, M. Truschzinski, P. Protzel, P. Ohler, and P. Rosenthal, "Emotionsmodell fr zuknftige mensch-technik-schnittstellen zur untersttzung von centerlotsen," in Proceedings der 57. Fachausschusssitzung Anthropotechnik der DGLR, M. Grandt and S. Schmerwitz, Eds. Bonn: Deutsche Gesellschaft fr Luft- und Raumfahrt - Lilienthal-Oberth e.V., 2015, pp. 219–232.

[10] N. H. Müller and M. Truschzinski, Human-Computer Interaction: Design and Evaluation: 17th International Conference, HCI International 2015, Los Angeles, CA, USA, August 2-7, 2015, Proceedings, Part I. Cham: Springer International Publishing, 2015, ch. Analytical Steps for the Calibration of an Emotional Framework, pp. 512–519.

[11] M. Truschzinski, H. . Dinkelbach, N. Mller, P. Ohler, F. Hamke, and P. Pretzel, "Deducing human emotions by robots: Computing basic non-verbal expressions of performed actions during a work task," in 2014 IEEE International Symposium on Intelligent Control (ISIC), Oct. 2014, pp. 1342–1347.

[12] N. H. Müller and M. Truschzinski, Human-Computer Interaction. Advanced Interaction Modalities and Techniques: 16th International Conference, HCI International 2014, Heraklion, Crete, Greece, June 22-27, 2014, Proceedings, Part II. Cham: Springer International Publishing, 2014, ch. An Emotional Framework for a Real-Life Worker Simulation, pp. 675–686.

[13] Y. Gao and W. Zhu, "Detecting affective states from text based on a multi-component emotion model," Computer Speech & Language, vol. 36, 2016, pp. 42 – 57.

[14] A. M. Zanchettin, L. Bascetta, and P. Rocco, "Acceptability of robotic manipulators in shared working environments through human-like redundancy resolution," Applied Ergonomics, vol. 44, no. 6, 2013, pp. 982 – 989.

[15] P. J. Lang, M. K. Gereenwald, M. M. Bradley, and A. O. Hamm, "Looking at pictures: Affective, facial, visceral, and behavioral reactions," Psychophysiology, vol. 30, no. 3, 1993, pp. 261–273.

[16] M. M. Bradley, B. N. Cuthbert, and P. J. Lang, "Startle reflex modification: Emotion or attention?" Psychophysiology, vol. 27, no. 5, 1990, pp. 513–522.

[17] E. W. Cook, L. W. Hawk, T. L. Davis, and V. E. Stevenson, "Affective individual differences and startle reflex modulation," Journal of Abnormal Psychology, vol. 100, no. 1, Feb. 1991, pp. 5–13.

[18] P. Ekman, Ed., Emotion in the human face: Guidelines for research and an integration of findings, 2nd ed. New York: Cambridge University Press, 1982.

[19] P. Sarkheil, A. Zilverstand, N. Kilian-Htten, F. Schneider, R. Goebel, and K. Mathiak, "fmri feedback enhances emotion regulation as evidenced by a reduced amygdala response," Behavioural Brain Research, vol. 281, 2015, pp. 326 – 332.

[20] T. Watanuki, K. Matsuo, K. Egashira, M. Nakashima, K. Harada, M. Nakano, T. Matsubara, K. Takahashi, and Y. Watanabe, "Precentral and inferior prefrontal hypoactivation during facial emotion recognition in patients with schizophrenia: A functional near-infrared spectroscopy study," Schizophrenia Research, vol. 170, no. 1, 2016, pp. 109 – 114.

[21] R. W. Picard, Affective Computing. MIT Press, 2000.

[22] R. P. Marinier and J. E. Laird, Encyclopedia of the Mind. SAGE Publications Inc., 2013, ch. Computational Models of Emotion, pp. 166 – 167.

[23] X.-F. Shi, Z.-L. Wang, A. Ping, and L.-K. Zhang, "Artificial emotion model based on reinforcement learning mechanism of neural network," The Journal of China Universities of Posts and Telecommunications, vol. 18, no. 3, 2011, pp. 105 – 109.

[24] J. L. Salmeron, "Fuzzy cognitive maps for artificial emotions forecasting," Applied Soft Computing, vol. 12, no. 12, 2012, pp. 3704 – 3710, theoretical issues and advanced applications on Fuzzy Cognitive Maps.

[25] R. W. Picard, Affective Computing. Cambridge: MIT Press, 1997.

[26] B. F. Yuksel, K. B. Oleson, L. Harrison, E. M. Peck, D. Afergan, R. Chang, and R. J. Jacob, "Learn piano with bach: An adaptive learning interface that adjusts task difficulty based on brain state," in Proceedings of the 2016 CHI Conference on Human Factors in Computing Systems, ser. CHI '16. New York, NY, USA: ACM, 2016, pp. 5372–5384.

[27] P. A. Schutz and R. Pekrun, Eds., Emotion in education. Elsevier Academic Press, 2007.

[28] M. Malekzadeh, M. B. Mustafa, and A. Lahsasna, "A review of emotion regulation in intelligent tutoring systems." Journal of Educational Technology & Society, vol. 18, no. 4, 2015, pp. 435 – 445.

[29] A. Mohanty, "Affective pedagogical agent in e-learning environment: A reflective analysis," Creative Education, vol. 7, 2016, pp. 586–595.

[30] B. Schuller, G. Rigoll, S. Can, and H. Feussner, "Emotion sensitive speech control for human-robot interaction in minimal invasive surgery," in RO-MAN 2008 - The 17th IEEE International Symposium on Robot and Human Interactive Communication, Aug. 2008, pp. 453–458.

[31] P. Rani and N. Sarkar, "Emotion-sensitive robots - a new paradigm for human-robot interaction," in Humanoid Robots, 2004 4th IEEE/RAS International Conference on, vol. 1, Nov 2004, pp. 149–167.

[32] W. Piechulla, C. Mayser, H. Gehrke, and W. Knig, "Reducing drivers mental workload by means of an adaptive manmachine interface," Transportation Research Part F: Traffic Psychology and Behaviour, vol. 6, no. 4, 2003, pp. 233 – 248.

[33] M. R. Endsley and E. O. Kiris, "The out-of-the-loop performance problem and level of control in automation," Human Factors: The Journal of the Human Factors and Ergonomics Society, vol. 37, no. 2, 1995, pp. 381–394.

[34] R. Parasuraman, M. Mouloua, and B. Hilburn, "Adaptive aiding and adaptive task allocation enhance human-machine interaction," Automation technology and human performance: Current research and trends, 1999, pp. 119–123.

[35] J. Langan-Fox, M. J. Sankey, and J. M. Canty, "Human factors measurement for future air traffic control systems," Human Factors: The Journal of the Human Factors and Ergonomics Society, vol. 51, no. 5, 2009, pp. 595–637.

[36] G. Borghini, P. Aricò, F. Babiloni, G. Granger, J.-P. Imbert, R. Benhacene, L. Napoletano, and S. Pozzi, "NINA : Neurometrics Indicators for ATM," SID 2013, SESAR Innovation Days, Nov 2013, sID 2013, 3rd SESAR Innovation Days, Stockholm, Suède.

[37] D. Zühlke, Nutzergerechte Entwicklung von Mensch-Maschine-Systemen: Useware-Engineering für technische Systeme. Berlin, Heidelberg: Springer Berlin Heidelberg, 2012, ch. Der Mensch und seine Fähigkeiten, pp. 5–34.

[38] C. Piepenbrock, S. Mayr, and A. Buchner, "Positive display polarity is particularly advantageous for small character sizes: Implications for display design," Human Factors: The Journal of the Human Factors and Ergonomics Society, vol. 56, no. 5, 2014, pp. 942–951.

[39] O. Vortac, M. B. Edwards, J. P. Jones, C. A. Manning, and A. J. Rotter, "En route air traffic controllers' use of flight progress strips: A graph-theoretic analysis," The International Journal of Aviation Psychology, vol. 3, no. 4, 1993, pp. 327–343.

[40] M. B. Edwards, D. K. Fuller, O. Vortac, and C. A. Manning, "The role of flight progress strips in en route air traffic control: a time-series analysis," International Journal of Human-Computer Studies, vol. 43, no. 1, 1995, pp. 1 – 13.

[41] C. A. Albright, T. R. Truitt, A. B. Barile, O. Vortac, and C. A. Manning, "How controllers compensate for the lack of flight progress strips." FAA Civil Aeromedical Institute, University of Oklahoma Dept. of Psychology, Tech. Rep., 1996.

[42] S. Fukushima, "Electronic flight progress strip prototype for en route atc," Aerospace and Electronic Systems, IEEE Transactions on, vol. 38, no. 1, Jan. 2002, pp. 119–127.

[43] N. A. Doble and R. J. Hansman, "Experimental evaluation of portable electronic flight progress strips," in Digital Avionics Systems Conference, 2004. DASC 04. The 23rd, vol. 1, Oct. 2004, pp. 51 – 11.

[44] W. E. MacKay, "Is paper safer? the role of paper flight strips in air traffic

control," ACM Trans. Comput.-Hum. Interact., vol. 6, no. 4, Dec. 1999, pp. 311–340.

[45] F. T. Durso and C. A. Manning, "Spinning paper into glass: transforming flight progress strips," Human Factors and Aerospace Safety, vol. 2, no. 1, 2002, pp. 1–31.

[46] C. Hurter, R. Lesbordes, C. Letondal, J.-L. Vinot, and S. Conversy, "Strip'tic: Exploring augmented paper strips for air traffic controllers," in Proceedings of the International Working Conference on Advanced Visual Interfaces, ser. AVI '12. New York, NY, USA: ACM, 2012, pp. 225–232.

[47] C. Letondal, C. Hurter, R. Lesbordes, J.-L. Vinot, and S. Conversy, "Flights in my hands: Coherence concerns in designing strip'tic, a tangible space for air traffic controllers," in Proceedings of the SIGCHI Conference on Human Factors in Computing Systems, ser. CHI '13. New York, NY, USA: ACM, 2013, pp. 2175–2184.

[48] T. Van Gog, F. Paas, and J. J. G. Van Merriënboer, "Uncovering expertise-related differences in troubleshooting performance: combining eye movement and concurrent verbal protocol data," Applied Cognitive Psychology, vol. 19, no. 2, 2005, pp. 205–221.

[49] S. Crevits, I.and Debernard and P. Denecker, "Model building for air-traffic controllers' workload regulation," European Journal of Operational Research, vol. 136, 2002, pp. 324–332.

[50] H.-C. She and Y.-Z. Chen, "The impact of multimedia effect on science learning: Evidence from eye movements," Computers & Education, vol. 53, no. 4, 2009, pp. 1297–1307.

[51] R. H. Mogford, "Mental models and situation awareness in air traffic control," The International Journal of Aviation Psychology, vol. 7, no. 4, 1997, pp. 331–341.

[52] S. T. Shorrock and A. Isaac, "Mental imagery in air traffic control." International Journal of Aviation Psychology, vol. 20, no. 4, 2010, pp. 309 – 324.

[53] B. Kirwan, L. Donohoe, T. Atkinson, H. MacKendrick, T. Lamoureux, and A. Phillips, "Getting the picture-investigating the mental picture of the air traffic controller," in Contemporary ergonomics 1998, M. A. Hanson, Ed. TAYLOR AND FRANCIS, 1998, pp. 404–408.

[54] C. Niessen and K. Eyferth, "A model of the air traffic controller's picture," Safety Science, vol. 37, no. 23, 2001, pp. 187 – 202.

[55] M. Tavanti and M. Cooper, "Looking for the 3d picture: The spatio-temporal realm of student controllers," in Human Centered Design, ser. Lecture Notes in Computer Science, M. Kurosu, Ed. Springer Berlin Heidelberg, 2009, vol. 5619, pp. 1070–1079.

[56] Y. Soraji, K. Furuta, T. Kanno, H. Aoyama, S. Inoue, D. Karikawa, and M. Takahashi, "Cognitive model of team cooperation in en-route air traffic control," Cognition, Technology & Work, vol. 14, no. 2, 2010, pp. 93–105.

[57] E. Brink, "The relationship between occupational stress, emotional intelligence and coping strategies in air traffic controllers," Master's thesis, University of Stellenbosch, 2009.

[58] R. M. Yerkes and J. D. Dodson, "The relation of strength of stimulus to rapidity of habit-formation," Journal of Comparative Neurology and Psychology, vol. 18, no. 5, 1908, pp. 459–482.

[59] B. Schaab, "The influence of ascending and descending levels of workload on performance," Automation Technology and Human Performance, 1999, pp. 218–220.

[60] K. Harper, S. Guarino, and G. Zacharias, "Adaptive human/machine interface for advanced air traffic management," in AIAA Guidance, Navigation, and Control Conference and Exhibit. American Institute of Aeronautics and Astronautics, 2000.

[61] M. S. Burnett and W. Barfield, "Perspective versus plan view air traffic control displays: Survey and empirical results," Proceedings of the Human Factors and Ergonomics Society Annual Meeting, vol. 35, no. 2, 1991, pp. 87–91.

[62] M. Lange, T. Dang, and M. Cooper, "Interactive resolution of conflicts in a 3d stereoscopic environment for air traffic control," in Research, Innovation and Vision for the Future, 2006 International Conference on, Feb. 2006, pp. 32–39.

[63] S. Bagassi, F. De Crescenzio, and F. Persiani, "Design and evaluation of a four-dimensional interface for air traffic control," Proceedings of the Institution of Mechanical Engineers, Part G: Journal of Aerospace Engineering, vol. 224, no. 8, 2010, pp. 937–947.

[64] D. J. Bergner and C. Schmand, "Fortschrittliche Visualisierungssysteme für Fluglotsenarbeitsplätze," TE im Fokus, 2012, pp. 7–13.

**International Academy, Research,
and Industry Association (IARIA)**
412 Derby Way
Wilmington, DE 19810

CURRAN ASSOCIATES INC.
proceedings.com

ISBN 978-1-5108-5623-3

9 781510 856233

G200 Youth Forum 2016 Conference

Garmisch-Partenkirchen, Germany
6 - 10 April 2016

G200
ASSOCIATION